Medieval History and Archaeology

General Editors
JOHN BLAIR   HELENA HAMEROW

Waterways and Canal-Building in Medieval England

MEDIEVAL HISTORY AND ARCHAEOLOGY

General Editors

John Blair   Helena Hamerow

The volumes in this series bring together archaeological, historical, and visual methods to offer new approaches to aspects of medieval society, economy, and material culture. The series seeks to present and interpret archaeological evidence in ways readily accessible to historians, while providing a historical perspective and context for the material culture of the period.

WATERWAYS AND CANAL-BUILDING IN MEDIEVAL ENGLAND

Edited by

JOHN BLAIR

OXFORD

UNIVERSITY PRESS

OXFORD

UNIVERSITY PRESS

Great Clarendon Street, Oxford OX2 6DP

Oxford University Press is a department of the University of Oxford.
It furthers the University's objective of excellence in research, scholarship,
and education by publishing worldwide in

Oxford New York

Auckland Cape Town Dar es Salaam Hong Kong Karachi
Kuala Lumpur Madrid Melbourne Mexico City Nairobi
New Delhi Shanghai Taipei Toronto

With offices in

Argentina Austria Brazil Chile Czech Republic France Greece
Guatemala Hungary Italy Japan Poland Portugal Singapore
South Korea Switzerland Thailand Turkey Ukraine Vietnam

Oxford is a registered trade mark of Oxford University Press
in the UK and in certain other countries

Published in the United States
by Oxford University Press Inc., New York

British Library Cataloguing in Publication Data

Data available

Library of Congress Cataloging in Publication Data

Data available

Typeset by Laserwords Private Limited, Chennai, India
Printed in Great Britain
on acid-free paper by
Biddles Ltd, King's Lynn, Norfolk

ISBN 978–0–19–921715–1

1 3 5 7 9 10 8 6 4 2

Preface

The starting point for this project was my growing conviction during the 1990s, as I worked on the medieval landscape of the upper Thames region, that I was encountering watercourses that were neither natural nor recent, and could only be understood as relict canals. Enquiries revealed various colleagues—working in different disciplines, in different parts of England, and mainly in isolation from each other—who had been moving equally hesitantly towards similar conclusions. An informal colloquium held in Oxford in December 1999 brought several of us together; by the end of it, we felt we had attained a clearer and more confident model of a well-managed network of waterways during c.950–1250, in which improved and artificial channels had an important place.

This book was completed six years later (a delay largely my fault). The authors include most of those who spoke in 1999, together with a few others. First drafts were exchanged between contributors, resulting in many cross-references and the removal of some overlaps. I have not, however, tried to eliminate duplication entirely, nor to make all statements mutually consistent: readers will find more than one account of, for instance, the Fens and the Somerset Levels. The contributors' range of expertise is diverse—history, archaeology, place-name and charter-boundary studies, geological science—and so are their approaches: it would be misleading to pretend that only one set of conclusions can be drawn from the data as we now have them. Rather, this book tries to give a synopsis of current methodologies and ideas, and of currently available evidence, in a field that is very young and will surely develop. If we are anywhere near the truth, the landscape contains numerous man-made and modified watercourses which are yet to be explored by the readily available tools of excavation, geomorphology, sedimentology, and radiocarbon dating. When more of this work has been done, we hope that the subject will not be seen as a blind lode, but as a useful tributary to the mainstream of medieval economic history.

The proofs of this book were not without problems, and I am especially grateful to Kate Hind and Lizzie Rowbottom for their care, support, and tolerance. Thanks too to Janet Lewendon for typing the index.

John Blair
The Queen's College, Oxford
December 2005

Contents

Contents

Figures

Abbreviations

ArchJ	*Archaeological Journal*
ASC	'Anglo-Saxon Chronicle' (cited with letter-symbol indicating version—except where the entry is from the common stock—and with corrected AD date), ed. C. Plummer, *Two of the Saxon Chronicles Parallel* (2 vols., Oxford, 1892–9).
BAR	British Archaeological Reports
BEASE	M. Lapidge, J. Blair, S. Keynes, and D. Scragg (eds.), *Blackwell Encyclopaedia of Anglo-Saxon England* (Oxford, 1999)
BL	British Library
Cal. Close R.	*Calendar of Close Rolls*
Cal. Pat. R.	*Calendar of Patent Rolls*
CBA	Council for British Archaeology
DB	Domesday Book. Citations are by folio in Great Domesday (DB i) and Little Domesday (DB ii).
EAA	*East Anglian Archaeology*
EcHR	*Economic History Review*
EPNS	English Place-Name Society
HRO	Hampshire Record Office
JBAA	*Journal of the British Archaeological Association*
JHG	*Journal of Historical Geography*
MA	*Medieval Archaeology*
OE	Old English
ON	Old Norse
PRO	Public Record Office
PWML	C. T. Flower (ed.), *Public Works in Mediaeval Law*, Selden Soc. 32 and 40 (2 vols., London, 1915, 1923).
RASC	A. J. Robertson (ed. and trans.), *Anglo-Saxon Charters* (Cambridge, 1956).
RCHME	Royal Commission on Historical Monuments (England)
S	Anglo-Saxon charters are cited by their 'S' number in P. H. Sawyer, *Anglo-Saxon Charters: An Annotated List and Bibliography* (London, 1968); the revised on-line version, currently available at *www.trin.cam.ac.uk/sdk13/chartwww/eSawyer.99/eSawyer2.html*, has been used.

SRO	Somerset Record Office
SxAC	*Sussex Archaeological Collections*
TNA	The National Archives
VCH	*Victoria Histories of the Counties of England* (cited by county and volume)
YAJ	*Yorkshire Archaeological Journal*

Introduction

JOHN BLAIR

The middle ages have rarely been seen as a dynamic era in the history of English water transport. Some common opinions were summed up in 2000 by Richard Holt:

> The limiting factor on water transport was the navigability and direction of rivers, as examples of canal-building and improvement of existing watercourses were extremely rare ... Rather than any development of water transport in medieval England ... there is ample evidence that a decline occurred as private, proprietorial rights outweighed perceptions of the public benefit to be derived from usable watercourses. Scholars have exaggerated the importance of water transport in the English economy; all too often assumptions of navigability depend on references to what can have been only occasional use. It is also apparent that whilst the medieval period saw a marked improvement in the conditions of road transport, with investment in the construction and repair of bridges and a significant move away from ox-hauling in favour of the horse, no such investment took place in improving watercourses.[1]

This comment, ostensibly on 'medieval England', actually refers to developments after 1200. The improvement of roads, bridges, and haulage in and around the thirteenth century, which recent research has demonstrated very clearly and convincingly,[2] raises in itself the possibility of an earlier stage when there had been more incentive to invest in water.

It is this possibility that the present essays chiefly seek to explore. The chronological focus of the volume is on the period 950–1250: some chapters range later, but all take the early to central middle ages as a reference point for assessing subsequent change. The division into two halves reflects a difference in emphasis between contributors whose main concern is the general economic picture, and those who expound the technology of canal-building and describe its physical remains. The second group of essays are, however, of fundamental

For comments on an earlier draft, I am very grateful to James Bond, Ann Cole, Mark Gardiner, Kanerva Heikkinen, John Langdon, Robert Peberdy, and Stephen Rippon.

[1] R. Holt, 'Medieval England's Water-Related Technologies', in P. Squatriti (ed.), *Working with Water in Medieval Europe: Technology and Resource-Use* (Leiden, 2000), 55–6.

[2] B. P. Hindle, *Roads and Tracks for Historians* (Chichester, 2001), 36–42; D. Harrison, *The Bridges of Medieval England* (Oxford, 2004); J. Langdon, *Horses, Oxen and Technological Innovation: The Use of Draught Animals in English Farming from 1066 to 1500* (Cambridge, 1986).

importance for the first, for they demonstrate the reality of widespread canal-building in a period which—largely preceding the thirteenth-century explosion in record sources—provides few references to it.

It may well be asked why, if canals were indeed common in late Anglo-Saxon and Norman England, their remains have not been more obvious. It was to answer this question that a geomorphologist, Ed Rhodes, kindly agreed to join a team of historians and archaeologists. His chapter shows how rapidly artificial drainage patterns will revert to nature, resulting in silting and the rapid onset of the meander process: 'Rivers tend to behave as if they have an unconscious desire to achieve what may be termed a minimum energy configuration' (below, p. 136). Canals that were not maintained will have metamorphosed into winding backwaters or disappeared entirely, while those that were will have been transformed by refashioning through the centuries. Archaeologists must now learn, as they have learnt for other historical features, to tease the exiguous traces of canals out of the landscape palimpsest.

This introduction thus starts with the basic evidence—the technical processes of canalization and their surviving remains—and then moves on to the broader patterns of transport and commercial freighting which the waterways served.

The European Background

Canal-building had been widely practised in the Roman empire, including Britain (see Bond, below pp. 158–69). After the fifth century it is likely to have been restricted by a decline in manpower, as also perhaps by a decline in the hydraulic knowledge needed to calculate and achieve viable water levels.[3] In many landscapes and with modest resources, however, the technology was straightforward and readily available: a gang of men with shovels working alongside wielders of mattocks and picks (Fig. 1).[4] It is a theme of this book that the determinants of medieval canal projects were more often economic than technological: where there was a strong enough will, the work-gangs could find the way.

Canal-building in post-Roman Europe seems to have started much earlier in the Germanic north than in the Mediterranean south.[5] At the pinnacle of all these endeavours stands Charlemagne's stupendous failure, the 30-metre-wide Karlsgraben designed to link the Rhine and Danube river systems.[6] Its

---

[3] It seems to have been failure to calculate water levels that made the Karlsgraben ultimately unviable: P. Squatriti, 'Digging Ditches in Early Medieval Europe', *Past and Present*, 176 (2002), 11–65, at 14–15.

[4] W. Jahn, J. Schumann, and E. Brockhoff (eds.), *Edel und Frei: Franken im Mittelalter* (Augsburg, 2004), 144–6, 181.

[5] In Italy, the construction of navigation canals on any scale does not seem to have resumed much before 1200: for Lombardy, for instance, see F. Menant, *Campagnes lombardes du Moyen Âge* (Rome, 1993), 174–5, 197–200. I owe this reference to Chris Wickham. See also Bond, below p. 174.

[6] Jahn, Schumann, and Brockhoff, *Edel und Frei*; Bond, below p. 172, for further references. Squatriti, 'Digging Ditches' argues that this and other earthwork projects were 'demonstrative acts' to show the power of rulers; I feel that he goes too far in minimizing their practical aims.

Fig. 1. Charlemagne's workmen digging the Karlsgraben to link the Rhine and Danube: an evocation by Lorenz Fries of Würzburg, 1546. (Stadtarchiv Würzburg, Ratsbuch 412, fo. 21^r; reproduced by kind permission of the Stadtarchiv Würzburg.)

astonishing aims and scale—at least three times the width of most of the canals discussed in this book—make all others look modest and realistic by comparison. It had several late- and post-Carolingian successors in northern France, including the river diversions around Caen in the 1060s supervised by the future Archbishop Lanfranc of Canterbury (Bond, below pp. 171–3). Against this background it is hard to believe that canals were not already being dug in the more watery Low Countries, although it seems difficult to identify specific cases before the late eleventh century.[7] It is an intriguing conjecture that St Dunstan, Abbot of Glastonbury from the early 940s but exiled in Flanders during 955–7, had first-hand experience of canals around Ghent, and that this experience had some impact on the late Anglo-Saxon canal at Glastonbury discussed below (pp. 235–8).

But an equally likely source for English canal technology may be the Scandinavian north. The Kanhave canal (Fig. 2) on Samsø, an island off the east coast of Denmark, can be dated by tree rings to precisely 726, more than fifty years before the Karlsgraben.[8] At 500 metres long and 11 metres wide, with slightly sloping sides and a shallow, almost flat-bottomed profile, it is similar to (though rather larger than) the Glastonbury canal and the upper Thames examples discussed by Blair (below, pp. 237, 283). It may also be relevant to some of the Anglo-Saxon cases that Danish archaeologists see the Kanhave canal as one element in a programme of eighth-century defensive

---

[7] The canalization of the rivers serving Ypres is indicated by the element *dic* in the place name Dixmude, mentioned in 1089: A. Verhulst, 'Les Origines de la ville d'Ypres', *Revue du Nord*, 81 (1999), 7–19, at 9. For other cases at around this time see A. Verhulst, *The Rise of Cities in North-West Europe* (Cambridge, 1999), 91, 94, 108. See also Bond, below p. 173.

[8] A. Nørgård Jørgensen, 'The Kanhave Canal on Samsø: New Investigations', *Château Gaillard: études de castellologie médiévale*, 18 (1998), 153–7. I owe this reference to Helena Hamerow and Julian Richards.

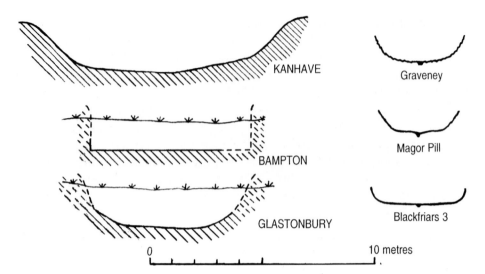

Fig. 2. Schematic cross-sections through the primary phases of three canals discussed in this book, compared with cross-sections through three English vessels of *c*.900–1400. (Note that the Bampton and Glastonbury canals are shown in relation to *modern* ground levels: in both cases any banks have been removed.)

and naval structures, ascribed to royal power.[9] Eastern England was always in touch with southern Scandinavia, and after the 870s Viking settlements would have created a more direct channel for technological dialogue, especially in the economically developing areas of the north-east midlands and the north. It is not credible that the later Anglo-Saxons were unfamiliar with the arts of channelling water.

Canals and Improved Rivers in England from Offa to Edward I

We have no evidence that King Offa of Mercia, Charlemagne's greatest English contemporary, built canals. But he did build his celebrated Dyke, a massive frontier earthwork which, as a statement of power, has been likened to the Karlsgraben.[10] As much recent work has shown, England in the age of Offa was powerfully influenced by Carolingian precedent in a range of political, legal, cultural, social, and economic fields:[11] could the Karlsgraben too have had English imitators? Graveney, on the north coast of Kent, was being negotiated by salt-laden boats by at least the early tenth century. Its name, first mentioned as *grafon eah* in a charter of 812, means 'dug river', and offers a strong hint that navigation of the salty coastal marshes was already being artificially facilitated when Kent was ruled by the Mercian kings.[12]

---

[9] A. Nørgård Jørgensen, 'The Kanhave Canal on Samsø: New Investigations', *Château Gaillard: études de castellologie médiévale*, 18 (1998), at 156–7.

[10] Squatriti, 'Digging Ditches'.   [11] e.g. J. Story, *Carolingian Connections* (Aldershot, 2003).

[12] S 169; cf. V. Fenwick, *The Graveney Boat*, BAR British Ser. 53 (Oxford, 1978), 179.

Explicit documentary references only come with the turn of the millennium, but then start to accumulate. In 1016 the Danes, attacking London, dug a big ditch ('ane mycele dic') on the south bank and dragged their ships around to the west side of London bridge:[13] a feat which—whether or not it reflects a tradition going back to the Kanhave canal of nearly three centuries earlier—shows that the rapid bypassing of a major river was within eleventh-century capabilities. Another episode of canal-digging in stressful conditions can be glimpsed vividly in 1038–40 in the narrative of a dispute between the communities of Christ Church and St Augustine's, Canterbury, over control of the port and fishery at Sandwich. Defeated hands down, the St Augustine's party retaliated by trying to establish a rival concern at Ebbsfleet, on their own side of the Wansum Channel:

Then Abbot Ælfstan turned up with a big gang of men, and had a big ditch (*an mycel gedelf*) dug at Ebbsfleet, with the intention that it should contain a ship-channel (*scipryne*) such as they had at Sandwich, but he was unable to achieve anything, for he who labours against the will of Christ labours in vain.[14]

More amicably, at Abingdon in the 1050s, the abbot cut a permanent bypass on a difficult stretch of the Thames for the benefit of Oxford traders (below, pp. 258, 266–8).

It is between the early twelfth and mid thirteenth centuries that references to the opening of Roman canals, the digging of new ones, and the improvement of rivers concentrate—unsurprisingly given the archival expansion of that period.[15] Many of these documented cases are monastic and especially Cistercian, though it is a grand royal project, Edward I's canalization of the Clwyd at Rhuddlan in 1277 (below, pp. 201–2), that marks the beginning of the end for English medieval canal-building. For whereas our lack of documents means that the level of activity in the eleventh century could, for all we know, have been as high as in the twelfth, the impression of a downturn after 1250 is certainly not a fiction of the sources. Except in the rather abnormal environment of the Somerset Levels, where small lodes were being built well after 1300 (Rippon, below pp. 217–19), construction projects tailed off well before the general fourteenth-century economic depression: it is here that Richard Holt's comments (above, p. 1) become apposite. As several authors in this book note, a cycle of decline set in: the diminishing attractiveness of water transport would have had an adverse effect on maintenance, and as the increasingly silted and weed-choked canals and improved streams gave way once more to natural drainage patterns, they would have become not merely uneconomic but impossible to negotiate.[16]

---

[13] ASC 'E' s.a. 1016 (p. 149).   [14] S 1467 (*RASC* 176–8).

[15] For details, below, Bond (pp. 188–202), Rippon (pp. 215–26).

[16] A good illustration of this is a presentment in 1360 that the stoppage of the river Ant (*Smallee*) (near Ludham, Norfolk) was by nobody's default because the river fell out of use during the plague and nothing was carried on it, so that it became weed-grown; and that it was not known who ought to clean it because nobody had cleaned it in living memory (*PWML* ii. 88–9). Cf. below pp. 106 and 176.

So much for the written evidence: it delineates the broad outlines of the phenomenon but is slight enough, especially for the earlier stages. Further knowledge will come mainly from the physical evidence to which much of this book is devoted, and to which the following comments are a brief introduction.

Artificial watercourses discussed here are of two main kinds: the feeder-stream (or closed-ended 'blind lode') which linked an important place or production centre to a navigable river; and the bypass or 'barge-gutter' which branched off and then rejoined a river to circumvent a difficult stretch or obstruction, enclosing an elongated oval island. Both of these, of course, occur in nature, as tributaries and braided channels, and piecemeal deepening and widening of existing streams may often have preceded and accompanied the digging of new ones (see Bond, below pp. 155–6). Sometimes navigation may have been an afterthought to the construction of other linear features, such as mill-leats and drains, or a consequence of the improved depth and control of water that some weirs afforded; thus it has been argued that the Cambridgeshire lodes are of late Anglo-Saxon origin and were primarily for drainage and water management, with transport perhaps 'a later, secondary function'.[17]

The physical sources, like the documentary ones, offer hints from the mid-Saxon period but solid evidence only from around 1000. At Old Windsor (Berks.) a ninth-century mill-leat 'over 6 m wide and 3–4 m deep', probably 'over 1100 m long and…dug across a bend in the Thames',[18] shows at least a technical capacity to bypass and canalize major rivers on a navigable scale. Predictably, it is in the Fens that a network of late Anglo-Saxon canals, extending the Romano-British system and developed by the wealthy monastic houses of the Benedictine reform, can most convincingly be proposed on topographical grounds (Bond, below pp. 180–2). The most impressive true canal excavated so far, however, is the blind lode linking Glastonbury to the River Brue, originally published by the Hollinrakes and discussed by them below on pp. 235–8, which produced radiocarbon dates in the late Anglo-Saxon period. At South Stoneham (Hants), on the Itchen above Southampton, Currie observes that the boundary-terms 'old Itchen' and 'new river' in a charter of 1012 had appeared since 990–2—when an earlier clause had described the same boundary without mentioning them—and identifies the 'new river' as a relict linear hollow (below, pp. 246–50). On the upper Thames a group of improved and artificial streams, also datable on charter-boundary and place-name evidence to the late Anglo-Saxon period, can now be set beside the documented Abingdon canal of 1052×1066 (Blair, below pp. 264–78).

---

[17] S. Oosthuizen, 'The Cambridgeshire Lodes', in T. Kirby and S. Oosthuizen (eds.), *An Atlas of Cambridgeshire and Huntingdonshire History* (Cambridge, 2000), section 32. Cf. Rippon, below, on how canals relate to other aquatic aspects of land management in the Severn region.

[18] G. G. Astill, *Historic Towns in Berkshire: An Archaeological Appraisal* (Reading, 1978), 70.

These examples show basic similarities. In their primary states they were relatively flat-bottomed, much wider than deep, and around 7 m wide.[19] The profiles suggest a maximum water depth of up to 2 m, though water-flows—the great imponderable factor throughout this discussion—must have varied greatly both geographically and seasonally, often only allowing a much shallower draft. Although the evidence is still very limited, we can perhaps start to recognize design norms in English tenth- to thirteenth-century canal construction.

How 'mycel' did 'an mycel gedelf' have to be if it was to meet contemporary freight needs? Figure 2 shows profiles of three canals to the same scale as three excavated ships, from the tenth-century Graveney Boat to the 'Blackfriars 3' shout of *c*.1400. These vessels were all in the width-range of *c*.3.5–4.4 m; Graveney and its successors, though not completely flat-bottomed like the shouts, had flattened profiles with shallow keels and would have required a draught of less than a metre.[20] Local trade must often have used logboats needing much less than this: the group from the Mersey near Warrington were typically 0.8–0.9 m wide, with a draft of only 0.3 m when loaded.[21] Other craft in general use that do not survive, such as rafts of logs or reed bundles, would have been equally slight. These factors suggest that freight vessels of the kind in normal coastal and estuarine use could—at least in the most favourable possible conditions—use the inland river and canal system of the time, albeit with much laborious negotiation of flash-locks and shallows, and occasional difficulties in passing one another. Certainly the smaller craft on which local trade and traffic are likely to have depended, such as punts and logboats, would have had no problems.

Who built and maintained the late Anglo-Saxon canals? Keeping major watercourses consistently open over long distances needed concerted activity by someone, as John Landers has recently described in relation to medieval Europe generally:

[T]he use of existing waterways for transportation, and especially their 'improvement', was prone to create serious conflicts with local communities who were likely to see 'their' watercourses primarily as fisheries and millstreams. Even purpose-built canals could be subject to obstruction and exactions by local landowners keen to seize on the chance of some additional income. The construction of commercially effective canal

---

[19] Below, pp. 237, 266, 272–7. Currie's earthwork at South Stoneham is 15–20 m wide (below, p. 248), but has never been excavated and could have been heavily altered. Rippon (below, p. 221) notes that medieval canals in the Somerset Levels are *c*.4–5 m wide in their present state.

[20] Fenwick *The Graveney Boat*; P. Marsden, *Ships of the Port of London: Twelfth to Seventeenth Centuries AD*, English Heritage Archaeological Rep. 5 (London, 1996); N. Nayling, *The Magor Pill Medieval Wreck*, CBA Research Rep. 115 (York, 1998). The shallow draught of the Magor Pill boat was *c*.60 cm (Nayling, *The Magor Pill Medieval Wreck*, 142); the draught of Blackfriars 3 with a safe load was *c*.40 cm. (Marsden, *Ships of the Port of London*, 97). See below, Langdon pp. 122–6 and Blair p. 284, for further discussion.

[21] S. McGrail, *Logboats of England and Wales*, BAR British Ser. 51 (Oxford, 1978), 287–98.

networks therefore required public authorities capable of protecting traffic as well as sufficient demand to justify the investment.[22]

Authorities and interests at all levels, from the national to the extremely local, might potentially have been involved. James Campbell has emphasized the importance of rivers to royal government and administration as the West Saxon kingdom took shape, and has noted the role of eleventh-century kings in protecting free passage along both the major roads and waterways and (by means of sheriffs and shire courts) the lesser ones.[23] Ann Cole's suggestion that *ēa-tūn* place names have a specific and functional meaning, denoting settlements responsible for servicing the passage along a river (below, pp. 78–82), encourages speculation that routes may not merely have been protected from obstruction, but actively improved through taxes similar to those which maintained forts and bridges. Duties like those of the Torksey men in the 1060s, who had to provide boats and boatmen to carry royal messengers down the Trent and then (presumably) up the Ouse to York,[24] might have been extended to the upkeep of minor waterways and the digging of new ones.

It does, however, seem likely that improvements in waterways were often initiated by their local users, or by owners of estates through which they passed. It is notable that some of the canals described in this book were on great estates with high-ranking proprietors, whether royal (Bampton and South Stoneham), Benedictine (Glastonbury and Abingdon), or Cistercian (Faringdon). Thereafter, either landlords, tenants, or river-users could have been responsible for maintaining them. A hint of the third arrangement occurs later in relation to Faversham creek (Kent), where shipowners using this inlet were responsible for finding men with iron rakes and shovels to keep it clear.[25] Presumably the smaller-scale the waterway the more local the interests involved: on inland systems one could well imagine logboat operators and smallholders taking their turns with a shovel.

As these possibilities illustrate, it may in fact be wrong to place too exclusive an emphasis on 'public authorities capable of protecting traffic'. Several studies in this book describe intensely local transport systems (the Fens; the Severn and Somerset Levels; the upper Thames; the trans-Pennine routes) which were

---

[22] J. Landers, *The Field and the Forge* (Oxford, 2003), 85.

[23] J. Campbell, 'Power and Authority 600–1300', in D. M. Palliser (ed.), *The Cambridge Urban History of Britain, I: 600–1540* (Cambridge, 2000), 51–78, at 53–5; id., *The Anglo-Saxon State* (London, 2000), 183. The main source is retrospective, the so-called 'Leges Edwardi Confessoris' of the 1130s: the king's protection extends directly over major rivers 'quorum navigio de diversis locis victualia deferuntur civitatibus sive burgis', whereas the shire and sheriff have responsibility 'de aquis minoribus, naves ferentibus cum eis que necessaria sunt civitatibus et burgis, scilicet ligna et cetera' (cc. 12, 12.11; ed. F. Liebermann, *Die Gesetze der Angelsachsen*, i (Halle, 1903), 637–40). The principle occurs in Domesday Book in Nottinghamshire, where anyone who 'impedierit transitum navium' on the Trent is to be fined £8 (DB i. 280). The strategic importance of rivers for burghal places, and thus for national defence, has often been noted.

[24] DB i. 337.

[25] E. Jacob, *The History of the Town and Port of Faversham* (London, 1774), 64–5. A 10-ton vessel provided a man working for six days every year, a smaller vessel one working for three days.

integrated within themselves and depended on intensive maintenance and expertise, but were linked to each other in ways that look rather less stable and more haphazard. If royal functionaries knew exactly whose job it was to move them around the kingdom, and premier suppliers of bulk goods used established links between the major arteries, ordinary traders and carriers are surely more likely to have fitted together permutations of local routes according to such variables as the season, the rainfall, and the attitudes of landlords and mill-owners.[26] As we turn to the wider picture, it is important to remember that there was probably no single transport system, but overlapping systems—national with local, local with local—distinguished by practices that were determined more socially and culturally than geographically.

Mills, Fish-Traps, and Bridges

There was a symbiotic relationship—normally contentious, occasionally creative—between the uses of flowing water for moving boats, for powering mills, and for harvesting fish. The last two endeavours, which are so central to the themes of this book that they justify a chapter (Hooke) devoted mainly to them, involved building barriers across the line of flow: to raise the water level high enough to drive mill-wheels, to attach nets and traps, and often to do both. Any dam was necessarily associated with a weir (on either the main channel or a bypass) for controlled release of the water, and this could take the form of a 'flash-lock' allowing craft to pass through (Fig. 3). Even if some weirs increased the draught on shallow rivers, there was an inherent conflict between people wanting to move along a river and people wanting to obstruct it; after 1000, as more and more dams and weirs were built, this conflict must have intensified.[27] The need to keep major waterways free of mills and fish-weirs was officially recognized by the early twelfth century,[28] and flooding blamed on excessively high mill-dams provoked countless lawsuits in the thirteenth.[29] On the other hand there were obvious similarities between mill-leats and navigable bypasses

---

[26] I owe this point to John Langdon, whose ideas have helped to clarify my own here.

[27] J. Langdon, 'Inland Water Transport in Medieval England: A Response to Jones', *JHG* 26 (2000), at 79, argues that there was 'a relatively small increase *at best* in water-mill numbers overall from 1086 to the beginning of the fourteenth century, but a much more substantial increase in mill weirs or dams and a notable movement of mills onto more major watercourses'. The standard work is now J. Langdon, *Mills in the Medieval Economy: England, 1300–1540* (Oxford, 2004).

[28] 'Leges Edwardi Confessoris', *c.* 12.8 (ed. Liebermann, 639); see R. H. C. Davis, 'The Ford, the River and the City', *Oxoniensia*, 38 (1973), 258–67 (reprinted in his *From Alfred the Great Stephen* (London, 1991), 264–5 for later royal pronouncements.

[29] See Blair, below pp. 263, 286–93. During the thirteenth and fourteenth centuries, legal redress for infringement of water-rights was essentially confined to nuisance actions alleging injury to property: see J. Getzler, *A History of Water Rights at Common Law* (Oxford, 2004), 46–116. For Bracton, flowing water was one of 'the things which are naturally everybody's' and which it was in the king's jurisdiction to maintain; the common law of individual economic rights in these public assets remained undeveloped (cf. Ibid. 67–9, 107, 110, 179). We badly need a more systematic study of 13th-century lawsuits over dams and weirs—recorded in abundance in the central legal records—focusing on evidence for the form of locks, and on the extent to which such suits were prompted by the obstruction of traffic as well as the flooding of land.

Fig. 3. A typical flash-lock: Hart's Weir on the upper Thames. In the upper picture the operator has just pulled up some of the 'paddles', allowing the punt to rush through on the 'flash' of suddenly released water. Details of the 'paddles' are shown in the lower picture. These Victorian engravings show the end of a technology that had probably remained largely unchanged for a thousand years. (Mr and Mrs S. C. Hall, *The Book of the Thames* (new edn. London, [*c*.1880]), 50–1.)

(leats could be of navigable proportions, as the early case at Old Windsor illustrates), and the spread of leat-building technology may have stimulated its application for other purposes.

Fishing and fish-weirs figure prominently in the late Anglo-Saxon sources; the primary reason for some bypass channels on major rivers was probably to maintain a breeding population by allowing enough migratory fish to circumvent the weirs and traps (Hooke and Currie, below pp. 53, 250). On the Severn the conflict between fishing and navigation was partly overcome by digging 'barge-gutters', through which boats could pass as well as salmon and eels.[30] This analysis—that the cutting of new channels was a reaction to blocking of the old ones—can be expressed more positively by saying that the weirs and mill-dams stimulated a dynamic approach to the engineering of waterways. At all events it explains what may be termed the *eald-ea* ('old-river') effect, seen on the Itchen above Southampton and the Thames and Cherwell above Oxford: the process by which the main course of a river is temporarily reduced to being the weir-encumbered backwater of a man-made navigation channel, but eventually recovers its status when the bypass is neglected and natural drainage patterns reassert themselves (Currie and Blair, below pp. 250, 266–8).

The investment that was now helping road transport to compete more strongly with waterways created another class of barriers across rivers. During the twelfth and thirteenth centuries, many fords and timber bridges were replaced by masonry arches and solid causeways (see Blair, below p. 262). If the effects were occasionally beneficial to river traffic, by encouraging a faster and deeper flow through the arches, they much more frequently limited vessel size and encouraged the formation of silty, static pools. A bridge built soon after the Conquest on the Adur at Bramber (Sussex) was quickly made an excuse for levying tolls and encouraging the offloading of ships there, to the detriment of Steyning minster's harbour further upstream; despite assurances in 1103 that 'the bridge will be constructed in such a way that ships can freely pass under it upstream and downstream', its continued presence meant the end of Steyning as a significant inland port.[31]

This case exemplifies an awareness of the specific problem, and a concern to mitigate it, which were still too short-term and localized to halt the progressively detrimental effect on the water transport system as a whole. The building of bridges and mills (like more recent engineering works: Rhodes, below pp. 143–8) were piecemeal activities, carried out with no thought to their broad environmental and economic consequences or to long-term change. If the improvement of navigable waterways had been purposeful and dynamic, the causes of their attrition were too fragmented and incoherent to admit any effective response.

Waterways in Trade and Transport

In considering whether there really was more water-borne transport in the earlier than the later middle ages, we can start by looking backwards from a

---

[30] D. J. Pannett, 'Fish Weirs of the River Severn with Particular Reference to Shropshire', in M. Aston (ed.), *Medieval Fish, Fisheries and Fishponds in England*, BAR British Ser. 182 (2 vols., Oxford, 1988), 371; Blair, below p. 270; cf. C. R. Salisbury, 'Primitive British Fishweirs', in G. L. Good, R. H. Jones, and M. W. Ponsford (eds.), *Waterfront Archaeology: Proceedings of the Third International Conference on Waterfront Archaeology*, CBA Research Rep. 74 (London, 1991), 76–87.

[31] *English Lawsuits from William I to Richard I*, ed. R. C. Van Caenegem, i, Selden Soc. 106 (London, 1990), 129, 132; E. W. Holden, 'New Evidence Relating to Bramber Bridge', *SxAC* 113 (1975), 104–17.

debate conducted in the *Journal of Historical Geography* during the 1990s. The initial article, by J. F. Edwards and B. P. Hindle, stated a 'maximalist' case for the use of major rivers, which John Langdon then countered with a more nuanced and cautious model derived mainly from royal purveyance accounts between 1294 and 1348.[32] Langdon drew attention to strong seasonal patterns in the use of the Thames: 'it appears that only in winter, when water levels were likely to be high, was it considered a safe option to use the river. It was also then that roads were most difficult.' He also observed that, given how much easier it must have been to go with the flow than to haul loaded boats upriver,[33] bulk water freight is likely to have been largely confined to moving goods from the interior of the country towards the estuaries and coasts. In conclusion he stressed both that there may have been a big difference between a *maximum possible* head of navigation and a *regularly effective* one, and that improvement of land transport during the twelfth and thirteenth centuries is likely to have reduced the cost differential,[34] and hence the advantages of water: 'Coupled with the increasing demands made upon rivers by mill and fishing weirs and problems like silting, this seems to have contributed to a slow degradation of the inland waterway system, which contemporaries were quick to recognize but did little about.'[35]

Edwards and Hindle made a perfunctory response[36] but then let the debate rest: it was left to Evan Jones, in 2000, to attempt mediation between their position and that of Langdon.[37] He did so mainly by developing Langdon's point about change over time, again stressing degradation in the extent and quality of waterways across a chronological span that Edwards and Hindle had treated as homogeneous: 'Edwards's most convincing evidence for extended navigation comes from the eleventh–thirteenth century, while Langdon's study is derived from fourteenth-century accounts, by which time the extent of navigation had already begun to decline.'[38]

There is thus a consensus that a more extensive, better-managed, and better-quality system of waterways is likely to have existed in the earlier than the later

---

[32] J. F. Edwards and B. P. Hindle, 'The Transportation System of Medieval England and Wales', *JHG* 17 (1991), 123–34; J. Langdon, 'Inland Water Transport in Medieval England', *JHG* 19 (1993), 1–11; he makes further use of these documents in his chapter below.

[33] This point is also made by A. C. Leighton, *Transport and Communication in Early Medieval Europe, AD 500–1100* (Newton Abbot, 1972), 126: in recent times flatboats could get down the Rhône from Lyon to Avignon in two to five days, whereas the same journey upstream took nearly a month and needed several animals to tow.

[34] J. Masschaele, 'Transport Costs in Medieval England', *EcHR* 46 (1993), at 270–3 calculates that freight along waterways in the later middle ages was often only half the cost of road transport, and could be cheaper still around the coasts.

[35] Langdon, 'Inland Water Transport', 6.

[36] J. F. Edwards and B. P. Hindle, 'Comment: Inland Water Transportation in Medieval England', *JHG* 19 (1993), 12–14.

[37] E. Jones, 'River Navigation in Medieval England', *JHG* 26 (2000), 60–75 (with an essentially supportive response by Langdon, 'Inland Water Transport in Medieval England: The View from the Mills').

[38] Jones, 'River Navigation', 72.

period: a consensus that can now incorporate the new evidence for canals, concentrated as it is in the eleventh to early thirteenth centuries. Our perceptions of this cycle of growth and decline must, however, be sensitive to other aspects of social and topographical change. The period with the better infrastructure of waterways was also the period of less developed contrasts between nodal points in the transport system, and of smaller-scale and more localized (though not always less intensive) activities. Before the evolution of sites of commerce and industry into a hierarchy of large and small towns during the twelfth and thirteenth centuries,[39] the multitudes of small hythes on the coasts, estuaries, and rivers probably had a greater relative importance (Gardiner, below). Boatmen in small-scale communities can make remarkably creative and flexible use of resources that developed economies reject as too limited, taking every chance to use small and even intermittent streams, and portaging where necessary.[40] Places at confluences of major with minor watercourses, or at intersections of watercourses with roads, could have been closely managed and maintained, with facilities for trans-shipment and portage: it was some of these intersection sites that became known as *ēa-tūns* (Cole, below, pp. 78–82). The 'break points' between different scales of commercial traffic, a notable feature of the English river system by the later middle ages (Langdon, below pp. 120–1), may have been correspondingly less important earlier.

This more localized world could have provided wider scope for small operators in navigation and freight, but this need not mean that it constrained the big ones. As Andrew Sherratt showed from archaeological evidence, rivers (rather than land routes such as 'ridgeways') were the major arteries of long-distance bulk trade in Britain by late prehistory:[41] the broad size and shape of early medieval river-traffic may have more to do with this largely undocumented *longue durée* than with the late medieval conditions delineated in such fine detail by archive sources. Ninth-century travellers, and even armies, could penetrate from the east coast to Tamworth and Repton—deep in the central midlands—along the Trent and Tame.[42] Rather than limiting long-distance operations, the intensely local nature of many riverine activities may have underpinned them, fostering a climate in which, as the economy grew during the tenth and eleventh centuries, improvement of waterways was a common goal.

Pottery was only one among many consumer products in local and regional trade, but it is a crucial one for us, since archaeologists have retrieved it in bulk and studied it intensively. One of the latest studies comes to the remarkable conclusion that 'the distances over which pottery was carried ... were actually as high or higher in the Middle to Late Anglo-Saxon Period as in the 13th

[39] M. F. Gardiner, 'Shipping and Trade between England and the Continent during the Eleventh Century', *Anglo-Norman Studies*, 22 (2000), 71–93, at 84–7.

[40] As observed by Leighton, *Transport and Communication*, 125.

[41] A. Sherratt, 'Why Wessex? The Avon Route and River Transport in Later British Prehistory', *Oxford Journal of Archaeology*, 15 (1996), 211–34.

[42] Cf. P. Sawyer, *Anglo-Saxon Lincolnshire* (Lincoln, 1998), 18, 197, and n. 23 above.

to 14th centuries'.[43] There is now abundant evidence that consignments of pots moved along the major English rivers in the ninth to twelfth centuries, especially *up*river from the Humber and Wash systems.[44] Far from operating in isolation from the trade in bulk goods, this trade in pottery was probably the visible half of a symbiotic process. As already noted, high-bulk, heavyweight loads (grain, wool, cloth, minerals, Droitwich salt) largely moved *down*river, from the interior of England to the coasts. If they were floated on log rafts, these could have been broken up and sold at the destination.[45] But properly built boats (such as Graveney or the later shouts) would have had to be returned upriver, either empty or with much lighter loads, and it was surely here that the pottery-sellers and shipowners could make deals to mutual advantage. A comparable rhythm can be identified in the 1050s on the Thames, where Oxford citizens shipping loads down to London returned upriver with smoked or salted herrings (Blair, below p. 259). There is thus some likelihood that regions to which pottery and other lightweight goods were regularly sent were also those from which bulk goods regularly came.

How accurately, then, can we reconstruct the geography of inland water transport in tenth- to twelfth-century England? Sherratt's analysis of river routes in late prehistoric south Britain identifies two broad systems—one focused on the Hampshire Avon and Severn, the other on the Thames—which at various stages competed for supremacy.[46] By our period the greater development of eastern England, and its orientation towards France, the Low Countries, and Scandinavia, had ensured the clear dominance of the Thames system, together with one centred on the Humber and Wash. The population recorded in Domesday Book (1086), the pattern of coin minting and circulation, and the distribution of metal small finds,[47] all point to more intensive economic activity in Yorkshire, Lincolnshire, and the south-east than in regions further west. This is the context for a pattern which was dominated by river

---

[43] A. Vince, 'Ceramic Petrology and the Study of Anglo-Saxon and Later Medieval Ceramics', *Medieval Archaeology*, 49 (2005), 219–45, at 219.

[44] Pottery is more likely to survive intact when carried by water than when carried by road: the growth of the Staffordshire potteries in the late 18th century was stimulated by canal-building.

[45] Cf. Leighton, *Transport and Communication*, 126, 134. A possible instance of two-way traffic is suggested at Kelvedon (Essex) in 1294, when tenants of Westminster Abbey owed a customary payment called 'ship-hire' (*schipur*) in lieu of carrying quantities of grain and malt from Kelvedon to Salcott, Heybridge or Maldon, presumably down-river along the Blackwater (Cambridge University Library, MS Kk. 5.29, fos. 114v, 115v; I owe this reference to Barbara Harvey). The estuary to which this grain was delivered—and whence it would have been shipped around the coast to Westminister in larger vessels—was well-known for producing coastal salt, and one wonders whether the grain boats carried loads of salt when they returned up-river.

[46] Sherratt, 'Why Wessex?'.

[47] D. Hill, *An Atlas of Anglo-Saxon England* (Oxford, 1981), maps 26 and 222, for population and mints; website of the Early Medieval Corpus of Coin Finds (**www.fitzmuseum.cam.ac.uk/coins/emc/ emc_notes.html**) for coins; website of the Portable Antiquities Scheme (**www.findsdatabase.org.uk/hms/ home.php?publiclogin=1**) for small finds. The only anomaly is that Yorkshire has a low recorded Domesday population.

routes to and from the east coast. It must at the same time be acknowledged that small-scale consumer durables need not adequately reflect the movement of bulk goods in the midlands and west (which may have been productive of raw materials despite their poorer material culture), or links to maritime trade in the Irish Sea region.

At present our best archaeological indicator is the diffusion of pottery from the Lincolnshire kilns at Stamford and Torksey, as studied nationally by Kilmurry and locally by Symonds.[48] Symonds's fine-grained analysis is especially useful in demonstrating a contrast between local distribution patterns, for which Roman roads were more important than rivers, and inter-regional transport in which waterways predominated.[49] The main long-distance axis for Torksey Ware was along the Trent, both downriver to the Humber and upriver via Newark into Derbyshire; the Witham, the Foss Dyke, and the Car Dyke were also regularly used for Torksey and Lincoln pottery.[50] Stamford Ware made heavy use of the Trent, Witham, and Wash systems, again probably as one element in a more diverse trade. Thus in the 1110s stone from Barnack—next to Stamford—was freighted down the Nene to its confluence with the Ouse, up the Ouse for more than 32 km to its tributary the Lark, then up the Lark for another 32 km to Bury St Edmunds;[51] Stamford Ware followed the same Nene–Ouse route, getting as far as Cambridge, Huntingdon, and Bedford. This kind of evidence again suggests a flexible system, which used roads for local journeys (and therefore smaller consignments?), but took advantage of the easier long-distance carrying capacity of water, and co-operation with transporters of other goods, for trade outside the region.

Figure 4 plots two categories of data discussed in this book—place names implying river traffic (after Cole, fig. 19) and average boat-loads recorded in purveyance accounts (after Langdon, fig. 23)—in relation to zones of extensive monetary circulation during 950–1180,[52] and selected Roman roads. The purveyance data are late and are biased towards larger loads and areas of royal activity, but they are probably reliable in highlighting the pre-eminent

---

[48] K. Kilmurry, *The Pottery Industry of Stamford, Lincolnshire, c. AD 850–1250*, BAR British Ser. 84 (Oxford, 1980); L. A. Symonds, *Landscape and Social Practice: The Production and Consumption of Pottery in 10th-Century Lincolnshire*, BAR British Ser. 345 (Oxford, 2003). Kilmurry's map as redrawn by S. E. James (D. A. Hinton, *Archaeology, Economy and Society* (London, 1990), fig. 6.4) has been used in compiling the present fig. 5.

[49] Symonds, *Landscape and Social Practice*, 135, 161, 168, 224.

[50] Ibid. 135, 140–2, 149, 161.

[51] This can be inferred from two writs of Henry I (C. Johnson and H. A. Cronne (eds.), *Regesta Regum Anglo-Normannorum; ii: 1100–1135* (Oxford, 1956), nos. 694 and 1410), protecting Ramsey and Bury abbeys from attempts by the monks of Peterborough to levy tolls on the stone as it passed along the Nene. I am grateful to Richard Sharpe for access to his forthcoming annotated edition.

[52] This is based on a map of single coin finds generated from the Early Medieval Corpus website (see n. 47). Although there are probably biases in recovery, the number of finds and the density of the clusters should by now give something approaching a reliable view of the zones in which money was in regular and widespread circulation.

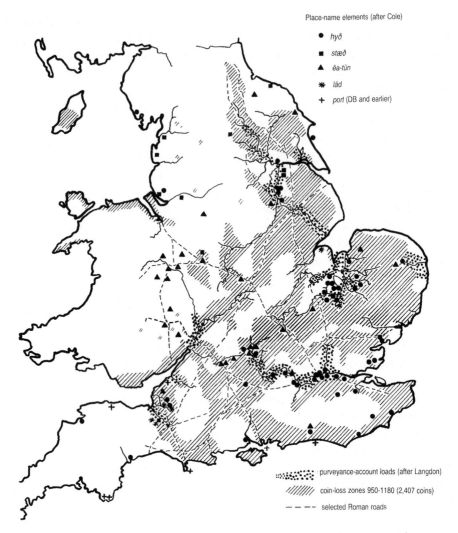

Place-name elements (after Cole)

● *hyð*

■ *stæð*

▲ *ēa-tūn*

✳ *lād*

+ *port* (DB and earlier)

············ purveyance-account loads (after Langdon)

////// coin-loss zones 950-1180 (2,407 coins)

— — — selected Roman roads

Fig. 4. Some evidence for inland water transport, tenth to thirteenth centuries. Place names (after Cole, below) and purveyance account loads (after Langdon, below) are plotted in relation to zones producing single finds of coins lost during 950–1180 and selected Roman roads.

importance of the Thames, the Fenland system, and the Ouse. The place names in *hȳð*, *stæð*, *ēa-tūn*, and *lād* are exempt from these biases but concentrate on precisely the same river routes. They also, however, hint at additional and more westerly zones of activity, such as the *ēa-tūns* in Shropshire and Staffordshire suggesting a system focused on the uppermost Severn (Cole, below p. 81), and the *stæðs* along the Lancashire coast. The coin distributions, which presumably bear at least a broad relationship to the intensity of local commercial activities, suggest two belts of monetary activity following the grain of central and southern England on a north-east to south-west axis: from Lincolnshire and the east midlands to the Warwickshire Avon and Severn; from the Wash and Fens to the upper Thames (perhaps via the Ouse, Ray, and Cherwell: see Blair, below pp. 269–270) and thence to the south coast of Dorset. As well as these lines of contact across England, the coins and place names highlight the 'trans-isthmian' route—a long-standing alternative to circumnavigating Devon and Corn-wall—from the Severn estuary via the Parrett and a Roman road to the south coast of Dorset.[53] In this picture the water routes look more consistently impor-tant over long distances than do the Roman roads, even if the latter were often locally important in the ways which Symonds has recognized in Lincolnshire.

Figure 5 is an attempt—perhaps rash, certainly broad-brush and specula-tive—to interpret the early medieval inland transport system in the light of the data in Fig. 4. Important rivers are shown as thick arrows, gradated according to hypothetical carrying capacity; the thin arrows give a schematic impression of minor routes linking them, whether by river, canal, or road. Also shown are the distribution pattern of pottery from Stamford, which followed the riverine system (and the coin distributions) closely; and the salt routes from Droitwich,[54] several of which supplied places en route to the upper Thames and shipment downriver. This map proposes a model in which the Severn and the Wash/Humber systems were regularly linked to each other along the line of the coin distributions, and were linked in turn to the Thames conduit towards London and the Continent: a model which therefore gives a special nodal importance to the uppermost Thames and Cotswold regions.

Even if this is a viable reconstruction in broad terms it may still be an incomplete one, especially in under-representing activity in the west. Here Edmonds's chapter is an important corrective and pointer to future work, revealing as it does a largely undocumented system of routes from the Irish Sea, up the main estuaries to roads leading over the Pennines and across into Yorkshire. Her argument that incentives, if they were strong enough, would have sustained traffic even in arduous conditions is surely to be preferred to a deterministic rejection (encouraged by absence of evidence rather than negative evidence) of trading systems based on the west coast. The recent unexpected discovery of a major Viking-age trading settlement at Llanbedrgoch, Anglesey, shows how little the written sources really tell us about western trade; there could well be more such sites along the Welsh coast.[55] Again, recent work on the so-called 'Chester Ware' raises the likelihood that its dissemination from Stafford utilized downriver transport on the upper reaches of the Severn and

[53] Sherratt, 'Why Wessex?', 212–13. Fresh sea-fish were moved up the Parrett to Langport, where bones recovered from 12th- to 14th-century contexts included, alongside the riverine eel, a range of maritime species of which some were unsuitable for salting: E. Grant, 'Marine and River Fishing in Medieval Somerset: Fishbone Evidence from Langport', in Aston (ed.), *Medieval Fish, Fisheries and Fishponds*, 409–16.

[54] After D. Hooke, 'The Droitwich Salt Industry', *Anglo-Saxon Studies in Archaeology and History*, 2 (1981), 123–69.

[55] M. Redknap, *Vikings in Wales: An Archaeological Quest* (Cardiff, 2000), 61–84.

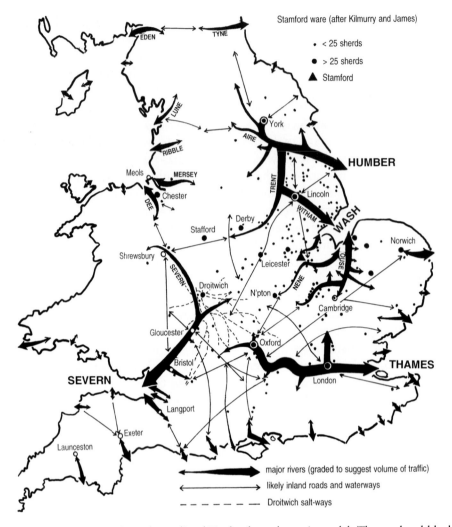

Fig. 5. Water transport in early medieval England: a schematic model. The gradated black arrows express an (obviously highly impressionistic) 'best guess', in the light of the various available indicators, at the relative volume of traffic along major waterways; the thin-line arrows suggest road and water links between trans-shipment points. The map also shows the distribution of excavated Stamford Ware (after Kilmurry) and Droitwich salt routes (after Hooke).

Trent and on the Dee.[56] For the study of traffic and transport in and around England before 1250, this book is a starting point rather than a summing-up: if it encourages others to explore further by showing how much we still do not know, it will have served its purpose.

[56] This conclusion is drawn from the map in ibid. 64.

PART I

WATERWAYS, GEOGRAPHY, AND ECONOMY

1

Barrier or Unifying Feature? Defining the Nature of Early Medieval Water Transport in the North-West

FIONA EDMONDS

Water both surrounds and divides north-western England (Fig. 6). The Irish Sea buffets the area's coast, and fast-flowing rivers, which run down from the Pennine watershed, break up the region's landscape. The question is whether water isolated this area or opened it up to outside influences. The Irish Sea is often characterized as a 'unifying feature', but rough weather frequently hinders maritime navigation. River valleys provide ways through upland areas of the north-west, but maps of major navigable rivers include few from the region. Indeed, unlike the southern and eastern English areas covered elsewhere in this volume, it is hard to postulate navigation beyond the tidal limit on north-western rivers. The same geographical factors which limited riverine navigation also rendered the rivers less susceptible to artificial improvements. However, the naturally navigable sections of the rivers can be viewed as an extension of the estuarine and marine water transport resources which were used by travellers during the early medieval period.

The area which will be considered in this chapter lies between the Mersey estuary and the Solway Firth. In the early medieval period this was not a coherent political entity; the southern part of the area has recently been described as a 'frontier landscape'.[1] All or parts of the region became incorporated into several kingdoms and territories—Northumbria, various areas of Scandinavian settlement, Strathclyde, the kingdom of the Scots, and eventually the Anglo-Norman realm. The area can appear peripheral when viewed in the context of kingdoms whose heartlands lay in the east of Britain, but the north-west has not often played a prominent role in discussions of the 'Irish Sea province'

I am grateful to Dr John Blair, Professor Thomas Charles-Edwards, and Dr David Griffiths for their comments on an earlier version of this chapter, and to Dr Andrew Bell for supplying some references. This piece was written during my tenure of an AHRB postgraduate studentship.

[1] N. J. Higham, *A Frontier Landscape: The North West in the Middle Ages* (Macclesfield, 2004).

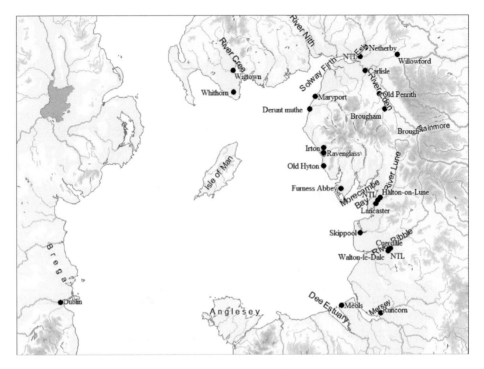

Fig. 6. Map of sites mentioned in the text. (Physical data © Collins Bartholomew.)

either.[2] The patchy nature of the source material available for the region during the early medieval period has no doubt contributed to this situation.[3] In the present chapter, many of the diverse categories of evidence employed elsewhere in this book, along with secondary works about the region,[4] are synthesized. From this evidence a twofold argument is tentatively constructed. First, a brief review of the geographical context and the Roman legacy compels the conclusion that it was possible to use some of the region's maritime and estuarine harbours during the early medieval period. Secondly, a review of historical events in the area suggests that various groups of people had reasons to exploit these water transport resources. Their journeys can be broadly divided into medium-distance voyages across the Irish Sea and localized movements around the coast and upriver.

---

[2] For example, the area enjoyed few mentions in D. Moore (ed.), *The Irish Sea Province in Archaeology and History* (Cardiff, 1969). However, the articles in J. Graham-Campbell (ed.), *Viking Treasure from the North-West: The Cuerdale Hoard in its Context* (Liverpool, 1992) set the north-west in its Irish Sea context.

[3] For a discussion of the challenges created by the source material see R. M. Newman, 'The Dark Ages', in R. Newman (ed.), *The Archaeology of Lancashire: Present State and Future Priorities* (Lancaster, 1996), 93–107.

[4] e.g. N. Higham, *The Northern Counties to AD 1000* (London, 1986); D. Kenyon, *The Origins of Lancashire* (Manchester, 1991).

The Landscape and the Roman Legacy

Whilst it is generally recognized that transport on the Irish Sea would have been limited by seasonal weather conditions, a pattern of maritime trade and cultural contacts between Ireland, Britain, and the Isle of Man is being reconstructed.[5] A recent book has gone so far as to link Irish Sea culture to an Atlantic-wide civilization in which 'geomorphology conspired to create a broad zone dependent on the sea'.[6] Yet the north-west's coastal landscape may have served to isolate the region from such maritime influences. Navigation on the Irish Sea is governed by tidal fluctuations, the prevailing westerly or south-westerly winds, and the availability of harbours. On the first count the north-west's coast seems particularly dangerous: the tidal range along the Lancashire coast is four to six metres greater than that of south-eastern Ireland. Thus rocks and sandbanks quickly become submerged and create hazards for sailors who are not acquainted with the locality.[7] A number of the region's rivers experience tidal bores, which are notorious for endangering travellers walking across the estuaries. Further inland, the surges of water that run off the Pennines during bad weather can be perilous for those hoping to travel by river.[8]

Even so, skilled navigators could use the swell of normal tides to propel them upriver, and similar tactics would have aided the return journey. There is some uncertainty about the nature of the relative sea level during the first millennium AD, and so the locations of contemporary tidal limits on the rivers are hard to ascertain. A survey of locations of Roman structures along the Wirral coastline and the Mersey estuary, amongst other coastal areas, has indicated that the relative sea level was considerably lower during Roman times than it is today.[9] On the other hand, a marine transgression detected in Lancashire may date either to the Roman period or to a later point in the first millennium.[10]

[5] J. M. Wooding, *Communication and Commerce along the Western Sealanes AD 400–800*, BAR Internat. Ser. 654 (Oxford, 1996) challenges the notion that long-distance traffic operated continuously in the Irish Sea region. For regional links in a later period see B. Hudson, 'The Changing Economy of the Irish Sea Province AD 900–1300', in B. Smith (ed.), *Britain and Ireland: Insular Responses to Medieval European Change* (Cambridge, 1999), 39–66. D. Griffiths's work has placed an important north-western site in the context of wider debates about Irish Sea communication: see below, p. 32 and n. 72.

[6] B. Cunliffe, *Facing the Ocean: The Atlantic and its Peoples 8000 BC–AD 1500* (Oxford, 2001), 542, 554.

[7] R. Buchanan, 'The Irish Sea: The Geographical Framework', in M. McCaughan and J. Appleby (eds.), *The Irish Sea: Aspects of Maritime History* (Belfast, 1989), 3.

[8] Kenyon, *Origins of Lancashire*, 12–13. For a survey of the geology of the coastline see J. A. Steers, *The Coastline of England and Wales*, 2nd edn (Cambridge, 1964), 70–111.

[9] A. C. Waddelove and E. Waddelove, 'Archaeology and Research into Sea-Level during the Roman Era: Towards a Methodology Based on Highest Astronomical Tide', *Britannia*, 21 (1990), 253–66.

[10] M. J. Tooley, 'Theories of Coastal Change in North-West England', in F. H. Thompson (ed.), *Archaeology and Coastal Change* (London, 1980), 84–6; G. D. B. Jones, 'Archaeology and Coastal Change in the North-West', ibid. 87–102; C. Wells, 'Environmental Changes in Roman North-West England: A Synoptic Overview of Events North of the Ribble', *Transactions of the Cumberland and Westmorland Antiquarian and Archaeological Society*, 3 (2003), 75–8.

In certain cases, there are indications that the tidal limit in the first millennium was similar to, or located further upriver than, its modern-day counterpart. Thus, at Walton-le-Dale, near Preston, a north–south Roman road crossed the River Ribble just below the modern-day Normal Tidal Limit (NTL) at River Side Farm, NGR SD 564 284.[11] Recent excavations at Walton-le-Dale have revealed the remnants of a settlement which was geared towards production, storage, and probably trans-shipment during the second and third centuries AD, and possibly later.[12] During the early modern period, the tide seems to have stretched further up the Ribble, although the notion that boats could reach Ribchester in antiquity remains controversial.[13] In the case of the River Esk, which flows into the Solway, Roman vessels seem to have travelled several miles upriver from the modern NTL (NGR NY 355 649) to the fort at Netherby. In the late sixteenth and early seventeenth centuries, antiquaries remarked upon the relics of a port which had existed by Netherby's Roman buildings.[14]

References to the Solway Firth in early modern documents indicate that the firth used to extend 2 miles further inland than it does today.[15] These indications may help to substantiate the notion that Carlisle sustained a Roman port, and it has been noted that barges were able to reach the city during the thirteenth century.[16] More localized processes of coastal change have also affected the degree to which the rivers and estuaries of the Solway and Galloway coasts were navigable. Wigtown's medieval port was situated at the mouth of the River Bladenoch (Dumfries and Galloway), but this harbour has since disappeared because the river's course shifted.[17] The processes of encroachment on and reclamation of mosses and marshland which were beginning in some areas of the north-west at the end of the first millennium AD have also greatly modified the north-west's estuarine and riverine environments.[18]

Nevertheless, the evidence cited so far suggests that the lower stretches of some rivers would have been just as navigable in the first millennium AD as they are now, if not more so. The notion that vessels passed far beyond the NTLs

[11] R. H. Critchley, 'The Lower Douglas and Ribble Navigation', *Ribble Archaeology*, 6 (1973) 20–2.

[12] R. M. Newman and A. Olivier, 'Walton-le-Dale: Proposals for the Publication of Excavations 1981–1983': **www.eng-h.gov.uk/archcom/projects/summarys/html98_9/cc2353.htm#top**, consulted 9 Nov. 2005. K. Buxton and D. Shotter, 'The Roman Period', in Newman (ed.), *Archaeology of Lancashire*, 78–81.

[13] D. Shotter, 'Introduction: The Study of Lancashire's Past', in Newman (ed.), *The Archaeology of Lancashire*, 6.

[14] E. Birley, 'The Roman Fort at Netherby', *Transactions of the Cumberland and Westmorland Antiquarian and Archaeological Society*, 53 (1953), 6–8, 28–9.

[15] G. Neilson, 'Annals of the Solway until AD 1307', *Transactions of the Glasgow Archaeological Society*, NS 3 (1899), 256–9.

[16] N. Higham and B. Jones, *The Carvetii* (Gloucester, 1985), 44; Higham, *Northern Counties*, 220; J. D. Anderson, *Roman Military Supply in North-East England: An Analysis of and an Alternative to the Piercebridge Formula*, BAR British Ser. 224 (Oxford, 1992), 68–9; C. Phythian-Adams, *Land of the Cumbrians* (Alde, shot, 1996), 13.

[17] A. Graham, 'Some Old Harbours in Wigtownshire', *Transactions of the Dumfriesshire and Galloway Natural History and Antiquarian Society*, 3rd ser. 54 (1979), 41, 66.

[18] Higham, *Frontier Landscape*, 5–6, 14–17, 43, 62, 77, 86, 90, 97; Steers, *Coastline*, 86, 91–2.

is less convincing. Selkirk has argued that Roman engineers constructed series of dams, which were negotiated by pound-locks, to render rivers navigable far upstream. The system has chiefly been researched in north-eastern England, but Selkirk also proposes that it enabled barges to reach a number of places in Cumbria: Brougham and Brough via the Eden and Eamont, Old Penrith on the Petteril, Willowford on the Irthing, and Netherby on the Esk.[19] These places are located between approximately 6 and 63 miles beyond the modern NTLs. However, the evidence for this programme of artificial improvements on the rivers of the north is drawn chiefly from aerial and ground surveys, which have not yet been substantiated by modern excavations.[20] Moreover, Anderson's detailed survey of a number of rivers in the north-east does not support the idea that Roman engineers constructed dams and pound-locks here. Anderson proposes instead that supplies were brought along maritime routes and naturally navigable sections of rivers and then trans-shipped to road-based vehicles.[21] A similar network of Roman roads linked estuaries, river valleys, and trans-Pennine passes in the north-west.[22]

From the perspective of the early medieval period, it is hard to believe that an extensive network of dams and lifting devices could have existed in the region. If the resources needed to maintain this system were hard to come by during the Roman period, it is unlikely that they were present during early medieval times, when the area was remote from the heartlands of the various kings who attempted to exercise power there. The more makeshift and less expensive methods used to improve navigation elsewhere in early medieval Europe, such as loosening the river bed, might have been available in the region.[23] However, the extent to which such measures would have facilitated navigation in the long term in northern England has been questioned.[24] Thus, the best guide to the water transport resources available at the start of the early medieval period may be found by plotting the sites of Roman harbours which lay on sections of rivers, estuaries, and coastal inlets that are likely to have been naturally navigable.

The Roman fortlets and milecastles along the northern Cumbrian coast were partly intended to debar intruders, and the defences were eventually consolidated at major coastal forts such as that at Maryport.[25] It is likely that the Roman

[19] R. Selkirk, *The Piercebridge Formula: A Dramatic New View of Roman History* (Cambridge, 1983), esp. 101.

[20] M. J. T. Lewis, 'Roman Navigation in Northern England? A Review Article', *Journal of the Railway and Canal Historical Society*, 28 (1984), 119.

[21] Anderson, *Roman Military Supply*.

[22] See e.g. Higham, *Northern Counties*, 216–21, esp. fig. 5.8, 218; I. D. Margary, *Roman Roads in Britain* (3rd edn. London, 1973), 371–5, 387–92, 433–6, 448–51. Some of these passes were used during pre-Roman times: Kenyon, *Origins of Lancashire*, 30–4, 47.

[23] M. Eckholdt, 'Navigation on Small Rivers in Central Europe in Roman and Medieval Times', *International Journal of Nautical Archaeology and Underwater Exploration*, 13 (1984), 3–10.

[24] Anderson, *Roman Military Supply*, 86.

[25] M. G. Jarrett, *Maryport, Cumbria: A Roman Fort and its Garrison*, Cumberland and Westmorland Antiquarian and Archaeological Society Extra Series 22 (Kendal, 1976), 83–9; Higham, *Northern Counties*, 215, 237, 239; D. Shotter, *The Roman Frontier in Britain* (Preston, 1996), 72–81.

complex at Maryport also sustained a port at which ships carrying supplies for Hadrian's Wall could berth.[26] Further south, at Ravenglass, a fortlet similar to those used in the coastal defences has been detected, and this feature was superseded by a fort.[27] Lancaster sustained Roman occupation for several centuries: the latest fort was constructed in the fourth century on the model of the Saxon Shore defences.[28] The notion that the shore forts of the western coast hosted a fleet, rather than providing land-based protection against piratical raids, is controversial.[29] However, an inscription on an altar found in the Lune valley mentions a 'numerus barcariorum' (unit of bargemen). Shotter suggests that the barges commanded by this unit could have engaged in assaults on the pirates that preyed on the local coastline as well as acting as lighters to larger vessels.[30] The barges would have been able to travel further upstream than seagoing vessels, and it may be relevant that the inscription in question was found at Halton-on-Lune,[31] a few miles upriver from the modern-day NTL at Skerton, NGR SD 483635.

The major ports, if not all of the natural harbours,[32] are likely to have been equipped with wharves and quays to facilitate the docking and embarkation of large ships. No certain examples have been excavated,[33] although possible traces of such structures have been found along an early waterfront at Walton-le-Dale and around the River Ellen at Maryport.[34] The upkeep of port facilities would have been assured while the region remained keyed into the Roman military network, but such amenities are unlikely to have remained as well maintained after the region fell out of the grasp of centralized Roman authority.[35]

However, it might have been possible for natural harbours to remain in use after their Roman amenities had decayed. Land-based and maritime traffic continued to pass through Ravenglass in the later medieval period,[36] and

---

[26] R. L. Bellhouse, 'Roman Sites on the Cumberland Coast', *Transactions of the Cumberland and Westmorland Antiquarian and Archaeological Society*, 70 (1970), 9–47; Jarrett, *Maryport*, 85–8.

[27] T. W. Potter, *Romans in North-West England: Excavations at the Roman Forts of Ravenglass, Watercrook and Bowness on Solway*, Cumberland and Westmorland Antiquarian and Archaeological Society, Research Series 1 (Kendal, 1979), 1, 14–18; J. Cherry, 'The Topography of the Site', ibid. 11.

[28] D. C. A. Shotter and A. White, *Roman Fort and Town of Lancaster* (Lancaster, 1990), 23–7; Higham, *Northern Counties*, 236–7.

[29] R. G. Livens, 'Litus Hibernicum', in D. M. Pippidi (ed.), *Actes du IXe Congrès international d'études sur les frontières romaines* (Bucharest, 1974), 333–9.

[30] D. Shotter, 'Numeri Barcariorum: A Note on RIB 601', *Britannia*, 4 (1973), 206–9.

[31] Ibid. 208; Anderson, *Roman Military Supply*, 33.

[32] H. Cleere, 'Roman Harbours in Britain South of Hadrian's Wall', in J. du Plat Taylor and H. Cleere (eds.), *Roman Shipping and Trade: Britain and the Rhine Provinces*, Council for British Archaeology Research Reports 24 (London, 1978), 36–40.

[33] J. Fryer, 'The Harbour Installations of Roman Britain', in D. J. Blackman (ed.), *Marine Archaeology; Proceedings of the Twenty-Third Symposium of the Colston Research Society* (London, 1973), 261–73, did not mention any examples from the region.

[34] See n. 12; Jarrett, *Maryport*, 6–8.

[35] For a survey of the late and post-Roman periods see Higham and Jones, *Carvetii*, 122–34.

[36] W. G. Collingwood, 'Ravenglass, Coniston and Penrith in Ancient Deeds', *Transactions of the Cumberland and Westmorland Antiquarian and Archaeological Society*, 29 (1929), 40.

the nearby place-name 'Drigg', first attested as *Dreg c.*1175–99, seems to be analagous to a Swedish name relating to a portage point. It is uncertain where the boats were being carried to, since the spit known as Drigg Point was much smaller during the early modern period than it is now. Then, it only sheltered the mouth of the Irt, which lay to the north of the present conjoined mouth of the Irt, Mite and Esk at Ravenglass.[37] Circumstantial evidence for the continued use of Ravenglass's port is provided by the siting of several important early medieval ecclesiastical settlements nearby, as discussed below. Another relevant place name, Skippool, is found on the Wyre estuary.[38] That this place name includes the Old Norse word for ship may indicate that the estuary sustained navigation during the early medieval period, although the port has since silted up.[39]

Medium-Distance Journeys

Thus, at the beginning of the early medieval period some of the region's ports and waterways were usable. In the next section, archaeological and historical evidence is invoked in order to argue that several groups of people and institutions had reasons to use these facilities during the period in question, and that these incentives proved sufficiently compelling for the constraints of the landscape to be overcome. Although there is little evidence for the participation of the inhabitants of the north-west in long-distance trade, they undertook trans-Irish Sea voyages and local water-borne communication.

One reason why the north-west has been excluded from discussions of early medieval communication across the Irish Sea is that the area has yielded no evidence for participation in the trade networks that brought Mediterranean, and subsequently continental, goods to western Britain and Ireland. The traders who plied these long-distance routes are likely to have dealt primarily in oil and wine, but their activities are evinced in the archaeological record by distinctive types of pottery.[40] The apparent absence of these ceramics in the north-west has been attributed to the deficiencies of the region's ports, or the incorporation of the area into the kingdom of Northumbria.[41]

Survival and discovery rates may have also played a part in preventing this pottery from being found in the north-west. Much of the continental E ware has been found during large-scale excavations at high-status sites

---

[37] A. M. Armstrong et al. (eds.), *The Place-Names of Cumberland*, vol. ii, EPNS 21 (Cambridge, 1951), 376–7, quoting Ekwall. For the relationship of the place-name to the changing coastline see the comments of E. Ann Cole, cited in D. Whaley, *A Dictionary of Lake District Place-Names* (Nottingham, 2006), 101, mentioning Steers, *Coastline*, 81–3.

[38] E. Ekwall, *The Place-Names of Lancashire* (Manchester, 1922), 140.

[39] For coastal changes on the Wyre Estuary, see Steers, *Coastline*, 94–6.

[40] It is impossible to cite all the scholarly discussions of this trade here, but the distinction between the Mediterranean and continental phases is drawn by C. Thomas, '*Gallici Nautae de Galliarum Provinciis*: A Sixth/Seventh Century Trade with Gaul Reconsidered', *MA* 34 (1990), 1–26.

[41] Wooding, *Communication*, 101; E. G. Bowen, *Saints, Seaways and Settlements in the Celtic Lands* (Cardiff, 1969), 17–18.

such as Lagore, in Ireland, whereas work of a similar nature has not taken place in the north-west.[42] The apparent absence of E ware at Meols, an early medieval trading site located on the Wirral, just to the south of the region considered in this chapter, may result from the fact that many of the finds from the site were accumulated by collectors, who were attracted to metalwork rather than dull-coloured pottery.[43] Indeed, there are signs that the region was not excluded from long-distance communication networks: Byzantine coins have been discovered around both Meols and Lancaster, where a sub-Roman community capable of consuming exotic goods may have survived.[44]

More positive results can be produced by examining medium-distance communication, especially routeways which linked Ireland and Northumbria. However, it was not necessary to travel through the north-west in order to communicate between these two places. Alternative routes linking Bernicia (north-eastern Northumbria) and Ireland had been forged through the Dál Riatan island of Iona, where several seventh- and eighth-century kings of Northumbria spent time in exile or in scholarly endeavours.[45] Iona maintained connections with Ireland not only because Dál Riata spanned territory in north-eastern Ireland and western Scotland, but also because many of Iona's abbots had family connections with Cenél Conaill, a northern Uí Néill group.[46] Moreover, Northumbrian territory stretched far enough north by the mid-seventh century to enable travellers to use the Firth of Forth. However, the upper stretches of the Forth passed through a no man's land between the rivers Carron and Avon, into which Pictish groups were pushing by the early eighth century.[47] The Firth of Clyde, which lay in a British kingdom, might have also been closed to Northumbrian travellers during turbulent times.

In political terms, therefore, it can be argued that the ports most favourable for communication between Bernicia and Ireland lay in modern-day Cumbria. By the mid- to late seventh century the Northumbrians were exercising a degree of influence in this area, which lay in the former British kingdom

[42] This point is also made by D. O'Sullivan, 'Cumbria before the Vikings: A Review of Some "Dark Age" Problems in North-West England', in J. R. Baldwin and I. D. Whyte (eds.), *The Scandinavians in Cumbria* (Edinburgh, 1985), 19–21.

[43] D. Griffiths, 'Coastal Trading Ports of the Irish Sea Region', in J. Graham-Campbell (ed.), *Viking Treasure from the North-West* (Liverpool, 1992), 67–8.

[44] S. Penney, 'Gazetteer', *Contrebis*, 5 (1977), 47–8; id., 'Gazetteer', *Contrebis*, 6 (1978), 43; Griffiths, 'Meols', 23. T. Potter, 'Recent Archaeology in the Northwest', *Current Archaeology*, 53 (1976), 187 reported that a sherd of African Red Slipware was discovered at Lancaster.

[45] Bede, *Historia Ecclesiastica Gentis Anglorum* (hereafter *HE*), ed. and trans. B. Colgrave and R. Mynors, *Bede's Ecclesiastical History of the English People* (Oxford, 1969; repr. 1979), III. 3; Bede, *Vita Sancti Cuthberti* (hereafter *VCB*), ed. and trans. B. Colgrave, *Two Lives of St. Cuthbert* (Cambridge, 1940), ch. 24; K. Hughes, 'Evidence for Contacts between the Churches of the Irish and the English', in P. Clemoes and K. Hughes (eds.), *England before the Conquest: Studies in Primary Sources Presented to Dorothy Whitelock* (Cambridge, 1971), 53–4.

[46] M. Herbert, *Iona, Kells, and Derry: The History and Hagiography of the Monastic Familia of Columba* (Oxford, 1998), 36–48.

[47] B. Hudson, *Kings of Celtic Scotland* (London, 1994), 12. I am grateful to Arkady Hodge for discussing this area with me.

of Rheged.[48] Textual evidence offers hints that Anglo-Saxon expansion into the more southerly regions of the north-west was motivated by a desire to establish routes to the Irish Sea. Bede states that Eadwine (reigned *c*.616–633) conquered the Isle of Man and Anglesey,[49] and that King Ecgfrith sent an army to attack the Irish region of Brega in 684.[50] In order to launch such military expeditions the Northumbrians would have needed to consolidate their control of the routes that led to the western estuaries. Moreover, during the eighth century the Northumbrian presence on the northern side of the Solway was growing stronger, as the development of an Anglian monastery and bishopric at Whithorn attests.[51] Although we cannot know how important such political considerations were to early medieval travellers, it can be imagined that they mattered to Anglo-Saxon nobles and high-ranking monks and bishops.

The accounts of foundation of ecclesiastical establishments in the relevant river valleys demonstrate that the Northumbrian kingdom had achieved mastery of these routes by the late seventh century, if not before. The communities based at these churches had incentives to use the north-west's rivers and ports to maintain communication with Ireland, as I explain in detail elsewhere.[52] For example, a number of churches were founded in the region from Lindisfarne, and under the aegis of St Cuthbert's community, during the seventh and eighth centuries; these establishments lay near harbours in Carlisle, on the western Cumbrian coast, and around Morecambe Bay. The communities had an incentive to use these ports in order to visit their counterparts in Ireland and Dál Riata in Britain, with whom they shared a common bond through Lindisfarne's mother-house, Iona.[53] The connection between Cumbria and Ireland is made explicit by the tenth- or eleventh-century work *Historia de Sancto Cuthberto* which notes that *Derunt muthe* (the mouth of the River Derwent, in Cumbria) was the location from which the Cuthbertine community attempted to dispatch its patron saint's relics to Ireland.[54]

Attention has also been drawn to the locations of the Anglian crosses at Waberthwaite and Irton on either side of the port at Ravenglass.[55] Waberthwaite lay by a stretch of the River Esk that is navigable for smaller boats at high

---

[48] Higham, *Northern Counties*, 256–74.   [49] *HE* II. 5, II. 9.

[50] *HE* IV. 26; S. MacAirt and G. MacNiocaill (eds. and trans.), *The Annals of Ulster* (hereafter *AU*) (Dublin, 1983), 685.2.

[51] *HE* V. 23.

[52] F. Edmonds, 'The Practicalities of Communication between Northumbrian and Irish Churches during the Seventh and Eighth Centuries', in J. Graham-Campbell and M. Ryan (eds.), *Anglo-Saxon/Irish Relations before the Vikings* (forthcoming).

[53] Ibid.; T. Johnson-South (ed. and trans.), *Historia de Sancto Cuthberto* (hereafter *HSC*), Anglo-Saxon Texts 3 (Cambridge, 2002), chs. 6 and 21 (pp. 48–9, 60–1). For these Cuthbertine houses cf. V. Tudor, 'St. Cuthbert and Cumbria', *Transactions of the Cumberland and Westmorland Antiquarian and Archaeological Society*, 84 (1984), 68–70 and Higham, *Northern Counties*, 292, 302.

[54] *HSC* ch. 20 (pp. 58–9); for indications of the location of a church at *Derunt muthe* see R. Cramp and R. Bailey (eds.), *Corpus of Anglo-Saxon Sculpture*, ii: *Cumberland, Westmorland and Lancashire North of the Sands* (Oxford, 1988), 11, 154–7.

[55] W. G. Collingwood, *Northumbrian Crosses of the Pre-Norman Age* (London, 1927), 111.

tide and fordable at low tide, while the Irton cross bears a number of motifs that are more at home in the Irish than the Anglian sculptural repertoire, although the cross resembles Northumbrian monuments in other respects.[56] Thus the communities that set up both of these crosses are likely to have benefited from the existence of an early medieval port at Ravenglass. The regularity and intensity of the contact engaged in by these ecclesiastical communities is debatable; movements to Ireland are likely to have been occasioned by notable events rather than day-to-day concerns. There is no evidence that the early medieval churches relied on Ireland for sustenance, as did the later medieval monastery of Furness.[57]

Viking raids, which occurred in the Irish Sea region long before Scandinavians settled in the region in the tenth century, must have sporadically interrupted the maritime connections of coastal monasteries. However, links between the kingdoms of Scandinavian York and Dublin perpetuated the incentive to use the north-western coast to maintain contact across the Irish Sea. Such communication was particularly desirable from the early tenth century, when a dynastic connection had been created between the two kingdoms following the Hiberno-Scandinavian Ragnall's capture of York,[58] although the association between the rulers of York and Dublin has been traced back before this.[59] There is no doubt about the pervasive nature of links between York and Dublin up to the mid-tenth century, and Downham has suggested that the final Scandinavian ruler of York, Erik, was also connected with the main Hiberno-Scandinavian dynasty.[60]

Evidence is compelling to suggest that the north-west sustained communication between York and the Irish Sea region at the very beginning of the tenth century. Higham has argued that a band of the Scandinavians of Dublin was based on the Ribble following their expulsion from Dublin in 902.[61] Higham suggests that the Ribble was attractive because the northern Cumbrian estuaries and the Isle of Man were experiencing incursions by the men of Strathclyde and hosting the camps of other seafaring bands respectively.[62] Such Viking residences on the Isle of Man must have been temporary since the full-scale

[56] M. C. Fair, 'The West Cumberland Group of Pre-Norman Crosses', *Transactions of the Cumberland and Westmorland Antiquarian and Archaeological Society*, 50 (1950), 101; Bailey and Cramp, *Corpus*, 117. The Esk has changed course slightly during the last few centuries: Steers, *Coastline*, 83.

[57] H. Sweetman (ed.), *Calendar of Documents Relating to Ireland 1171–1251* (London, 1875), 79, 235, 297, 307, 424.

[58] A. Smyth, *Scandinavian York and Dublin* (2 vols., Dublin, 1975–9), i. 100–1,108–10; *ASC* 'D', 'E' s.a. 923 (p. 105) notes Ragnall's presence in York. However, the 12th-century work *Historia Regum Anglorum et Danorum* places the event in 919. Cf. C. Downham, 'Britain and Scandinavian Ireland: The Dynasty of Ívarr and Pan-Insular Politics to 1014' (Ph.D. thesis, Cambridge, 2003), 88. For the relevance of these developments to the north of the area under consideration here, see N. Higham, 'The Scandinavians in North Cumbria: Raids and Settlement in the Later Ninth to Mid Tenth Centuries', in Baldwin and Whyte (eds), *Scandinavians in Cumbria*, 39–43.

[59] Smyth, *Scandinavian York*, i. 15–92, now modified by Downham, 'The Dynasty of Ívarr', 57–88.

[60] C. Downham, 'Eric Bloodaxe—Axed? The Mystery of the Last Scandinavian King of York', *Mediaeval Scandinavia*, 14 (2004), 51–77.

[61] *AU* 902.2.

[62] N. Higham, 'Northumbria, Mercia and the Irish Sea Norse 893–926', in *Viking Treasure*, 27.

Scandinavian settlement of the island seems to have begun later than was once assumed, possibly not until well into the tenth century.[63]

A key element of Higham's argument is the suggestion that the residence of a band of Hiberno-Scandinavians on the Ribble in the early tenth century may explain the deposition of the Cuerdale hoard by this river *c.*903–5. Much of the hack-silver in the hoard originated in Ireland, but a recently minted package of coins from York had also been added to the hoard shortly before it was buried. It is more likely that such additional material was arriving while the band was waiting on the Ribble than that the whole hoard had been accumulated in York. Graham-Campbell points out that it would have been more convenient to have the bullion minted if the hoard had been collected in that city.[64] Thus, the Ribble estuary's advantage of linking water transport to the trans-Pennine Ribble–Aire road to York explains why it was a useful route between exiled bands of Dublin Norsemen and their supporters in York even at the very start of the tenth century, as Edwards notes. Cuerdale is located just under a mile upriver from the current NTL.[65]

It cannot be taken for granted that the dynastic links between kings of Scandinavian York and Dublin promoted the continuing use of the north-west's water transport resources. War-bands might have moved into eastern England directly by boat; the name *Dyvelinstanes*, which was once applied to a street that ran down to York's waterfront, implies that ships from Dublin had once berthed in the vicinity.[66] In this circumstance, the fleets might have sailed around the northern or southern tips of Britain, but Smyth has argued that a much shorter route was followed: boats were transferred between York and Dublin by way of the Forth–Clyde isthmus. The expanse of land between these two estuaries apparently sustained a portage, along the lines of those which operated in Russia.[67] It is likely that the Forth–Clyde route was sometimes followed by kings of Scandinavian York and Dublin and their armies, who seem to have come to arrangements with the kings of Strathclyde and Alba.[68] However, it is not clear that the isthmus was regularly used as a portage point, as opposed to a route which involved some transport overland. It may have been possible to drag ships over the isthmus, taking advantage of lakes on the way, but the practice may not have been viable in all weathers, or under

---

[63] J. Graham-Campbell, 'The Early Viking Age in the Irish Sea Area', in H. B. Clarke et al. (eds.), *Ireland and Scandinavia in the Early Viking Age* (Dublin, 1998), 116–20.

[64] J. Graham-Campbell, 'Some Archaeological Reflections on the Cuerdale Hoard', in D. M. Metcalf (ed.), *Coinage in Ninth-Century Northumbria*, BAR British Ser. 180 (Oxford, 1987), 340, 344. For the origins of the bullion see also idem, 'The Cuerdale Hoard: A Viking and Victorian Treasure', in *Viking Treasure*, 10–11; idem, 'The Cuerdale Hoard: Comparisons and Contrasts', in ibid., 113–14, and idem, 'The Northern Hoards: From Cuerdale to Bossall/Flaxton', in N. J. Higham and D. Hill (eds), *Edward the Elder 899–924* (London, 2001), 220–3.

[65] B. J. N. Edwards, *Vikings in North-West England* (Lancaster, 1998), 65–7.

[66] Smyth, *Scandinavian York*, ii. 236–7, 256.   [67] Ibid. i, 22, 301–3.

[68] Ibid., 35–6, 63–4, 94–5, 108; ii. 43–4, 272, 278–82; cf. B. Crawford, 'The "Norse Background" to the Govan Hogbacks', in A. Ritchie (ed.), *Govan and its Early Medieval Sculpture* (Stroud, 1994), 109–10.

turbulent political conditions.[69] Thus, the Forth-Clyde route need not have had many advantages over its more southerly land-based counterparts. When travelling overland it would have been necessary for the kings of York and Dublin to keep affiliated fleets on either side of Britain. Indeed, since Erik was killed on Stainmore in 954 it seems likely that he was intending to escape to a western harbour.[70] The roads from this trans-Pennine pass led towards the harbours of the Cumbrian coast or the Solway Firth.[71]

The north-west is not served well by historical evidence, so much of the above discussion as to whether water transport separated or connected the area in the early medieval period is tentative. However, as other contributors to this book have found, some archaeological evidence supports the picture drawn from written sources. Griffiths's work on the site at Meols on the Wirral has demonstrated that the north-western coast occupied a place in Irish Sea communication during early medieval times and beyond. The abundant arte-factual evidence yielded by eroding sand dunes at Meols includes continental and Byzantine coins and artefacts indicative of links with Ireland, such as Hiberno-Scandinavian ringed pins. This coastal trading site was located just to the south of the area discussed in this chapter. It may therefore have drawn its importance from links between Mercia and Ireland, rather than the contacts emanating from Northumbria which have been described above. However, the discoveries made at Meols highlight the fact that the changing coastal landscape may have concealed other, smaller, medieval trading places along this coast. Finds began to be made at the site when the sands of a promontory on which Roman and medieval activity may have occurred began to be washed into the sea in the nineteenth century.[72]

Even so, the distribution of coin finds in the north-west and the Irish Sea region does not support the notion that the area was engaged in communication with Ireland. The ninth-century Northumbrian coins that are known as *stycas* have been found at several coastal sites in the region, but they did not filter across the Irish Sea. In contrast, a few Mercian Offan pennies of the later eighth century have been discovered in Ireland.[73] The westward transmission of tenth-century coins from England's Irish Sea coast can also be accounted

---

[69] The regularity with which the Forth–Clyde portage was used has also been questioned by J. Graham-Campbell and C. Batey, *Vikings in Scotland: An Archaeological Survey* (Edinburgh, 1998), 98, and C. Phillips, 'Portages in Early Medieval Scotland: The Great Glen Route and the Forth–Clyde Isthmus', in C. Westerdahl (ed.), *The Significance of Portages: Proceedings of the First International Conference on the Significance of Portages*, BAR Internat. Ser. 1499 (Oxford, 2006), 196. I am grateful to Alex Woolf for discussing the Forth–Clyde route with me.

[70] *Rogeri de Wendover Chronica, sive Flores Historiarum*, ed. H. Cox (5 vols., London, 1841–4), i. 402–3; Smyth, *Scandinavian York*, ii. 173–5.

[71] A. Woolf, 'Eric Bloodaxe Revisited', *Northern History*, 34 (1998), 192–3 tentatively suggests that Erik had supporters in the Solway area.

[72] David Griffiths has worked extensively on Meols and is preparing a publication about the site. See in the meantime D. Griffiths, 'Great Sites: Meols', *British Archaeology*, 62 (2001), 20–5, and Griffiths, 'Coastal Trading Ports', 67–8.

[73] D. M. Metcalf, 'The Monetary Economy of the Irish Sea Province', in *Viking Treasure*, 93–4, 96.

for to a substantial extent by the activities of the Mercian mints and ports, especially Chester until its decline at the end of the tenth century.[74] Yet sufficient numbers of York-minted Anglo-Scandinavian coins have been found amongst the hoards deposited at this time in Ireland to suggest that coins were travelling across the sea along more northerly routes.[75] The Ribble estuary may have been a more likely candidate for such an outlet than the Solway Firth, since the latter lay in territory that belonged to Strathclyde, rather than Northumbria, by this period.[76]

Political changes were also occurring in the more southerly region of the north-west, but incentives to make the most of trade along the western seaboard clearly still existed there. Griffiths has argued that the line of burghal fortresses built in north-western Mercia was not solely intended to ward off Viking attacks. Rather, it represented an attempt to stamp royal power on this edge of the Mercian kingdom, and to exploit its economic possibilities.[77] Higham has argued that the site of at least one of these burhs was connected with navigation along the Mersey. He suggests that the burh at Runcorn not only oversaw the lowest ford over the Mersey, but also protected navigation through a relatively narrow section of the river, and perhaps also overlooked a mooring place that served the Mercian navy.[78] Higham further notes that similar proximity to a fordable and navigable section of the Ribble may have entailed the construction of a burh at Penwortham after Mercia's frontier was pushed northwards.[79] Further north, the region's estuaries and rivers remained in a politically turbulent zone.

Localized Movements

Finally, it is necessary to assess the extent to which the harbours, estuaries, and rivers of the north-west sustained transport within the region during the early medieval period. The distributions of some artefact types that have already been discussed seem to reflect the courses of localized traffic. Metcalf thought that *stycas* were used in the north-west during the course of 'coastwise shipping and local trade'.[80] Indeed, coins have been found in locations that were difficult of access overland, for example, on the southern Cumbrian coast. The *stycas* that were discovered here seem to have come from hoards, however, and so need not have been lost in the course of exchange.[81] Bailey has shown that whilst the distributions of certain motifs found on sculptured stones do not attest

---

[74] For Chester see Higham, *Frontier Landscape*, 167–70.

[75] Metcalf, 'Monetary Economy', 97.   [76] Phythian-Adams, *Land of the Cumbrians*, 112.

[77] D. Griffiths, 'The North-West Frontier', in N. Higham and D. Hill (eds.), *Edward the Elder 899–924* (London, 2001), 178–81.

[78] N. J. Higham, 'The Cheshire Burhs and the Mercian Frontier to 924', *Transactions of the Lancashire and Cheshire Antiquarian Society*, 85 (1988), 199–200, 203–4.

[79] Ibid. 213–14; id., 'Irish Sea Norse', 28; id., *Frontier Landscape*, 142.

[80] D. M. Metcalf, 'The Monetary Economy of the Irish Sea Province', in *Viking Treasure*, 96.

[81] Higham, *Northern Counties*, 303–4.

links between northern England and Ireland, such material reveals coastal links between areas of the Wirral, the Lune valley, the Cumbrian coast, Galloway, and the Clyde.[82] However, such connections say as much about the conditions of sculptural patronage that obtained in the territories bordering the Irish Sea as they do about the capabilities of maritime routes and rivers to sustain traffic.[83]

The extent to which boats moved goods from place to place along inland stretches of river during the early medieval period also bears consideration. As discussed above, it is unlikely that the region's larger rivers were actively managed by dredging, canalization, and straightening during Roman times, or that small rivers were rendered navigable by such means. Neither does textual evidence deriving from the later medieval period suggest that the smaller rivers of the region sustained much traffic. Of the north-western rivers, Edwards and Hindle placed only the Mersey, Lune, Derwent, and Eden into their 'major river' category.[84] Langdon has argued that their method of analysis produced overly optimistic results. He uses purveyance accounts to produce a more limited picture of the extent of river transport,[85] but such evidence does not exist for the north-west.

Nevertheless, the evidence advanced in this paper does not contradict the conclusion of other chapters that rivers were sometimes navigable further inland in the early medieval period than they were later. Thus, Edwards and Hindle consign the Ribble to the 'minor river' category, which indicates that it was navigable for less than 10 miles from the sea.[86] This concurs with the fact that Preston docks, which is located slightly downstream from the current NTL, was maintained only with a major dredging operation.[87] However, such measures were necessary for modern boats, rather than early medieval vessels with a shallow draught. Indeed, it has been suggested above that some boats navigated the Ribble, as well as a number of other rivers in the region, up to and beyond the current NTL. There are also indications that some of the region's smallest rivers were navigable during the early medieval period. Thus, whilst 'Old Hyton' in southern Cumbria is nowadays situated on a small river, its name (from *hȳð*: cf. Cole, below p. 64) indicates that this location was used as a landing place for goods brought inland during the early medieval period.[88]

Indeed, small boats could have been used to carry portable luxuries along marshy stretches of the north-western rivers. Few early medieval vessels have been excavated in the north-west, but those that have are of the logboat

---

[82] R. Bailey, 'What Mean these Stones?', *Bulletin of the John Rylands Library*, 78 (1996), 31–2.

[83] R. Bailey, 'Irish Sea Contacts in the Viking Period: The Sculptural Evidence', in G. Fellows-Jensen and N. Lund (eds.), *Tredie Tvaerfaglige Vikingesymposium*, (Copenhagen, 1984), 12–13, 28–30.

[84] J. F. Edwards and B. P. Hindle, 'The Transportation System of Medieval England and Wales', *JHG* 17(1991), 131.

[85] J. Langdon, 'Inland Water Transport in Medieval England', *JHG* 19 (1993), 1–11, and his chapter below.

[86] Edwards and Hindle, 'Transportation System', 128, 131.

[87] J. Barron, *A History of the Ribble Navigation* (Preston, 1938), 261–321.

[88] Phythian-Adams, *Land of the Cumbrians*, 14.

variety. Of thirteen boats and fragments found at Warrington and nearby on the Mersey and Irwell rivers, the eight that were analysed are likely to date from the earlier medieval period. The dates for the felling of the logs which were used in their construction, obtained through radiocarbon analysis, cluster around the twelfth century AD. Even though these specimens were in the smaller range of logboat sizes they could be put to a variety of uses, including the carrying of personnel and goods. Analysis of 'Warrington 2' suggests that it could have been used for transporting bulky cargo such as hay and that it could bear up to four men. The logboat found at Barton lent itself more to a high-density load.[89] Thus these boats would have been suitable for the transport of agricultural or luxury products such as hides from relatively upland regions; it is likely that these basic canoe-style boats could have travelled further upstream than larger vessels.

Little information is available about the larger boats which would have been active in the waters around the north-west. The text *Míniugud Senchasa Fher nAlban*, which records genealogical and military details about Dál Riata in Britain, reveals that twenty 'houses' were able to produce two seven-bench ships, which were predominantly used for military activities.[90] Several developments made on Scandinavian ships would have enabled their seagoing boats to navigate into the estuaries of the north-west, and perhaps some way upriver. These changes seem to have been general, although it should be remembered that these ships were not of uniform type or size, and indeed their form changed over time. Nevertheless, the ability of the ships to operate in shallow water and land on open beaches as a result of their shallow draught is well known.[91] Scandinavian ship technology filtered down to other territories neighbouring the north-west: dendrochronological analysis and dendroprovenancing carried out on one of the ships recovered from a submarine blockade at Skuldelev, Roskilde, revealed that it had been constructed from trees felled in Ireland in the mid eleventh century.[92] No such boats have been recovered from the north-west: a mound that once stood on the north side of the Solway Firth at Graitney Mains was once thought to have contained a boat burial, but it was never investigated and has since been destroyed.[93]

Localized navigational considerations were not merely affected by ship technology; the quality of the landscape surrounding the rivers determined the

[89] S. McGrail and R. Switsur, 'Medieval Logboats of the River Mersey: A Classification Study', in S. McGrail (ed.), *The Archaeology of Medieval Ships and Harbours in Northern Europe*, BAR Internat. Ser. 66 (Oxford, 1979), 93–112.

[90] J. Bannerman, *Studies in the History of Dalriada* (Edinburgh, 1974), 148–54. Cf. Wooding, *Communication*, 8–21.

[91] P. Sawyer, *Age of the Vikings* (2nd edn. London, 1971), 76–7.

[92] O. Crumlin-Pedersen and O. Olsen (eds.), *The Skuldelev Ships*, i: *Topography, Archaeology, History, Conservation and Display* (Roskilde, 2002), 185, 326–30.

[93] Neilson, 'Annals of the Solway', 265; J. Graham-Campbell, *Whithorn and the Viking World*, Whithorn Lecture (Whithorn, 2001), 17.

ease with which boats could be beached and debarked.[94] In other regions of medieval England, non-navigational use of rivers impeded the passage of boats. The only indications of such activities in the north-west are found in Domesday Book, which records half of a fishery at Penwortham and alludes to others in West Derby hundred.[95] That the fisheries were few in number may reflect the fact that the Domesday account is exceptionally patchy for this region, but there is no reason to think that the region's rivers were unduly encumbered with weirs and traps. It is also unlikely that obstructions associated with mills were frequently present during the early medieval period, since the more intense exploitation of the region's landscape which led to a growth in milling happened during the second millennium AD.[96]

Conclusion

In this chapter, various types of evidence, and the views of earlier commentators, have been accumulated in order to shed light on the network of water transport resources which was available in the north-west during the early medieval period. It appears that the region's rivers and coasts acted as much as unifying features as they did as barriers. This is not to underestimate the obvious limitations of the area's water transport: the Irish Sea is frequently rough during the winter and the rivers often flow dangerously fast following bad weather. It is impossible to determine the frequency with which the sea and rivers were travelled in this area, but it is likely that the poverty of the region meant that there was little incentive artificially to improve the upper sections of the rivers in order to facilitate bulk transport of goods. However, there are some indications that small boats plied inland stretches of rivers in order to move goods locally. Moreover, the naturally navigable stretches of the region's rivers were part of a network of estuarine and coastal water transport which enabled armies, craftsmen, monks, and traders to pass through the region on their journeys between the Irish Sea and the east of Britain.

[94] For example, Higham suggests that the mossy landscape surrounding the Mersey would have made debarkation from Viking ships problematic: 'The Context of *Brunanburh*', in A. Mills and A. Rumble (eds.), *Names, People and Places: An Onomastic Miscellany in Memory of John McNeal Dodgson*, (Stamford, 1997), 153.

[95] H. C. Darby, *The Domesday Geography of Northern England* (Cambridge, 1962), 415, 442–4; Higham, *Frontier Landscape*, 47–8; however, a number of fisheries existed in later medieval Cumbria: see A. Winchester, *Landscape and Society in Medieval Cumbria* (Edinburgh, 1987), 107–13.

[96] Higham, *Frontier Landscape*, 47, 138–9; Winchester, *Landscape*, 117–119.

2

Uses of Waterways in Anglo-Saxon England

DELLA HOOKE

Rivers were the lifelines of early and high medieval England. They provided water for men and animals, and their seasonal flooding helped to grow the meadow-grass that provided the hay essential for winter fodder. While there may be little direct documentary or literary evidence for the use of rivers for trade and transport in early medieval England, the circumstantial evidence is overwhelming, backed up by the evidence of place names. The seas themselves were highways of navigation and trade, rather than impediments to movement, and river estuaries were gateways to the early medieval kingdoms.

When early medieval lines of communication are plotted, whether known from archaeological or documentary evidence, a dense network of functioning roads is revealed, with a clear hierarchy of routes ranging from lanes to major highways. These ran for long distances across the country, and they have been the subject of many studies.[1] For an indication of routeways combining both road and river transport one has only to look at the lines of the early medieval saltways which radiate, either directly or indirectly, from the inland salt-producing centre of Droitwich in the Hwiccan kingdom. These can be reliably reconstructed from charter and place-name evidence and it is clear that several routes which ran south-eastwards through Gloucestershire ran to the Thames (Figs. 5, 62).[2] In particular, two routes can be traced which ran south-south-east: the first, after leaving the Vale of Evesham, followed the River Isbourne and crossed the high Cotswolds near Hawling before dropping down to the Colne and the Thames; the second crossed the Cotswolds to Stow-on-the-Wold and continued southwards to cross the Windrush, probably linking with the first route before reaching the Thames at Lechlade. The route

I am grateful to Tim Grubb, Sites and Monuments Officer for Gloucestershire County Council Archaeology Service, and to Elizabeth Townsley, University of Bristol, for information on fisheries in the River Severn.

[1] F. M. Stenton, 'The Road System of Medieval England', *EcHR* 7 (1936), 1–21 (reprinted in D. M. Stenton (ed.), *Preparatory to Anglo-Saxon England* (Oxford, 1970), 234–52). D. Hooke, *Anglo-Saxon Landscapes of the West Midlands: The Charter Evidence*. BAR British Ser. 95 (Oxford, 1981), 300–14.

[2] D. Hooke, *The Anglo-Saxon Landscape: The Kingdom of the Hwicce* (Manchester, 1985), 125, fig. 31; cf. Blair, below, pp. 255–6.

continued on in a south-easterly direction towards the royal estate of Wantage in the Vale of the White Horse, but it seems highly likely that the salt was being transported from Lechlade by barge to London.

Lechlade, *Lecelade* 1086, is probably a river name derived from OE **læc(c)*, **lece* 'boggy stream' with OE *gelād*, which is generally interpreted as 'river crossing',[3] to be distinguished from *lād* 'a watercourse'.[4] Lechlade is one of several such sites in this area: Cricklade lies some 15 km upstream where the Roman road from Gloucester and Cirencester crossed the Thames and close to another branch saltway from the north-east.[5] Evenlode (now in Gloucestershire) lies on a much smaller river, also now called the Evenlode but in the early medieval period known as the *Bladen*.[6] There are similar *gelād* names along the River Severn, as at Framilode, Abloads Court, Wainlode, Lower and Upper Lode in Gloucestershire and, in Worcestershire, at Clevelode and Worcester itself. Forsberg[7] notes cognates like *lid* 'a ship' and *līðan* 'to travel by sea or water', and concludes that the most likely meaning of *gelād* was 'river crossing by boat', i.e. 'ferry', but Smith considers this interpretation to be a meaning established by usage, 'for many places on the Severn ferries happen to be found at places called "lode"'.[8] Gelling has recently suggested that use of this term implied a 'difficult water-crossing' or even a 'water-crossing liable to be rendered impassable by flooding'[9] but, at least on the Severn, most of these places could never have been forded. In literary sources *gelād* is used for a seaway,[10] and there remains the possibility that the term was related to passage along, rather than across, a river[11] — salt could have been transported by river to London from these places on the Thames—but all the locations seem to be definite points, as at Worcester.

It is also interesting to note how, at least on the River Severn, many of the crossings are close to minsters and major churches, and could have been on roads used to transport building materials or produce to these centres. In Gloucestershire, Framilode lies on a route that runs from the Frome valley and the Cotswolds across the river towards Westbury-on-Severn (a ferry crossing); Upper and Lower Lode are on routes leading towards Tewkesbury (the latter

[3] A. D. Mills, *Dictionary of English Place-Names* (Oxford, 1991), 207.

[4] A. H. Smith, *English Place-Name Elements* (2 vols., 2nd impression: Cambridge, 1970), i. 8–9. For discussion and further suggestions, see Cole, below, pp. 77–8.

[5] Hooke, *The Anglo-Saxon Landscape*, 125, fig. 31.

[6] S 109; mapped in D. Hooke, *The Landscape of Anglo-Saxon England* (London, 1998), 88.

[7] R. Forsberg, *A Contribution to a Dictionary of Old English Place Names* (Uppsala, 1950), 22.

[8] Smith, *English Place-Name Elements*, ii. 9.

[9] M. Gelling, 'The Landscape of *Beowulf*', in M. Lapidge (ed.), *Anglo-Saxon England* (Cambridge, 1972), 31, 10–11.

[10] S 1280.

[11] Watts appears to prefer to interpret the name of Lechlade as indicating 'channel, watercourse of the River Leach' although, paradoxically, he continues to translate OE *gelād* in the names of Cricklade and Evenlode and other 'lode' names as 'a river crossing' adding, in the case of the former, Gelling's rider of 'particularly one liable to be difficult owing to flooding as happens here on the Thames': V. Watts (ed.), *The Cambridge Dictionary of English Place-Names* (Cambridge, 2004), 366, 168, 220.

also a ferry crossing) and at Worcester the lode noted in a pre-Conquest charter may have lain on the site of the later medieval bridge;[12] Clevelode and Wainlode were not far distant from the minsters of Kempsey and Deerhurst.

That river travel and transport may be further implied by place-name evidence has recently been suggested by Bryony Coles, who argues that *trisantona* river names such as the Trent, Tarrant, etc. meant 'a way through', in reference to non-local routes which may have been followed for long distances.[13] However, Ekwall[14] claims that the common base for such names, *Trisantōn-* or *Trisantonā*, is British, emphasizing the implication of 'to cross, to pass over', and even considers the term as appropriate to 'one who goes across, a trespasser': hence 'the great traveller' referring to a river liable to floods. Nevertheless, Coles's suggestion is an attractive one and could apply very effectively to the midland Trent, which was clearly navigable for considerable distances in medieval times, to the Dorset Trent (now Piddle), which flows into Poole Harbour near the early medieval borough of Wareham, and to the Sussex Tarrant (now the Arun), which provides a route northwards towards the valley of the Wey and hence to the Thames. Two other Dorset rivers, the Trent Brook and Tarrant, and the *trentan*, a tributary stream of the Avon which flows north-north-west across the Vale of Evesham in Worcestershire, are rather less convincing as routeways. Coles identifies the *trentan* as the Badsey Brook, but a boundary clause for Evesham holdings[15] clearly shows this identification to be wrong, for the name refers to a stream flowing southeastwards through Bretforton and Willersey; there is, however, a possible route here up the Cotswold scarp towards the Ryknield Street and a major saltway, but one that only runs close to the headwater section of the stream.

In later periods, it is clear that rivers were much used for the transport of building stone, especially for oolite from the south Cotswolds up the River Severn and its tributaries, perhaps combined with further transport by road where the roads were adequate for ox-drawn carts,[16] although even minor rivers could be used in some cases. Stone from the Taynton quarries in Oxfordshire reached Newent and Acton Beauchamp in Herefordshire and Tenbury in north-west Worcestershire. At Tenbury the stone was used for an Anglo-Saxon cross.[17] Jope comments upon the 'watermarks' found in the fluted piers of Deerhurst church and the 'Lechmere' stone (part of a tombstone) at Hanley Castle that suggest barge transport of stone up the Severn from the

[12] D. Hooke, 'The Hinterland and Routeways of Anglo-Saxon Worcester: The Charter Evidence', in M. O. H. Carver (ed.), *Medieval Worcester: An Archaeological Framework*, Trans. Worcestershire Archaeol. Soc. 3rd ser. 7 (1980), 47–9.

[13] B. J. Coles, '*Trisantona* Rivers: A Landscape Approach to the Interpretation of River Names', *Oxford Journal of Archaeol*, 13/3 (1994), 295–311.

[14] E. Ekwall, *English River-Names* (Oxford, 1928), 417–18.

[15] S 1591a; D. Hooke, *Worcestershire Anglo-Saxon Charter-Bounds* (Woodbridge, 1990), 377–82.

[16] T. Eaton, *Plundering the Past: Roman Stonework in Medieval Britain* (Stroud, 2000), 42–4.

[17] E. M. Jope, 'The Saxon Building-Stone Industry in Southern and Midland England', *MA* 8 (1964), 106–7.

Bath area.[18] The River Wye was navigable past Chepstow at least as far as Hereford as late as the fifteenth century. As noted by David Pelteret (pers. comm.), Hwiccan minster sites were frequently located beside rivers, and many of their buildings are likely to have used stone transported along the rivers Severn and Avon. Similarly, Blair[19] notes the location of early minsters at regular intervals along the Thames. On Romney Marsh, Pearson and Potter[20] argue for the use of the rivers and marshland watercourses (even minor lodes only a few metres wide) for the transport of beach boulders from the Ashdown sands and Hythe Beds which were used in the construction of some of the earliest churches of the area.

There were many landing places along estuaries, and the *Liber Llandavensis* notes how *c.*895 King Brochfael granted Bishop Cyfeiliog two churches and 'free landing rights for ships at the mouth of the Troggy', a river now know as the Nedern that empties into the Severn estuary near Caldicot in Monmouthshire, and at about the same time returned *Yscuit Cyst* with its free landing rights at the mouth of the Meurig a little further north.[21] Along several estuaries, seagoing boats could have been beached on the shore, as suggested along the west bank of the Itchen near *Hamwic* and beside the Thames at Aldwych, the 'old' *wīc* of London.[22] In some of these locations, double tides would have assisted the launching of boats. Although the Roman quayside in London consisted of massive oak timbers, Anglo-Saxon traders appear to have preferred at first to beach their boats upon the foreshore, perhaps influenced by the lowering of water levels during the Roman period. A pre-Conquest survey of the estate of Tidenham (Gloucs.)[23] refers to 'scipwealas' ('sailors') who held some land at 'Kingston' for rent, apparently a reference to the Welsh sailors who plied up and down the Wye.[24]

Place-name evidence contributes the term OE *hȳð* for a 'landing place or harbour'. In 898/9, the Bishop of Worcester and the Archbishop of Canterbury were both granted the right to moor ships along the width of their properties at *Æðeredes hyd*, the later Queenhithe, on the Thames: the earliest evidence, if dated correctly, for commercial activity in the newly refounded city of London.[25] The authenticity of the charter is not in doubt, although the document is only available in later copies, the earliest dated to the twelfth century. The grant consisted of two *iugera* divided by the public road from the River Thames, one given to Archbishop Plegmund and to Christ Church, the other to Bishop Wærferth and the Church of Worcester. Both plots extended as

---

[18] E. M. Jope, 'The Saxon Building-Stone Industry in Southern and Midland England' .

[19] J. Blair, 'The Minsters of the Thames', in J. Blair and B. Golding (eds.), *The Cloister and the World: Essays in Medieval History in Honour of Barbara Harvey* (Oxford, 1996), 5–28.

[20] A. Pearson and J. E. Potter, 'Church Building Fabrics on Romney Marsh and the Marshland Fringe: A Geological Perspective', *Landscape History*, 24 (2002), 87–107, esp. 93–6, 104–5.

[21] W. Davies, *An Early Welsh Microcosm: Studies in the Llandaff Charters* (London, 1978), 183; Eaton, *Plundering the Past*, 44–5.

[22] M. Welch, *Anglo-Saxon England* (London, 1992), 118–19.   [23] S 1555.

[24] *RASC* 205, no. 109.   [25] S 1628.

far as the town wall, and outside the wall were wharves, *navium staciones*, of the same width as that of the *iugera* within the wall:

'... duo jugera ad locum qui dicitur *ÆÐEREDES HYD* ... Est autem via publica a flumine Tamis. dividens hæc duo jugera. Et tendens in aquilonem. ambo autem jugera in murum protelantur. Et extra murum navium staciones tante latitudinis quante et jugera sunt intra murum. Habet vero jugerum ecclesie Christi. artam semitam. in occidente. Jugerum Wygornacensis ecclesie viam artam ab oriente. Caput amborum jugerorum semita ad orientem dirimitur.

'... two *iugera* at the place which is called *ÆÐEREDES HYD* ... The public road from the River Thames is, however, separating these two *iugera* and, extending to the north, both *iugera* are driven forth within the wall. And there are wharves outside the wall of the same width as the *iugera* within the wall. The *iugerum* of Christ Church has, in truth, a narrow path on the west, the *iugerum* of the Church of Worcester a narrow way on the east; the end of both *iugera* is separated by a path [leading] to the east.

A second, less trustworthy version of the charter names the location as *At Eredyshythe*.[26] The identification of *Æðeredes hyd* with Queenhithe is not in doubt as *Edredeshede* is called *Ripa Regine anglice Quenhyth* in a twelfth-century charter.[27] The site has been identified as Bull Wharf, and excavations here have revealed lines of insubstantial timber trestles, which may have supported a walkway to vessels moored alongside, and possible evidence for barge beds where the barges had been beached. However, timber wharves consisting of low revetments were soon to be erected, utilizing old timbers, in the late Anglo-Saxon period. Only in the eleventh century, between 1021 and 1045, were these to be replaced by more substantial embankments, a response perhaps to serious flooding and tidal changes.[28]

The *hȳð* term is found in other Thameside locations in the early recorded place names Erith, Stepney, and Chelsea, sometimes indicating the kind of goods being landed: perhaps chalk or limestone at Chelsea or lambs at Lambeth (noted in the later place name *Lamhytha* in 1088, 'landing place for lambs').[29] It is also found on other rivers: Lakenheath (*Lacingahið* eleventh century), 'landing place of the people living by the streams', is on the Little Ouse in Suffolk; the Little Ouse joins the Great Ouse and flows into the Wash. The term was not confined to river wharves but also occurs in coastal locations, as at Hythe in Kent (*Hede* 1086), but was most common along navigable rivers such as the Thames, lower Trent, and Ouse (for other names, and the reasons for the choice of these locations, see Cole, below, pp. 61–74). A similar OE word is *stæð*, but the sense 'landing place' is not, according to Smith, evidenced before the fourteenth century. He interprets its earlier meaning as 'the bank of a river, a shore',[30] but Mills[31] claims that the name of Stafford, the borough founded

---

[26] Again S 1628.
[27] M. Gelling, *The Early Charters of the Thames Valley* (Leicester, 1979), 188.
[28] [R. Wroe-Brown], 'Bull Wharf Queenhythe', *Current Archaeology*, 158 (1998), 757.
[29] Place names after Mills, *Dictionary*.
[30] Smith, *English Place-Name Elements*, ii. 142.   [31] Mills, *Dictionary*, 305.

by Æthelflæd in 913, means 'ford by a landing place', the borough being sited on the River Sow, a tributary of the Trent (discussed further by Cole, p. 75).

The use of lesser rivers can only be estimated from the type of circumstantial evidence discussed above, but there is also evidence of new channels being deliberately cut in the early medieval period to ease navigational problems. Perhaps the best-known example is a new channel dug by the abbey of Abingdon for river traffic in the mid-eleventh century when a stretch of the Thames was silting up, greatly inconveniencing Oxford traders (see Bond and Blair, below, pp. 179–80, 258, 266–8). Haslam has also suggested a similar man-made channel to aid river-borne trade and the construction of a mill in the Edwardian borough of Cambridge.[32]

While rivers were probably arteries for trade, they were also increasingly used in ways that had the propensity to create impediments for navigation. Rights to the use of running waters are a common component of the scribal formulae which relate to the appurtenances of estates in pre-Conquest charters, and are rarely specified, but fishing rights are also mentioned in some charters. The construction of various features associated with fisheries and mills may have increasingly hindered navigation.

Watermills were becoming common by late Anglo-Saxon times and begin to appear in place names and charters in the ninth century. A place name *Mylentun*, referring to an unidentified place near Kemsing in Kent, is recorded in 822[33] in a charter known to be authentic, and a 'mylen pul' is recorded upon the boundary of Stoke Bishop (Gloucs.), in 883, again in a charter that is probably authentic.[34] By the tenth century, mills were becoming increasingly common, especially, at first, upon royal estates. There may have been three pre-Conquest mills at the *villa regalis* of Calne in Wiltshire[35] but some of the best known are the two mills of the Mercian borough of Tamworth. The first was a horizontal-wheeled watermill located just outside the Anglo-Saxon defences. This was subsequently abandoned, to be replaced by a second mill, also of a horizontal-wheeled design and dated by dendrochronology to the mid-ninth century, which was destroyed by fire before the Norman Conquest.[36]

In East and West Woolstone in the Vale of the White Horse (Oxon.) a mill appears to have been built on the River Ock there between 856 and 958, probably after 944.[37] By the later ninth and tenth centuries, mills not infrequently appear in charters, found among the appurtenances of estates (e.g. in 822 at Milton in Seal, Kent; in 968 at Bemerton, Wilts.; in 963×975

[32] J. Haslam, 'The Towns of Wiltshire', in J. Haslam (ed.), *Anglo-Saxon Towns in Southern England* (Chichester, 1984), 281 n. 22.

[33] S 186.   [34] S 218.   [35] Haslam, 'The Towns of Wiltshire', 106.

[36] P. A. Rahtz and R. Meeson, *An Anglo-Saxon Watermill at Tamworth*, CBA Research Rep. 83 (London, 1992).

[37] S 317 and S 575; Hooke, *Anglo-Saxon Landscapes of the West Midlands*, 267–8.

at Downton, Hants),[38] in boundary clauses, and as additions to a charter grant or lease. Examples of the latter are recorded at Holborough (Kent) in 838 where the charter notes 'et unam molinam in torrente qui dicitur Holan beorges burna' ('and a mill in the torrent [river] which is called Holborough's bourne'),[39] and at Longstock, (Hants) in 982 where 'se mylenham and se myln ðærto', ('the mill ham (loosely interpreted as meadow) and the mill belonging [to it]') are added to the grant.[40] By the tenth to eleventh centuries, mills are recorded in charters as far distant as Cornwall (despite the small number recorded for that county in Domesday Book) on the ecclesiastical estates of Burnewhall in West Penwith, of Tinnell in Landulph ('the king's mill' beside the Tamar), and at Trerice in St Dennis on the Fal.[41]

The church was quick to encourage features which would be of economic benefit on its estates. If 'mylen-steall' may rightly be interpreted as 'the site for a mill'[42] then the charters may catch the provision for the building of mills on certain estates, often as they were leased out. In 871 × 877 Bishop Ealhferth and the community at Winchester leased out an estate at Easton (Hants) the boundary of which ran down to 'ðone mylensteall' on the Itchen.[43] In Warwickshire, 'þreo æcras benorðan afene to mylln stealle' ('three acres north of the Avon as a site for a mill') were included in a lease of three hides at Alveston, Upper Stratford, and *Fachanleah* by Bishop Oswald (of Worcester) in 966.[44] On the Warwickshire Stour at Blackwell in Tredington, another 'mylen stall' ('site for a mill') was added to another of Oswald's leases in 977.[45]

But were these mills located upon the main channels of streams and rivers? In most cases, the answer must surely be no, for most sites continued in use for centuries and the locations of later, medieval, mills are well known. In most cases a leat drew water from the main channel in order to obtain a maximum head of water to feed the wheel which was placed on such a side channel; the water was then returned to the main stream by another leat. Even when the mills were located close to the river, as at Bath, they lay upon side channels separated from the main flow of the river.[46] The mill-pool was often a source of fish and eels. Such leats were man-made and could be of impressive dimensions. At Old Windsor, beside the Thames in what is now Berkshire, the mill, at first consisting of three vertical wheels, was fed by a leat over 6 m wide and 3–4 m deep; this was cut across a bend in the Thames and may have been over a kilometre long. The mill and a nearby stone building were destroyed in the early tenth century when the leat was filled in, but the latter was then recut to drive another horizontal-wheeled mill which continued in use until the early

---

[38] S 186, S 767, S 821.   [39] S 280.   [40] S 840.
[41] D. Hooke, *Pre-Conquest Charter-Bounds of Devon and Cornwall* (Woodbridge, 1994), 26, 59–60, 66.
[42] *RASC* 88–99.   [43] S 1275.
[44] S 1310; D. Hooke, *Warwickshire Anglo-Saxon Charter-Bounds* (Woodbridge, 1999), 52.
[45] S 1330; Hooke, *Warwickshire Anglo-Saxon Charter-Bounds*, 89–94.
[46] B. Cunliffe, 'Saxon Bath', in Haslam, *Anglo-Saxon Towns*, fig. 116.

eleventh century.[47] At Totnes in Devon, too, streams flowing into the Dart appear to have been canalized to feed the town mill at the time the borough was established.[48]

A mill-leat was usually referred to as a 'mylendic' ('a mill-dyke')[49] but occasionally there are references to a 'mylen wær', 'mylenwaru', 'myle(n)wer' 'a mill weir, mill-dam'. In the late sixth century, Gregory of Tours relates how the Abbot of Loches built a mill on the River Indre to relieve his monks from the labour of hand-grinding flour, and made a weir

defixisque per flumen palis, adgregatis lapidum magnorum acervis, exclusas fecit atque aquam canale collegit, cuius impetu rotam fabricae in magna volubilitate vertere fecit.

He had piles driven into the river, and collected a mass of great stones and made a dam and a channel for the water, whose force would turn the mill-wheel with great speed.[50]

The weir could, therefore, be a dam directing the flow of water to the mill. Mills and weirs are included (but not necessarily closely in association), as 'mid milnan … mid waterum. 7 mid werum', in the appurtenances of several writs issued by Edward the Confessor declaring that he has granted estates at Pershore (Worcs.), Deerhurst (Gloucs.), and Staines (formerly Middlesex, now Surrey) to Westminster Abbey,[51] but these documents are not necessarily in their original form.[52] In Warwickshire charters, the weir appears in association with the mill in the appurtenances of an estate at Ruin Clifford (in Stratford-upon-Avon) in 988 where the appurtenances include 'et aquis … mid were 7 mid mylene'.[53] In boundary clauses, a mill-weir is noted as a landmark on the boundary of an estate beside the River Nadder in Wiltshire in the mid-tenth century (956 for 959), the bounds running 'and lang streames on þa mylen ware' ('along the stream to the mill-weir'),[54] and in a charter for Witney (Oxon.), dated to 1044, where the bounds run 'to ðam mylewere ðe hyrnð into duceling dune' ('to the mill-weir that belongs to Ducklington').[55] Another term appearing as *mylengeares* in a Winchester prayer-book c.900 and as *mulenger* in the bounds of Padworth (Berks.), in 956[56] is less easy to understand. At Padworth it seems, like the *myle[n]wer*, to have described a stream-dam holding back the mill-race until it was released down the *mylendic*, thus 'holding back or diverting a watercourse to form a head of water for the mill'.[57]

[47] G. Astill, *Historic Towns in Berkshire: an Archaeological Appraisal* (Reading, 1978), 70–1; G. Astill, 'The Towns of Berkshire', in Haslam, *Anglo-Saxon Towns*, 81; D. M. Wilson and J. G. Hurst, 'Medieval Britain in 1957', *MA*, 2 (1958), pp. 183–5.

[48] J. Haslam, 'The Towns of Devon', in Haslam, *Anglo-Saxon Towns*, 261.

[49] e.g. S 308, S 620, S 842.

[50] Gregory of Tours, 'Liber Vitae Patrum', c. 18, ed. B. Krisch, *Monumenta Germaniae Historica: Scripta Rerum Merovingicarum*, i (Hanover, 1884), 734–5; trans. E. James, *Gregory of Tours, Life of the Fathers* (Liverpool, 1985), 122.

[51] S 1146, S 1142. See, too, similar scribal formulae in other writs of Edward such as S 1142, in which Westminster Abbey receives an estate at Staines, (Middx) and land at *Stæningahaga* in London in which the appurtenances include rights 'on waterin and on weren'.

[52] See comments in S. [53] S 1356. [54] S 586. [55] S 1001.

[56] S 1560, S 620. [57] Rahtz and Bullough, 'The Parts of an Anglo-Saxon Mill', 23–6.

In many cases, however, the weir was also used as part of a fishery. A Carolingian capitulary of 802–13 lays down that the good steward of a royal estate should properly maintain 'vivaria cum pisces, vennas, molina' ('fishponds, fish-traps, ponds closed by a weir, mills') and 'vivaria cum pisces' in royal forests.[58] One of the duties of 'the discriminating reeve' in *Gerefa*, a text produced in tenth- or eleventh-century England, was 'fiscwer 7 mylne macian' ('to construct a fish-weir and mill').[59]

Weir place names are in evidence in pre-Conquest contexts at a number of places. Wareham in Dorset, 'homestead or river-meadow by a weir', is recorded from Asser, in the late 9th century, onwards: William of Malmesbury claims that Aldhelm visited the place, where he built a church, on his way to Rome *c.*698—the place name is given in a marginal note of the twelfth-century manuscript[60]—and five memorial stones from an earlier Celtic burial ground were built into the piers of the nave of the ninth-century church.[61] The weir was probably a fish-weir across one channel of the River Frome, in the same location as the twelfth-century weir. Here the waters were tidal and initially reserved for the Crown.[62] Other manors recorded in Domesday Book include Ware (Herts.) 'the weirs', Warham (Norfolk) 'homestead or village by a weir', and Warwick (Warwicks.) 'dwellings by the weir or river-dam' (like Stratford, on the Avon), while Great and Little Warley (Essex) may be 'wood or clearing near a weir', as is Wardley (Leics.) in 1067.[63]

In addition to the weirs noted above, fisheries are noted among the appurtenances in some of the earliest Anglo-Saxon charters, especially those of Kent. Here, rights in *piscaris/piscariis* are recorded as early as the later seventh century on the Isle of Thanet,[64] and 'capturam piscium quae habetur in ostio fluminis cuius nomen est Limenea. ... cum domibus piscatorum' ('the catching of fish which is in the mouth of the river whose name is *Limen* ... with fish-houses') is recorded in the eighth century, donated to Abbot Dunn and Christ Church, Canterbury, in 732 and 741.[65] Sea fishing obviously took place and a charter of *Hwitanclife* (Kent) refers to the toll on the catch of one fishing boat ('navis piscationem') in 962.[66] While some fishing may also have been carried out in salt-water coastal channels, as at Graveney, Kent, where the rights in 811 included 'piscuosis ac maritimis fretibus paludibus uallibusque

[58] 'Capitulare Aquisgranense' (801–13) c.19 (ed. A. Boretius, *Capitularia Regum Francorum*, i: *MGH Legum*, ii (Hanover, 1883), 172; 20); Rahtz and Bullough, 'Parts of a Mill', 20; for observations on the meaning of *venna* and OFr *vanna*, see their n.3.

[59] F. Liebermann, *Die Gestetze der Angelsachsen* (Leipzig, 1903), Gerefa 9: 1. 454; P. D. A. Harvey, 'Rectitudines Singularum Personarum and Gerefa', *English Historical Review*, 108 (1993), 1–22.

[60] *Willelmi Malmesbiriensis Monachi de Gestis Pontificum Anglorum*, ed. N. E. S. A. Hamilton, Rolls Ser. 52 (London, 1870), 363–4.

[61] B. Yorke, *Wessex in the Early Middle Ages* (London, 1995), 69–72.

[62] H. J. S. Clark, 'The Salmon Fishery and Weir at Wareham', *Proc. Dorset Nat. Hist. & Archaeol. Soc.* 72 (1950), 99–110.

[63] Mills, *Dictionary*, 346, 347.   [64] S 58, S 515.   [65] S 23, authentic charter; S 1611.
[66] S 701.

dulcis' ('sea-bays abounding in fish, marshes, and sweet vales'),[67] others in
the same kingdom were associated with salt-water marshlands. But most fisheries noted in charters seem to have been in rivers or estuaries, although
those near Little Thetford (Cambs.), must have been in freshwater marshlands
beside the Ouse.[68] At Ripple in Worcestershire, an estate on the River Severn, for instance, the appurtenances included 'fluminales piscationes' ('riverine
fisheries'). In central Warwickshire at *Ufera Stretford* on the River Avon,
allegedly in 845, they again included 'piscationes', and at Clifford Chambers
mention is made in 922 of profits from 'fixnoð' ('fisheries') on the River
Avon or the River Stour.[69] There were breeding pools (OE *tēam pōl*) in the
estuary of the Exe below Topsham[70] and later evidence indicates rich salmon
fisheries.

A ninth-century charter relating to land at Bromhey in Frindsbury (Kent),
adds a fishery in the Thames called *Fiscnæs*.[71] When, in 983, King Æthelred
granted another fishery on the Darent in Kent to Bishop Æthelwold, it was
referred to as 'a trap (*captura*) constructed in the River Darent, called *Ginanhecce* in common speech, and a small parcel of land useful to the fisherman
(*piscator*) at the said trap and to the ?bailiff (*procurator*) for necessary purposes'. Moreover, it seems that the fishing rights were protected by a curse:

Haec sunt nomina gurgitum qui ad piscandum extraneis cum anathemati prohibiti sunt.
Cytala pol. Lympol. Wyllen muðe.

Here are the names of the waters which strangers are forbidden with a curse from
fishing: Cytwala pool, Lym pool, Wylles mouth.[72]

Other fisheries granted by charter include 'captura piscium' ('fish-traps') at
Twinam (Christchurch, Hants), which is on the south coast at the mouths of
the rivers Stour and Avon (this is a lost charter of 954 noted in Glastonbury
Abbey records),[73] and another fishery at *Bræge* restored to the Old Minster at
Winchester in 996.[74]

The archaeological evidence for fish as part of the diet of early medieval
England has recently been assessed by Barrett, Locker, and Roberts.[75] Sea-fish
were traded inland to urban centres: herring, in particular, has been noted in
the environmental evidence available from the wics of York, Ipswich, London,
and *Hamwic* (Southampton) between the late seventh and tenth centuries. The
food preferences of Scandinavian immigrants may have led to the intensification

---

[67] S 168.   [68] C. R. Hart, *The Early Charters of Eastern England* (Leicester, 1966), 80, no. 28.
[69] Ripple: S 52 (but this is a charter of doubtful authenticity); *Ufera Stretford*: S 198, ('suspicious');
Clifford Chambers: S1289 (authentic).
[70] S 433; Hooke, *Pre-Conquest Charter-Bounds of Devon and Cornwall*, 122–6.
[71] S 157.   [72] S 849.
[73] S 1741, H. P. R. Finberg, *The Early Charters of Wessex* (Leicester, 1964), 45, no. 76; L. Abrams,
Anglo-Saxon Glastonbury: Church and Endowment (Woodbridge, 1996), 234–5.
[74] S 889.
[75] J. H. Barrett, A. M. Locker, and C. M. Roberts, '"Dark Age Economics" Revisited: The English
Fish Bone Evidence AD 600–1600', *Antiquity*, 78 (2004), 618–36.

of fishing in the ninth and tenth centuries, as noted in northern Scotland. A shift to the consumption of marine fish is evident at the end of the first millennium: the importance of herring increased fourfold on urban sites in southern and eastern England in the eleventh to twelfth centuries with codlike fish appearing as a significant component only at this time. It is not yet possible to know if this merely reflects increased demand for fish in general but, after 1000, Barret et al. note that fishing appears to have become more regulated and intensively practised through the use of fish-traps (for eels and salmon) and the construction of fishponds, perhaps as the proliferation of such obstacles as mill-dams hampered the availability of migratory and freshwater fish. Monastic fishponds may have become increasingly important as dietary requirements encouraged fish consumption, arguably after the Benedictine reforms at the end of the tenth century.

How much, then, is known about the early medieval art of fishing? Ælfric's fishermen took their boats and cast their nets into the river, using baited hooks and baskets; however, they rarely ventured into the open sea in their rowboats, presumably catching 'hæringas 7 leaxas, mereswyn 7 stirian, ostran 7 crabban, muslan, winewinclan, sæcoccas, fagc 7 floc 7 lopystran 7 fela swylces' ('herrings and salmon, porpoises and sturgeon, oysters and crabs, mussels, winkles, cockles, plaice and flounders and lobsters, and many similar things') in coastal waters, but Ælfric's fisherman notes that 'I can't catch as many as I can sell'. Other fish named (from inland waters) were 'ælas 7 hacodas, mynas 7 æleputan, sceotan 7 lampredan, 7 swa hwylce swa on wætere swymmaþ. Sprote', ('eels and pike, minnows and turbot, trout and lampreys, and whatever swims in the water. ?Small fish').[76]

Charter references occasionally shed further light on the question of fishing methods. Grants of estates which included Ickham and Palmstead (Kent), dated 785 and 786, included a fishery, adding 'atque unius hominis piscatum in ðæm pusting uueræ/were', 'fishing for one man in the *pusting* weir'.[77] The 'pusting were' may have been similar to, or the same feature as, the 'cytwere' described below. Land at Ombersley beside the Severn in Worcestershire was allegedly associated with two weirs:

in captura etiam piscium que terre illi adiacet. ubi sunt scilicet [duo] quod nostratim dicitur weres. id est alter ubi fontanus qui nominatur Ombreswelle deriuatur in fluuium qui dicitur Saberna. alter qui est ad uadum qui nuncupatur Leuerford.

also in the fish-trap which adjoins the land, namely where there are two of what amongst us are called *weres* (weirs), the one, that is, where the spring which is called Ombreswell flows into the river which is called Severn, the other which is at the ford which is called Leverford.[78]

---

[76] G. N. Garmonsway, *Ælfric's Colloquy*, Exeter Medieval Texts & Studies (Exeter, 1991), 2–9; M. Swanton, *Anglo-Saxon Prose* (London, 1975), 110. Old English *sprott* is literally 'sprat'.

[77] S 123 (known to be authentic); S 125.

[78] S 46; D. Hooke, *Worcestershire Anglo-Saxon Charter-Bounds* (Woodbridge, 1990), 36–40.

Again, at Fenstanton (Cambs.), fishing (*fixnað* for *fiscað* 'fishing', but perhaps confused with *fiscnett* 'a fishing-net') was carried out at weirs in the Great Ouse, probably in the eleventh century: 'æt holanwere. 7 æt deopanwere. 7 æt suðan ea. 7 æt niwanwere. 7 æt dinde. 7 æt biscopes were. 7 æt bradan were. 7 æt niwanwere. [sic] 7 æt merbece'.[79] Also amongst the appurtenances of estates, other weirs include one on the River Itchen which is added to a grant of Candover (Hants) allegedly in 900. This is in a document which may be a later forgery, but that added to an estate of Upton-on-Severn (Worcs.), in a lease of 962, is authentic.[80] Several of the ninth- and tenth-century charters recorded in the *Liber Llandavensis* also grant estates with weirs either on the Severn itself or on its tributaries: ('in/cum coretibus') at *Cairnonui*, St Julians at Caerleon, *Yscuit Cyst* with 'its weirs on the Severn and on the Meurig on both banks', and Caldicot where the 'Troggy' (Nedern) flows into the Severn.[81]

Weirs do not appear frequently in boundary clauses, but one for *Hysseburnan* (St Mary Bourne and Hurstbourne Priors, Hants) notes that the bounds begin at *twyfyrde* on the River Test, a ford where the western boundary of Hurstbourne Priors leaves the river, and eventually return to the Test at 'þone syþeran stēð. þonne 7 lang steþes 7 be neoðan beamwǣr on þone norþere stēþ' ('the southern bank/shore; thence along the bank/shore that is below the wooden weir to the northern bank/shore') before returning to *twyfyrde*. Unfortunately this charter, ascribed to 900, is not an entirely reliable document.[82] Other more reliable charters referring to weirs in their bounds include one of 801 which refers to 'bregedeswer' on the boundary of Butleigh (Somerset); one of 961 referring to 'eadmundes wer' on the boundary of Easton near Winchester (Hants); 'caluwan wer' on the boundary of Olney (Bucks.), in 979; and 'ægces wer' at *Lothers leage* (Middx.) in 972–8.[83]

Few charters are as detailed, however, as the documents for an estate at Tidenham (Gloucs.), one of which provides further details of the fish-traps themselves. In 1061 × 1065 Archbishop Stigand received a life-lease from the community at Bath which included an annual rent of 1 mark of gold, 6 porpoises ('merswun') and 30,000 herrings ('hæringys') annually,[84] apparently referring to a sea catch, but a further survey provides much fuller details of the fishing methods employed in the rivers themselves by noting the rights of individual estates within the capital manor:

To stræt synd .XII. hida ... 7 on sæuerne .xxx. cytweras. To middel tune .V. hida ... XIIII cytweras on sæuerne. 7 .II. hæcweras on wæge. To cinges tune .V. hida sind ... on

---

[79] Hart, *The Early Charters of Eastern England*, 33, no. 30, treats this as part of the 11th-century charter of Æthelred to Bishop Godwine, but S 1562 separates the bounds and rights from the main grant.

[80] S 360; S 1300.

[81] 60. J. Gwenogvryn Evans, *The Text of the Book of Llan Dâv Reproduced from the Gwysaney Manuscript* (Oxford, 1893; repr. National Library of Wales, 1979), 221, 225, 234, 236; see also Davies, *An Early Welsh Microcosm*, 181, 183.

[82] S 359.   [83] S 270a; S 695; S 834; S 1451.   [84] S 1426.

sæuerne .XXI. cytwera. 7 on wæge XII. To bispes tune synd .III. hida. 7 .XV. cytweras. on wæge. On land cawet synd .III. hida. 7 .II. hæcweras on wæge. 7 .IX. cytweras. .... Æt ælcum were þe binnan þam .XXX. hidan is. ge byreð æfre se oðer fisc þam land hlaforde. 7 ælc seldsynde fisc þe weordlic byð. styria. 7 mere swyn. healic oðer sæfisc. 7 nah man nænne fisc wið feo to syllane þone hlaford on land byð ær man hine him gecyðe.

Se gebur sceal his riht don. ... to wer bolde .XL. mæra oðde an foþer gyrda. oðde .VIII. geocu byld .III. ebban tyne.

At Stroat there are 12 hides ... and 30 basket weirs on the Severn ... At Milton 5 hides ... 14 basket weirs on the Severn and 2 hackle weirs on the Wye. At *Kingston* there are 5 hides ... 21 basket weirs on the Severn and 12 on the Wye. At Bishton there are 3 hides and 15 basket weirs on the Wye. In Landcaut there are 3 hides and 2 hackle weirs on the Wye and 9 basket weirs. ... At every weir within the 30 hides every alternate fish belongs to the lord of the manor and every rare fish which is of value—sturgeon or porpoise, herring or sea fish; and no one has the right of selling any fish for money, when the lord is on the estate, without informing him about it ...

For weir-building the *gebur* must supply 40 larger rods (?), or a fother of small rods, or he shall build 8 yokes for 3 ebb tides ...[85]

The *cytweras* appear to have been the basket weirs employed on both rivers as late as the nineteenth century, then used to catch salmon, as described by Seebohm: 'wattled basket-hedge weirs' rather than any solid structure. They had been devised 'to meet the difficulty presented by the unusual volume and rapidity of the tidal current'. In the nineteenth century the 'cytweir' consisted of

rows two or three deep of long tapering baskets arranged between upright stakes at regular distances. These baskets are called *putts* or *butts* or *kypes*, and are made of long rods wattled together by smaller ones, with a wide mouth, and gradually tapering almost to a point at the smaller or butt end. These *putts* are placed in groups of six or nine between each pair of stakes, with their mouths set against the outrunning stream; and each group of them between its two stakes is called a 'puttcher' [from *putts weir*, i.e. a weir made of *putts*].[86]

The *hæcwer*, the 'hackle weir', in contrast, was a barrier or fence of wattle set across the current to produce an eddy in which the fish could be caught from a boat with a stop-net. The Tidenham charter refers to a smaller number of such weirs. In the nineteenth century the hackle weir was used almost exclusively for the capture of salmon and, according to Seebohm, both these fishing methods were then peculiar to the Wye and the Severn (today, local terminology for the various features may be different from that used by Seebohm). Numerous weirs are also mentioned in a post-Conquest document noting the rents due to Thorney Abbey from its fenland estates. These also include a 'tynadwere' ('fenced weir').[87]

[85] S 1555; *RASC* 204–7.

[86] F. Seebohm, *The English Village Community* (4th edn. London, 1890), 150–3. 'Putcher' is a west midland dialect term (*OED* 1655).

[87] *RASC*, appendix 2, 252–7, no. 9.

The fishery at *Ginanhecce* in Kent takes its name from OE *hæc(c), hec(c)* 'a hatch, a grating, a half-gate, a gate', *∗hæcce* 'a fence'. This term may have described a sluice, a flood-gate, or a trap, presumably a small-scale fixture which operated in a similar way to a weir but may also have been used for other kinds of traps in non-riverine locations as well—Wollage in Kent appears to have referred to a wolf-trap.[88] Like the Leverford associated with one of the Ombersley fisheries, Hackforth in North Yorkshire and Hackford in Norfolk may each refer to a 'hatch' by a ford, and other names include Hatch Warren in Hampshire (*Heche* 1086) and Hatch Beauchamp in Somerset (*Hache* 1086).

Fish-weirs appear, therefore, to have been relatively fixed structures, consisting of barriers across the stream to direct fish into a trap in which they would be caught, or to produce an eddy in which the fish would become trapped and netted. Such features have now been identified archaeologically. A pre-Conquest fishing weir at Colwick (Notts.) was uncovered in a gravel pit beside the River Trent in 1978. It consisted of a double row of oak, holly, and hawthorn posts set about 1 m down into the river bed. One sample post was dated by radiocarbon dating and dendrochronology to *c.*662–764. The posts were interlaced with hurdles of hazel and willow which have themselves been dated to 810–80. These would have needed constant renewal, suggesting that the weir was in use between the eighth and ninth centuries. Earlier, in 1973, a later weir had been uncovered in the same parish. This again consisted of oak and holly posts driven into the gravel some 50 cm apart, and was again interlaced with wattle hurdles (of hazel and alder) standing vertically against the posts and kept in place with a packing of clay containing bundles of branching twigs bound with knotted withies. This weir was dated to the eleventh century, and may be related to the fishery recorded in 1086 on the manor of Colwick. The 15 acres (6 hectares) of small wood recorded on the manor may have provided the coppiced timber necessary for the construction of the weir.[89] One arm of the weir seems to have started in the shallows, running out into deeper water to join a longer arm, thus producing an angled barrier with the 'V' pointing downstream to make a funnel-shaped weir. The excavators suggest that coarse fish such as lampreys and eels might be thus driven towards gaps in the wattle where nets or baskets would have been placed. The excavated examples at Colwick are but a few of the possible sites along the Trent where stake alignments associated with abandoned channels of the river have been noted, although most of these may have been bank revetments.

The concept was not a new one in early medieval times, for rows of wooden stakes joined by wattling found in an ancient river bed at New Ferry, Lough Begg, N. Ireland, have been dated to 1000 BC,[90] their appearance identical

---

[88] Smith, *English Place-Name Elements*, i. 213.

[89] P. M. Losco-Bradley and C. R. Salisbury, 'A Saxon and a Norman Fish Weir at Colwick, Nottinghamshire', in M. Aston (ed.), *Medieval Fish, Fisheries and Fishponds in England*, BAR British Ser. 182 (2 vols., Oxford, 1988), 329–52.

[90] N. C. Mitchel, 'The Lower Bann Fisheries', *Ulster Folk Life*, 11 (1965), 1.

to that of eel-weirs still in use on the River Bann. Rows of pointed stakes have also been found in a gravel pit alongside the Thames at Shepperton and radiocarbon dated to the fifth century[91] while lengths of wattle fencing have been found in the river silts of the Witham at Lincoln, interpreted as fish-weirs dating from the second and tenth centuries AD.[92]

In Domesday Book, fisheries were common along English rivers (Fig. 7), and Darby notes how 'the mention of a fishery in Domesday Book may imply some kind of fixed contrivance such as a weir or fish-trap'. He also notes that weirs in eastern England were referred to as *gurgites*.[93] Boats plied on Whittlesey and Soham meres and elsewhere in the Fens, nets were used at Swaffham in Cambridgeshire, and boats and nets on the Dee, while on the Thames seines and dragnets ('De sagenis et tractis in aqua Temisiae') were in

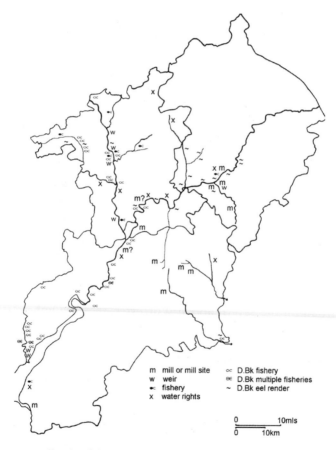

m   mill or mill site   oc D.Bk fishery
w   weir   œ D.Bk multiple fisheries
◄ fishery   ~ D.Bk eel render
x   water rights

0 _____ 10mls
0 _____ 10km

Fig. 7. The west midlands: fisheries and mills in pre-Conquest charters with additional Domesday Book fisheries.

[91] Bird, cited in Losco-Bradley and Salisbury, 'A Saxon and Norman Fish Weir', 345.
[92] B. Gilmour, 'Brayford Wharf East', *Tenth Ann. Rep. Lincoln Archaeol. Trust* (1982), 22–4.
[93] H. C. Darby, *Domesday England* (Combridge, 1977), 280–6. Cf. below, p. 263 n. 42.

use. There are only a few references to a 'vivarium piscium' ('a fish-stew'), as in eastern England on the holding of Osbern the fisherman at Sharnbrook on the River Ouse in Bedfordshire, on the monastic land at Bury St Edmunds, at St Albans in Hertfordshire, and at Caversfield in Oxfordshire; an earlier fishpond may be represented by the 'styrian pol', 'stirigan pole' (OE *styria*, used of various fishes) in two Berkshire charters of allegedly mid-tenth-century date describing landmarks on the boundary between Besselsleigh and Cumnor,[94] but neither is a reliable document. Although most fisheries were in rivers, sea fishing took place off the Dorset coast and the coast of East Anglia: herring renders are recorded at eighteen places in the three Suffolk hundreds of Blything, Lothingland, and Wangford, including 68,000 herrings due from Dunwich, 60,000 from Beccles, and 25,000 from Southwold. Fishing also took place along the shore, for the abbey of Bury St Edmunds had a moiety of a sea-weir and a fourth part of another at Southwold. The 'heia de riseburc' which belonged to the manor of Blythburgh was also some kind of fixed net or dam along the shore. Coastal fisheries of this nature were common in medieval England, and on the north Somerset coast were composed of lines of beach boulders forming drystone walls erected to form a V-shape with the apex pointing out to sea. A net would be stretched across the neck of the 'V' in which fish would have become trapped as the tide receded.[95] Of the sea-fish, herrings are most prominent but porpoises are noted as well, as at Southease and Stone in Kent (where porpoises are also recorded in the Domesday Monachorum at Gillingham).[96] These are two of the creatures also named in the Tidenham survey discussed here.

It seems, however, that eels were the main catch from the Domesday fisheries and millponds, although 1,000 lampreys in addition to 1,000 eels were rendered by the manor of Petersham in Surrey and salmon were caught in the Severn in Gloucestershire (sixteen rendered to the church of St Peter in Gloucester), in the Dee near Chester, and in south Devon in the rivers Dart and Avon.[97] On Worcestershire manors eel fisheries are recorded at Grimley, Ombersley, and Hallow, all on the Severn above Worcester, that at Ombersley rendering 2,000 eels. At Martley two weirs rendered 2,500 eels and 5 'stitches' (an additional 625 eels), probably from weirs on the River Teme. The Stratford-on-Avon fisheries, first noted in the appurtenances of a charter relating to *Ufera Stretford* in 845 as 'fluminibus uel piscationibus' ('rivers or fisheries'),[98] still rendered 1,000 eels in 1086. Tidenham remained the outstanding fishery in Gloucestershire in 1086, with 65 fisheries of which 55 were in the Severn, 8 in the Wye, and 2 in unspecified locations.[99]

[94] S 673, S 757.
[95] M. Aston and E. Dennison, 'Fishponds in Somerset. Appendix 1', in Aston (ed.), *Medieval Fish, Fisheries and Fishponds*, 401–3.
[96] Darby, *Domesday England*, 285–6, fig. 101.   [97] Ibid. 283.   [98] S 198.
[99] DB i. 164, 166[v], 167[v].

It is difficult to ascertain how far fisheries impeded navigation in early medieval times. It is clear that semi-permanent weirs were present in most major rivers. In the medieval period, however, a space had normally to be left at one end of the weir to create a free passage in order to preserve fish stocks, and on the River Severn it was customary to maintain a bypass around the weirs in order to allow migrating fish, both salmon and eels, to pass upstream. Use was often made of natural braided channels; or an artificial 'gutter' could be cut on the adjoining floodplain to create an island and 'bylet'.[100] Such channels could also be used by river craft. The twelfth-century weir at Wareham made use of one channel close to the castle in the braided River Frome, although this was gradually to silt up; later another weir was to be built across the second channel to the south of the 'fishing island'. This, which made use of a hoop-net, was to decimate salmon stocks by the nineteenth century until use of the net was banned.[101]

Even when weirs were constructed across rivers, however, they need not necessarily have hindered navigation; rather, on the contrary, there are instances of navigation being facilitated by weir-building. Davis notes how the deepening of a channel by the construction of a weir to serve a mill might actually provide deeper water for barges: in the seventeenth century, bargemen paid the miller a fee to open the weir to allow a rush or 'flash' of water to carry their barges over the shallows (Fig. 3).[102]

It is clear, however, that the proliferation of weirs during the middle ages was creating problems for river navigation, and it was the stone weir at Strata Marcella Abbey in Montgomeryshire which proved the ultimate barrier to craft moving up the Severn.[103] In Magna Carta it was instructed that weirs should be 'removed from Thames and Medway and throughout England, except upon the sea-shore'.[104] This led to the cutting of new channels to avoid the edict, despite the fact that it was not strictly enforced.[105] Losco-Bradley and Salisbury[106] note how a royal commission was appointed in 1378 to investigate the 'many weirs, mills, dams, pales and kiddles fixed or raised in the waters of the Trent, impeding the passage of ships', and another only five years later. (The kiddle consisted of a semicircular hedge of stakes, wattles, and nets which were covered at high tide so that the fish were trapped as the tide receded.[107]) The latter led to the Byrons of Colwick being prosecuted for obstructing the river.

---

[100] D. J. Pannett, 'Fish Weirs of the River Severn with Particular Reference to Shropshire', in Aston (ed.), *Medieval Fish, Fisheries and Fishponds* 371–5, fig. 2. Cf. above, pp. 10–11.

[101] Clark, 'The Salmon Fishery and Weir at Wareham'.

[102] Davis, 'The Ford, the River and City', 263; cf. R. Peberdy, 'Navigation on the River Thames between London and Oxford in the Late Middle Ages: A Reconsideration', *Oxoniensia*, 61 (1996), 311–40, and Blair, below pp. 254–5.

[103] Davis, 'The Ford, the River and the City', 371.   [104] Magna Carta, cap 33.

[105] Clark, 'The Salmon Fishery and Weir at Wareham'.

[106] Losco-Bradley and Salisbury, 'A Saxon and a Norman Fish Weir', 344.

[107] 'Kiddle' is a word of Anglo-Norman origin *kidel, kydel*, OFr *quidel*, later *quideau* meaning 'a wicker engine whereby fish were caught', recorded by the 13th century, which acted as a dam or barrier with openings for nets or baskets in which to catch the fish: *OED* 690.

Pannett notes how similar complaints are heard on the Severn by the thirteenth century: in 1286 Henry de Ribbesford and others were appointed to investigate the narrowing and heightening of weirs on the river between Gloucester and Shrewsbury 'so that vessels cannot pass through as they were wont, and to pull the same down where necessary',[108] and again in 1425 commissioners were appointed to view the Severn and to pull down any mills or weirs obstructing passage. In particular, the Abbot of Lilleshall had been accused in 1415 of causing obstruction to barge traffic at Bridgnorth. Some weirs in Shropshire were ordered to have enlarged openings or were closed down in the sixteenth century, especially in locations where gutters could not be created, as in the Ironbridge gorge.[109]

Seebohm[110] shows the nineteenth-century 'puttcher' weir extending from the shore only a short way into the river near Tidenham, and it is difficult to see how any greater length could have withstood the currents of this particular river. Linear features which were probably associated with post-medieval putts and putchers have been observed at many places on the Severn: a long irregular wooden structure at right angles to the foreshore in Ham and Stone extends for some 500 m, but others were only about 100 m in length (the estuary is over 2 km across in this area) and might each be associated with several hundred basket-like features.[111] Putcher ranks in recent times usually extended from low-water mark up to the mean high-water mark to ensure maximum coverage by the tidal flow, but the mesh size was such as to catch only sizeable salmon. It is unlikely that these would have hindered navigation, although they may have caused problems with landing or entering the pills draining into the estuary (E. Townley pers. comm.). It is far more likely that navigation, on this particular river, would be hindered further upstream, where it is much narrower and free of tidal surges, especially as other types of traps such as kiddles were in use here in the medieval period.

It seems unlikely that other users of river water, such as mills and fisheries, had become sufficiently sophisticated or ubiquitous in the early medieval period to create enormous problems for shipping. The ability to construct artificial channels to aid navigation, as described in Part II of this volume, marks a new development in technology that further illustrates how important river transport was to trade in this period.

---

[108] Pannett, 'Fish Weirs of the River Severn'; *Cal. Pat. Rolls. 1281–92*, 257; M. E. Simkins, 'Ribbesford with the Borough of Bewdley', in *VCH Worcs.* iv: 307.
[109] Pannet, 'Fish Weirs of the River Severn', 379.
[110] F. Seebohm, *The English Village Community* (1890), fig. facing 152.
[111] Gloucestershire SMR 9520, 9521, 9523.

3

The Place-Name Evidence for Water Transport in Early Medieval England

ANN COLE

The vocabulary of place names includes terms for sheltered anchorages and for sites by rivers at which goods could be loaded and unloaded, and these terms have an obvious relevance to the study of water transport in post-Roman times. The words *port*, *hȳð*, *stæð*, and *stoð* refer to sea and river traffic; also relevant are *lād*, the place-name term for an artificial watercourse, and the compound *ēa-tūn* 'river settlement'. The evidence comes from throughout England, although it is very sparse in Cornwall and not very plentiful in northern counties. Most of the names considered are settlement names, many appearing in DB, but some additional *hȳð* names are boundary marks culled from Anglo-Saxon charters; their distribution will give a basic picture of the country's water transport network at the time of the Norman Conquest. Other names first appearing in later records may be genuinely new coinages, although most of them are of a type which could have been in use in Anglo-Saxon times.

The changes to the coastline, and to a lesser extent along the rivers, in the last 1,000–1,500 years may have altered the configuration of a site very considerably since it was named. In order to understand the significance of some names, it is necessary to know the configuration as it was as near to the time of naming as possible.

The place names referring to sheltered anchorages and landing places, namely *port*, *hȳð*, *stæð*, and *stoð*, will be considered first.

The term *port* (see Fig. 8) was borrowed by the Anglo-Saxons from the Latin *portus* in the early days of settlement. It occurs sparingly along the south coast and Bristol Channel. It must not be confused with the later use of 'port' meaning market town (e.g. Newport) or Mod E 'harbour', nor Cornish/Welsh 'porth'. The examples of *port* are Portslade, Portsmouth, Portland, Portlemouth, Porlock, and Portishead. The use of a Latin loan suggests that some or all of these names were formed at a relatively early date.

Old Portslade (Sussex) is at GR TQ 255 064 (see Fig. 9). It first occurs as *Porteslage*, *Porteslamhe* in Domesday Book (1086) and as *Porteslad(e)* in

Ann Cole

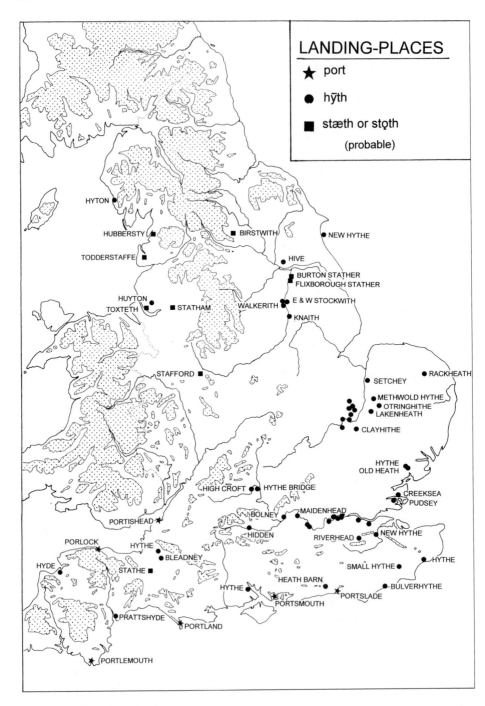

Fig. 8. Landing places in place names.

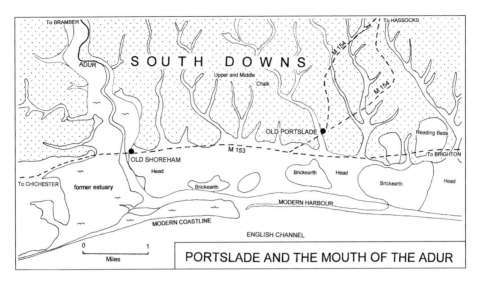

Fig. 9. Portslade and the mouth of the Adur.

1179.[1] The second element in the name is either *slæd*, a small, damp valley, or *gelād*, a difficult water crossing: the spellings are inconclusive. Old Portslade is built in a dry valley in the South Downs leading towards the coastal plain, however, so *slæd* is not a very likely second element. The configuration of the coast in late Roman/early Anglo-Saxon times is unknown because it has been destroyed by erosion. The eustatic rise in sea level accompanying the melting of the ice sheets would have flooded the Sussex coast valleys of the Arun, Adur, Ouse, and Cuckmere, and the longshore drift of material would have begun to build up spits across the mouths of these rivers. Behind the spits were expanses of sheltered tidal water. Just when the spits began to form is not known, but certainly by the fifteenth century there was a spit or bar extending eastwards from Lancing across the mouth of the Adur deflecting its outfall towards Southwick and Aldrington. It is very likely that the spit was already in existence in Roman times. The marshes fringing the estuaries behind the spits were being reclaimed by the thirteenth century, reducing the volume of the tidal flow and hence the scouring effect of the water. The rate of silting of the estuaries therefore increased, making it more difficult for ships to penetrate far upstream and reducing the size of the sheltered anchorage behind the spit, until today there is no wide expanse of water to which the Anglo-Saxons would

---

[1] The spellings are taken from the EPNS county volumes as appropriate, or E. Ekwall, *The Concise Oxford Dictionary of English Place-Names* (4th edn. Oxford, 1960), unless otherwise stated. In a few instances interpretations have been updated by reference to M. Gelling and A. Cole, *The Landscape of Place-Names* (Stamford, 2000).

have applied the term *port*. Portslade is linked to the Roman road system northwards by twin routes over the South Downs (Margary 154)[2] to Hassocks via Pyecombe or Poynings, and southwards to a coastal road running from Chichester to Brighton (Margary 153) crossing the Adur at Old Shoreham. If the second element of Portslade is *gelād*, then this is likely to have been the difficult crossing referred to. The mouth of the Adur is known to have been fordable at low water in 1680.[3]

Portsmouth, happily, presents fewer problems of interpretation. The *port* is the area now called Portsmouth Harbour, and it is referred to in the names *Portesmouth* (SU 62 03, *Portesmutha* ASC s.a. 501 *c*.890), 'mouth of the harbour'; *Portchester* (SU 625 044, *Porceastra* 904 (twelfth century), *Portcestre* DB), 'Roman settlement by the harbour'; Portsea Island (SU 66 01, *Portesīeg* 982 (fourteenth century)), 'island at the port'; and *Portsdown* (SU 63 07, *Portesdone hdr* DB), 'hill overlooking the harbour'.[4] The extent of the harbour would have been similar in early Anglo-Saxon times, since this stretch of coast is sheltered by the Isle of Wight from the vigorous wave action which produced the spits of the Sussex coast, and it would have suffered less silting as there are fewer, smaller rivers contributing sediment. Roman Portchester must have been linked into the Roman road system, probably to a ridgeway on Portsdown or possibly to an extension of Margary 420, Winchester to Wickham.

In the case of Portland (Dorset) (SY 68 76, (*on*) *Port* ASC s.a. 837, *Porland* DB) 'tract of land by the harbour' the *port* was the sheltered bay formerly called Portland Roads and now Portland Harbour. To the south lay the Isle of Portland (*insulam de Portland(e)* ASC s.a. 982) ending in the Bill of Portland (1649), its name stemming from its profile resembling the shape of a bird's bill when seen from the side. Portland Harbour is linked to the Roman road system by Margary 48 leading from Weymouth to Dorchester.

East and West Portlemouth (Devon) are at SX 749 385 and SX 710 391 respectively. The name probably means 'mouth of the harbour stream' (*Portlamuta* DB, *Portelemuthe* 1308) from OE *portwellanmutha*. The *port* would have been the sheltered waters of the ria (drowned river-valley) leading past Salcombe to Kingsbridge at the northern end. Kingsbridge (*cinges bricge* 962) could have been reached by smaller ships and was an important crossing place. It had road connections to Halwell, a place listed in the Burghal Hidage, along a *weg*. Halwell is at the hub of a series of ridgeways, one of which leads to Totnes by way of Harbertonford.[5] After an 8½-mile gap a Roman road, traced as far south as Newton Abbot, is reached, and leads to Exeter and the rest of the Roman road network.

---

[2] The Roman road numbering system is taken from I. D. Margary, *Roman Roads in Britain* (3rd edn. London, 1973).

[3] H. C. Brookfield, 'The Estuary of the Adur', *SxAC* 90 (1952), 153–63.

[4] R. Coates, *The Place-Names of Hampshire* (London, 1989).

[5] T. R. Slater, 'Controlling the South Hams: The Anglo-Saxon Burh at Halwell', *Reports and Transactions of the Devonshire Association*, 123 (1991), 57–78.

Porlock (Somerset) (SS 886 466, *Portloca* ASC s.a. 918, *Portloc* DB) means 'enclosure by the harbour', and it lies on the south side of the Bristol Channel (Fig. 10). The present entrance is at Porlock Weir, sheltered by hills either side. A storm beach encloses an area of saltmarsh which can even now be flooded for several weeks at a time when the storm beach is breached.[6] The preliminary results of borings show that there was once an embayment at the western end of the marsh and a period (after 3000 BC) when marine clay was being deposited there.[7] This suggests that there could have been sheltered, open water in Porlock Bay, where the marsh now is, such as would have attracted the term *port* in the late Roman/early Anglo-Saxon times—a welcome haven along a steeply cliffed coast. Porlock village is situated above the marsh on a deposit of Head, and is linked by a steep ascent of 1300 ft (400 m) to the ridgeway system on Exmoor and more particularly to the Brendon Hills ridgeway.[8]

The only other known example of *port*, on the Severn estuary, is represented by Portishead (ST 465 760, *Portesheve* DB) and Portbury (Somerset) (ST 503 757, *Porberie* DB), 'the headland' and '*burh* of the harbour' respectively (Fig. 11). They overlook the Gordano valley, now an alluvial area, but very difficult to traverse before it was drained. It is not known whether the area of

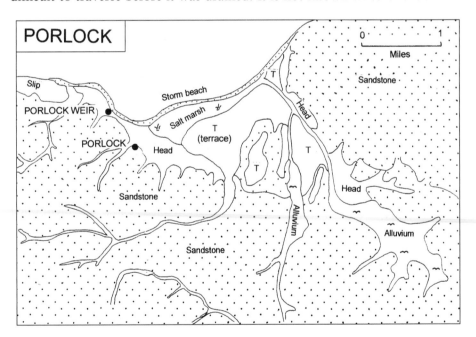

Fig. 10. Porlock.

[6] R. A. Edwards, *The Minehead District: A Concise Account of the Geology* (London, 1999), 103.

[7] M. Canti, V. Heal, R. McDonnell, V. Straker, and S. Jennings, 'Archaeological and Paleoenvironmental Evaluation of Porlock Bay and Marsh', *Archaeology of the Severn Estuary,* 6 (1995), 59, 64.

[8] H. Eardley-Wilmott, 'New Light on Old Travel Routes: Combwich Causeway and the Harepath', *Somerset Archeological and Natural History Society,* 134 (1990), 187–91.

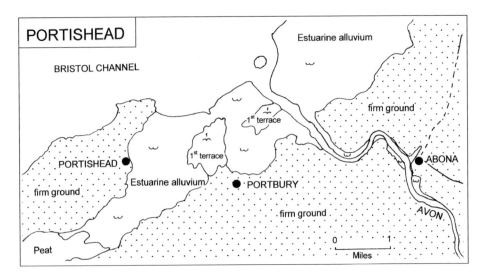

Fig. 11. Portishead.

marine alluvium was open water in late Roman/early Anglo-Saxon times, but at this period of marine transgression the sea was encroaching on the low-lying lands beside the Severn estuary, so the area was likely at least to have been inundated at high tide and criss-crossed by salt-water creeks.[9] The place names, however, suggest that there was a haven here: the western part forming the end of the Gordano valley, overlooked and sheltered by the 'headland of the *port*' (Portishead), and the eastern part forming part of the mouth of the Bristol Avon on the shore of which stood 'the stronghold of the *port*' (Portbury). Sea Mills nearby is the site of the Roman port, Abona, which linked into the road system by way of Margary 541.

The *port*s were, then, large stretches of sheltered water connected to the road system of the time. Portsmouth, Portland, and Porlock, in company with Avonmouth the other side of the harbour from Portishead, are mentioned in connection with seafaring in the *Anglo-Saxon Chronicle*, showing these harbours to have been in active use in early medieval times. Portsmouth was the reputed site of a landing s.a. 501; Portland saw a skirmish with the Danes s.a. 837, a raid by three Viking ships s.a. 982, and a raid by Earl Godwine in 1052. Porlock was subject to a raiding party from Brittany s.a. 915D/918A, and Harold, coming from Ireland with nine ships, raided Porlock s.a. 1052E/1051F. Avonmouth was mentioned under 915D/918A in connection with positions being arranged from Cornwall to Avonmouth to protect the coast. In 1052D severe weather was noted outside the mouth of the Avon.[10] These *port*s would have served both coastal and cross-channel shipping.

---

[9] S. Rippon, *The Severn Estuary: Landscape Evolution and Wetland Reclamation* (London, 1997), 42–3.

[10] *ASC*, years as stated (pp. 15, 63, 124, 99, 178, 176).

Another term sometimes used of sheltered inlets is *pōl*, for instance Poole (Dorset), Poulton (Cheshire, Lancs.), Hartlepool (Durham). Its primary meaning cannot be 'coastal haven' because it is also used of small creeks and places inland, and so the term is not pursued any further here.

The OE term *hȳð* means a landing place (Fig. 8). Hythes are to be found in sheltered coastal sites and along rivers well inland. Like the previous group, many of the sites of the coastal hythes have changed since Anglo-Saxon times. These settlements will be considered first, starting in north-east England and working clockwise round the coast.

Erosion along the Holderness coast has been so severe that New Hythe in Skipsea (*c.* TA 1955, *le Neuwe Hithe* 1260), which would have served the boulder clay lands between the Hull valley and the coast, disappeared altogether sometime after 1416.

Old Heath (*c.* TM 016 229, *Hetha(m)* 1158–1237) and Hythe (TM 013 247, *la Newheth* 1311) (Essex) are on the estuary of the River Colne just seawards of Colchester. Drained marshland lies between Old Heath and the river, suggesting that early silting triggered its replacement by a new hythe nearer Colchester. It would have been within easy reach of the Roman road system there.

Creeksea and Pudsey lie on the north and south banks respectively of the River Crouch. Creeksea (TQ 930 969, *Criccheseia(m)* DB, *Krikesheth* 1240) is where the river briefly abuts firm land allowing easy landing—indeed it still has a small ferry service. It would have served the Dengie peninsula, though not by any known Roman road. Pudsey Hall (TQ 881 951, *Puteseiam* DB, *Podes(h)ethe* 1361) is now one mile from the river on firm ground behind reclaimed marshland. In earlier days it would have served villages like Canewdon on a ridge of drier ground between the Crouch and the Roach.

Since the time when the Roman port at Lympne flourished the coasts of south-east Kent and north-east Sussex have been subject to great changes, with the further development of the large cuspate foreland of Dungeness and the reclamation of marshland to the south and west affecting the sites of Hythe, Small Hythe, and Bulverhythe. At the time when Lympne flourished the River Rother reached the sea by a northerly route past Appledore and along the foot of the North Downs, but when silting and reclamation made Lympne less useful as a port and the course of the Rother had shifted to the Rhee Wall/New Romney course, a new landing place developed a little seawards of Lympne and the dunes at Santon, at what is now West Hythe (TR 130 342, *Huthe* 1052, *Hede* DB).[11] By the time of Henry VIII shingle growth had reduced Hythe Haven to a narrow gut running east towards Folkestone, and even this had disappeared by Elizabeth I's time.[12] The area was linked by Roman roads north

[11] J. K. Wallenberg, *The Place-Names of Kent* (Uppsala, 1931).

[12] J. G. O. Smart, G. Bisson, and B. C. Worssam, *Geology of the Country around Canterbury and Folkestone* (London, 1966), 258–9.

to Canterbury (Margary 12) and west to the Weald (Margary 131), and within easy reach of an ancient downland track sometimes called the Pilgrims' Way.

Figure 12 shows the geology of the area around the Isle of Oxney and the neighbourhood of Small Hythe near Tenterden (Kent) (TQ 893 302, *Smalide* thirteenth century), an area that has undergone much change in the last two millennia. Small Hythe appears in the records at a time when the Rother is thought to have flowed round the northern side of the Isle of Oxney and out to sea via the Rhee Wall and/or a nearby channel. The great storm of 1287, which so damaged Old Winchelsea, caused the mouth of the Rother to shift southwards to the vicinity of Rye. The marshes of the Rother valley have been gradually reclaimed over a long period, but mainly since the fourteenth century. It was in the seventeenth century that the Rother was diverted to flow south of Oxney.[13] Small Hythe was known to be functioning as a ship repair yard in the fifteenth century,[14] at which time ships must have come up the Rother past Rye and round to the north of Oxney, but since Small Hythe is mentioned in the thirteenth century it is also possible that vessels used the Rhee Wall route in earlier times. It is now beside the Reading Sewer some 8½ miles from the open sea.

Bulverhythe (TQ 765 083, *Burewarehethe* 1229) means the 'hythe of the burghers [of Hastings]'. Anglo-Saxon Hastings was on the western side of the mouth of the Priory valley, but because of coastal erosion had declined by the eleventh century; the haven itself was probably useless for navigation by the early twelfth century. Later medieval Hastings grew up a little to the west in the Bourne valley,[15] but the hythe to serve it grew up on the western side of a larger inlet—the Combe valley. The site of Bulverhythe deserted medieval village is at TQ 768 082, and the ruins of a medieval chapel are nearby. Since the coast is eroding and the estuaries silting up, Bulverhythe had a short life as a haven and port, being first mentioned in the thirteenth century and being already in decline at the end of the fourteenth.[16] Medieval Bulverhythe and Hastings were on opposite sides of Combe Haven, so that travellers and goods would have had a circuitous route from the hythe to Hastings and thence to the Roman road system at Ore (Margary 13).

A Roman road (Margary 423) led down the western side of Southampton Water to a small inlet used as a landing place close to Stone Farm (SZ 458 994). However, as it silted up a new landing place was needed. Hythe (SS 425 078) occurs as *Portmonna hyth* in AD 962—the 'landing place of the men of

[13] E. R. Shephard-Thorn, J. G. O. Smart, G. Bisson, and E. A. Edmonds, *Geology of the Country around Tenterden* (London, 1966), 97–100.

[14] T. Taylor, *Behind the Scenes at Time Team* (London, 1998), 126–7.

[15] E. M. Ward, 'The Evolution of the Hastings Coastline', *Geographical Journal*, 56 (1920), 110–23.

[16] G. R. Burleigh, 'An Introduction to the Deserted Medieval Villages in East Sussex', *SxAC* 113 (1973), 65.

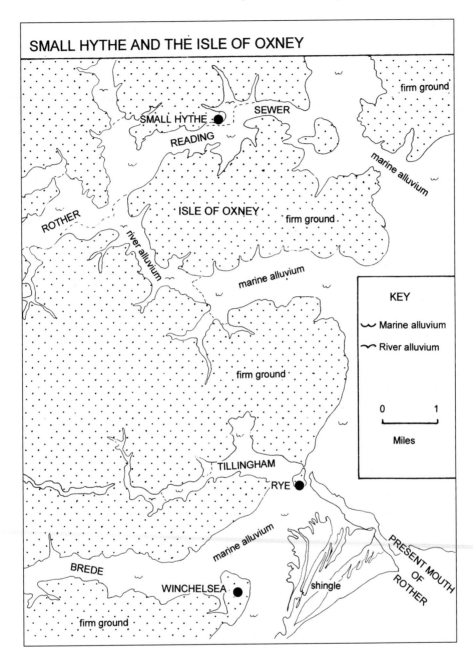

Fig. 12. Small Hythe and the Isle of Oxney. Compare Fig. 21.

Southampton'.[17] It was within 1 1/2 miles of the Roman road, but across the estuary of the Test from the busy trading centre of *Hamwic*.

[17] A. Mawer, F. M. Stenton, with J. E. B. Gover, *The Place-Names of Sussex* (Cambridge, 1929–30), 535; S 701.

Prattshide (Devon) (SY 000 825, *Pratteshithe c.*1250) is a name no longer surviving in use. It was on the Exe estuary in Withycombe Raleigh, 4 miles south of the nearest Roman road, the M490 Exeter to Topsham.

Hyde near Bideford (Devon) (SS 461 290) was recorded as *West Hede* in 1520 but appears in the Lay Subsidy Rolls as the home of John *atte Hithe* in 1333. It lies on the west bank of the Torridge where deep water approaches the bank on the outside of the bend. The nearby fourteenth-century bridge over the Torridge would have given access to lands east of the river.

The last coastal hythe is in Cumberland—Old Hyton in Bootle (SD102 875, *Hytona c.*1210). It is on a rise beside Annaside Beck, another stream whose little estuary has silted up and which has been deflected to the north by the growth of a storm beach.

The riverside hythes are concentrated in the drainage basins of the Humber, Wash, and Thames, and once again are dealt with clockwise from the north–east.

Hive in the East Riding of Yorkshire (SE 821 310, *Hythe* 959 *c.*1200, DB) now lies 4½ miles from the Humber and 1 mile from the little River Foulness (also called Foulney): an unlikely looking site for a hythe, but the countryside was very different a thousand years ago (Fig. 13). The north side of the Humber was bordered by a strip of saltmarsh up to about 3 miles wide; beyond this were waterlogged clays and sands extending north to beyond the Pocklington Beck 8 to 9 miles away, this wet area being drained by the Derwent and the Foulness. The latter flowed into Wallingfen, which was normally a lake, and on to an inlet of the Humber called Skelfleet.[18] An east–west deposit of sand a few feet above the marshes was the site of Eastrington, Portington, Cavil, and Hive, all recorded in DB. These settlements, surrounded by marshland, were difficult to reach from any direction, but the easiest, safest route (because the Humber estuary is dangerous for shipping) must have been up Skelfleet, through Wallingfen and into the Foulness, and thence to a mile-long ditch, perhaps the Delph Drain, through the waterlogged clays to a landing place at Hive. It served some small marshland communities.

Knaith, Walkerith, and East and West Stockwith are beside the lower Trent. Knaith (*Cheneide* DB, *Knayth c.*1225 (fourteenth century)) is at SK 828 847, on a river bend to which the first element *cneo* 'a knee' probably refers. It is on rising ground on the east bank, about 1½ miles north of the point where Margary 28 crosses the Trent en route to Littleborough from Marton. The significance of the locations of West Stockwith (SK 793 950, *Stochith'* 1226) (Notts.), and East Stockwith (SK 788 945, *Stokhede* 1188) and Walkerith (SK 788 931, *Walkerez* late thirteenth century, *Walkreth* 1300) (both Lincs.) is best understood by looking at the map showing the pre-Vermuyden drainage of the area (Fig. 14). The Don formerly reached the lower Trent just before its

---

[18] June A. Shephard, 'The Draining of the Marshlands of South Holderness and the Vale of York', *East Yorkshire Local History Series*, 20 (1966).

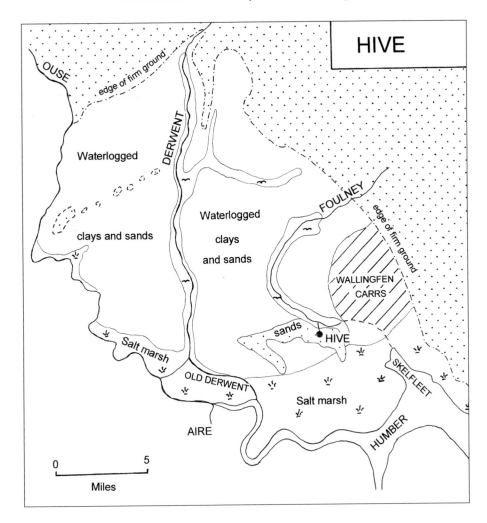

Fig. 13. Hive.

confluence with the Humber. The Idle and Torne were tributaries of the Old Don. At some stage before 1344 (perhaps even in Roman times)[19] a channel was cut to take the Don northwards, cutting off a loop of the Went and joining the Aire near East Cowick.[20] Another channel called Bykers Dyke was cut, before the Domesday Survey, to link the Idle to the Trent.[21] These two/three channels created a through waterway from the Trent to the Ouse, avoiding the dangerous eagre, tides, and shifting sandbanks of the Humber. It would have been an important part of the water route between Lincoln and York,

[19] J. A. Steers, *The Coastline of England and Wales* (Cambridge, 1964), 418.
[20] G. D. Gaunt, 'The Artificial Nature of the River Don North of Thorne, Yorkshire', *YAJ* 47 (1975), 15–21.
[21] D. Gaunt, *Geology of the Country around Goole, Doncaster and the Isle of Axholme* (London, 1994), 128–30.

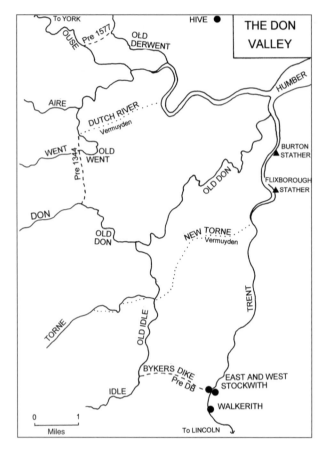

Fig. 14. The Don valley.

for instance. East and West Stockwith and Walkerith (the fuller's hythe) were strategically placed at the confluence of Bykers Dyke and the Trent. Although the situation was very good the sites of these three places were poor, as they are among the few hythes built on alluvium (rather than on firm ground such as river terraces) and therefore liable to flood.

Around the edges of the Fens and on the islands within the Fens are numerous hythes (Fig. 15). The area in Anglo-Saxon times was mostly untamed marshland through which rivers such as the Ouse, Nene, and Cam wound their sluggish way, their courses considerably different from today. There was usually little option but to travel by boat. Accordingly, a series of landing places grew up around the eastern and southern margins of the Fens where the higher, firmer ground of the East Anglian Heights fell away to the marshlands, and where small but navigable streams gave access to the rising ground. Setchey (Norfolk) (TF 635 136, *Seche* 1202, *Sechithe* temp. Henry III) is on a distinct rise some 10 ft above the Nar which today, though a small river, is big enough for a quanted (i.e. poled) boat. Setchey is not at the head of navigation but at the

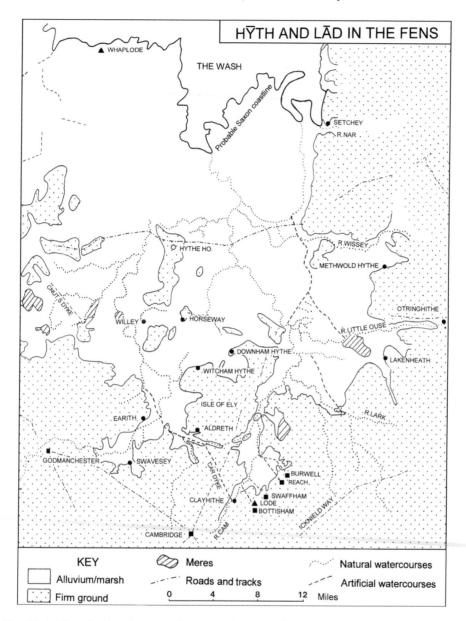

Fig. 15. *Hȳð* and *lād* in the Fens. Compare Figs. 20 and 37.

Fenland/upland junction, and, as so often happens where two areas with different resources and products meet, trading developed: Setchey had a market by 1258, and its road links south were improved by the construction of a causeway across the Nar valley sometime before 1271 and by a bridge by 1413. The landing place must have been substantial, as the inquisition of 1274–5 refers to the 'great hythe of Secheth' where an accumulation of wreck, rubbish, and silting was impeding the flow of the Nar.[22] Methwold Hythe (Norfolk) (TL 712 948, *Methelwoldehythe* 1277), 1 1/2 miles west of Methwold, is similarly placed at the Fenland/upland junction. In medieval times it was a small port, and a large enough settlement to have its own church. As the hythe is 2 miles from the River Wissey there must have existed a man-made waterway linking the two by the second half of the thirteenth century. Lakenheath (Suffolk) (TL 713 827, *aet Lacingahith* 945, DB) overlooks the Fens and lies on rising ground about 2 miles south of the Little Ouse, to which it was linked by Lakenheath Old Lode or its predecessor. This waterway must have been in existence by 945 for Lakenheath to have been named after its hythe ('hythe of the dwellers by streams'). It is surrounded by peat on three sides, so it would have been relatively easy to modify an existing stream, or cut a new channel, to link the settlement to the Little Ouse. The Little Ouse itself has been straightened downstream of Hockwold, and so one cannot now be certain of the layout of the waterways when the hythe at Lakenheath developed in the Anglo-Saxon period. Fortunately there are no such problems associated with *Otringehythe* (Norfolk) (TL 801 876, *Otringheia*, DB), which was mistakenly equated with Methwold Hythe in the *Dictionary of English Place-Names*. Dymond[23] has shown it to be a mile upstream of Brandon, at a point where the phragmites swamp each side of the river is briefly replaced by firm ground on the north bank, and where the Little Ouse ceased to be navigable for heavy goods. It is a little further into the uplands than the three previous hythes, and close to a Roman road, Margary 332.

Clayhithe on the River Cam at TL 501 644 was originally just called Clay (*Cleie* 975 (twelfth century)). It had become Clayhithe (*Cleyheth*) by 1268. It probably developed as a landing place because it was at the junction of the Car Dyke and the Cam. Swavesey (Cambs.) and Earith (Hunts.) are one each side of the Great Ouse near the fen-edge. Swavesey (TL 362 692, *Suauesheda* 1086, *Suauishith* 1290) is on a river terrace, the nearest dry point to the Great Ouse three-quarters of a mile away, and 2 miles from the Cambridge to Godmanchester Roman road (Margary 24). Earith (TL 385 748, *Herheth* 1244) is also on a river terrace (hence the qualifying element *ear* 'gravel') where it abuts the river for a quarter of a mile and where the Great Ouse and Car Dyke meet. Several of the 'islands' in the Fens have their own hythes: ports of entry for goods from the mainland, or for boats exploiting the resources of the marshes. Aldreth (Cambs.) (TL 445 735, *Alrehed(a)* 1169–72), on the Isle of Ely, is built on Kimmeridge Clay a quarter of a mile from the Great Ouse (Old West River): the two are linked by the Aldreth Causeway across the intervening peat and alluvium. Earith is only 3 1/2 miles away along the Great Ouse. Witcham Hythe (TL 459 816, *Wichamhythe* 1251) and Downham

---

22 R. J. Silvester, *The Fenland Project 3: Marshland and the Nar Valley, Norfolk*, EAA 45 (1988), 141.

23 D. Dymond, 'A Misplaced Domesday Vill: Otringhithe and Bromehill', *Norfolk Archeology*, 43 (1998), 161–8.

Hythe (TL 500 837, *Dunham hythe* 1251) also served the Isle of Ely, giving access to the streams in the heart of the Fens and perhaps to Horseway (TL 425 870, *Hors(e)hythe* 1238), a landing place serving both Honey Hill and the island upon which Chatteris stands. Chatteris also had access to the Car Dyke and thence to the Nene and Great Ouse through the hythe at Willey Farm (TL 382 875, *Wyliethe* 1240). There is a record of a *hethelod* in 1221 in March (Cambs.), perhaps now Hythe House, which was probably the landing place for the island on which March stands. It cannot be far from the Roman road Margary 25 from Peterborough to Denver.

Rackheath (Norfolk) (TG 270 150, *Racheitha* DB) is on a short tributary of the Bure which meanders down to the Great Estuary. The first element *hraca* 'a throat' refers to the deep little valley that the stream by Rackheath church flows through. If the second element of Rackheath is indeed *hȳð* (it could also be *hæð*) the stream would need to have been canalized if the landing place was nearby (no stretch of the Bure is in the present parish of Rackheath). It would appear to have served quite a limited area, since nearby Norwich would have had its own landing places on the Yare.

The Thames has more *hȳð* place names than any other river, running from the estuary to just above Oxford. The river offers a means of access to the heart of southern England from across the English Channel and for local water-borne traffic (see also Blair, above p. 18, below p. 255). Although the estuary east of Lambeth is mostly bordered by alluvium and saltmarshes, the sites for all the hythes but Rotherhithe were carefully chosen where little patches of firm ground—usually a gravel terrace—formed the riverbank, and often where a Roman road was close by. Greenhithe (TQ 585 753, *Grenethe* 1264), being on chalk, was less liable to poaching by the trampling of many feet than some other hythes and perhaps had a covering of grass and other plants, hence the qualifier 'green'. It lay 1½ miles from the Roman road from London to Rochester (Margary 1c). Erith (TQ 515 781, *Earhyth* 695, *Erhede* DB) stands on Thanet Sands, chalk, and the gravel to which the name refers; it is 2 miles from Margary 1c. Stepney (TQ 367 810, *Stybbanhyth*, *c*.1000, *Stibenhede* DB) is on floodplain gravel, and a mile from Margary 3a, the London–Colchester road. The first element might be a personal name *Stybba, or alternatively *stybb*, a tree stump, + *hȳð*, perhaps referring to some part of the hythe's construction or to a mooring post (cf. Stockwith; *stocc*; a tree stump + *hȳð*). Queenshithe (TQ 323 807), formerly *Aetheredeshyd* (898), on river gravels, was within the old city, by London Bridge and close to the hub of the Roman road network.[24] Rotherhithe (TQ 359 804, *Rederheia c*.1105), referring to cattle, is on the south bank on alluvium opposite Stepney—and very muddy it must have become with hooves churning up the ground! It was 2 miles from Margary 1c. Endiff (Middx.), now lost, (*c*. TQ 30 80, *Anedehea* thirteenth century) was close to the Houses of Parliament on a little gravel patch at the western end of the old ford

[24] See Hooke, above pp. 40–1

at Westminster, and judging by its name a place much frequented by ducks, whilst across the river, at the eastern end of the ford, lambs were to be seen at Lambeth (TQ 307 790, *Lambhyth* 1041), again on gravel. Upstream of Lambeth and Endiff the gravel patches are more extensive, giving a wider choice of site. Chelsea (TQ 271 776, *Cealchyth* 785, *Chelched* DB) on the north bank is 1½ miles from Margary 40 (London–Silchester) and was possibly a place where chalk was landed, to use for sweetening and lightening the rather heavy soils of the neighbourhood. Putney (TQ 242 755, *Putelei* DB, *Puttenhuth(e)* 1297), like Chelsea, is also on the First Terrace; the High Street links the landing place to the London–Kingston–Guildford road (present A3), an old route though not Roman.

Upstream of this cluster, around London, the hythes are more dispersed. Where the London to Silchester Roman road, Margary 4a, crosses the Thames at Staines, on the south bank lie Hythe (c. TQ 030 715, *Huthe* 675 (thirteenth century)) and Glanty (c. TQ 020 718, *Glenthuthe* 675 (thirteenth century), 'the hythe frequented by hawks'), both of them on small gravel patches. Maidenhead (SU 888 815, *Maideheg* 1202, *Maydehuth* 1241), earlier called Elington, is on the Flood Plain Terrace about 3½ miles south of the crossing of the St Albans to Silchester road (Margary 163) over the Thames. Bolney Court (SU 777 807, *Bollehede* DB, *Bulehethe* 1176) is on a gravel terrace deposit 1½ miles south of Henley-on-Thames. The Roman road from Dorchester via Nettlebed crossed the Thames near Henley to link with Margary 163, and would be readily accessible from Bolney. This bulls' hythe is likely to have been linked by way of the long Harpsden valley to an area of rough grazing on the Chiltern plateau where cattle were raised at Rotherfield (*hrȳðer* + *feld*, cf. Rotherhithe above and Riverhead below). Hythe Bridge Street (SP 507 064, *Hithe* 1233–44) in Oxford not only ties in with the important north–south route through the city but also to Henley and Bolney by way of an old route over Shotover and through the Chilterns called Knightsbridge Lane. The use of this overland route could shorten the journey and cut out a difficult stretch of the river. Further up the Thames in Eynsham is High Croft Lodge (SP 446 093, *Huythecroft* 1328) which would have been associated with Eynsham Abbey and the nearby river crossing at Swinford. Nearly 2 miles further upstream is Bablock Hythe (SP 434 043, *Babbelack* 1277): *hȳð* was added to this name sometime before 1581–2,[25] and a ferry service still operated here into the mid-twentieth century. As the number of fish-weirs and watermills along the river increased in later medieval times, so the usefulness of the river for transport decreased, particularly from Henley up to Oxford (see Blair, below pp. 285–6). The presence of these *hȳðs*, and of *ēa-tūns* (below) upriver as far as Cricklade, suggests that the upper Thames was a busy thoroughfare in pre-Conquest times.

---

[25] W. H. Stevenson and H. E. Salter, *The Early History of St John's College, Oxford*, Oxford Hist. Soc. NS 1 (Oxford, 1939), 236.

The only other *hȳð* in the Thames basin is referred to in Hidden (*on Hyddene* 984) 'valley with a hythe'. It is a dry chalk valley running down to the Kennet at Kintbury (*c.* SP 385 675) where the landing place would have been. The valley leads up past Margary 53 to Mildenhall and on towards Margary 41 (Silchester to Cirencester).

Two rivers flowing into the Thames estuary have *hȳðs*: Riverhead (Kent) (TQ 515 561, *Reddride* 1278, *Reydrythe* 1292), 'cattle hythe', is on a dry point where the floodplain of the Darent is locally at its narrowest, and can be crossed at Langford should one wish to reach the Pilgrims' Way. New Hythe, Kent (*c.* TQ 711 600, *La Newehethe* 1254)[26] is on the Medway near Aylesford, and is one of the rare hythes built on alluvium. The Roman road Margary 13 (Rochester to Bodiam) and the Pilgrims' Way are both the far side of the river. Both Riverhead and New Hythe are located in the Vale of Holmesdale, an area of Kent considered by Everitt to have been favourable to relatively early settlement and occupation.[27] The Darent and Medway linked Holmesdale to the coastlands of Kent, also well settled by the early Anglo-Saxons.

Heath Barn (Sussex) (TQ 190 146, *atte Huth* 1327), like Portslade, is on the Adur, once an estuary stretching inland at least as far as Bramber and Steyning and probably navigable up to Bines Bridge (Fig. 16). St Cuthman's port had been established at Steyning in the time of Edward the Confessor; boats reached it by using the eastern branch of the Adur adjacent to Upper Beeding. This was bridged by 1086 when DB records the collecting of tolls from passing ships. The western stream adjacent to Bramber had been bridged by 1230, but the continued silting of the estuary rendered both bridges impassable by boats by the late fifteenth century, so that port activities shifted seawards to Shoreham.[28] The hythe at Heath Barn 3 miles upriver from Steyning was well placed for access to the east–west Roman road (Margary 140, Hardham–Hassocks) crossing the Adur at TQ 201 134. The hythe's usefulness would have decreased as the thirteenth-century drainage works accelerated the silting of the estuary, and it became inaccessible to larger vessels.

Unlike the Fens, the islands and periphery of the Somerset Levels have only two settlement names in *hȳð*. Bleadney (ST 486 453, *Bledenithe* eighth century)[29] is where the Axe breaks through the ridge running from Wells to Wedmore. This makes it a good trans-shipment point for goods or people going by river between Glastonbury and the coast at Uphill with those going along the ridge between Wells and Wedmore. Hythe (Somerset) (ST 457 523 *Huthe* 1212)[30] is on a tributary of the Axe, called the Yeo, which rises in Cheddar, which estate Hythe would have served (Fig. 17).

[26] Wallenberg, *The Place-Names of Kent*, 149.
[27] A. Everitt, *Continuity and Colonisation* (Leicester, 1986).
[28] E. W. Holden, 'New Evidence Relating to Bramber Bridge', *SxAC* 113 (1975), 104; J. Blair, 'St Cuthman, Steyning, and Bosham', *SxAC* 135 (1997), 173–92, at 182–6.
[29] Gelling and Cole, *The Landscape of Place-Names*, 84.
[30] *Book of Fees*, i: 1198–1242, 82.

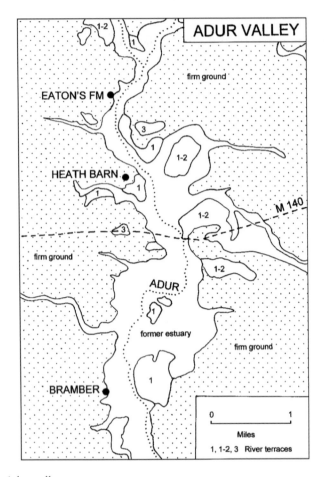

Fig. 16. The Adur valley.

The last place, Huyton (Lancs.) (SJ 443 912, *Hitune* DB), is a problem. It is on a low watershed between the River Alt and the Ditton Brook. It has no access to a navigable stream, nor would it appear to have done prior to the drainage and reclamation of the Lancashire mosses. The significance of the name remains a puzzle at present.

The now-lost examples known to us from Anglo-Saxon charter-boundaries reinforce the pattern shown by the present-day settlements called *hȳð*, even though their exact positions are not always known. Three are estuary-side sites: *Hyðe* in Stoke near Ipswich must have been beside the River Orwell (970, in S 911); *ða Hyðe* is in an estate believed to be Brede (Sussex), and would have been beside the River Brede, at that time a long estuary leading inland from Winchelsea (1005, in S 911); *Wichyðe* in South Stoneham (Hants) is believed by Currie to refer to *Hamwic*, the forerunner of Southampton, and lay at about

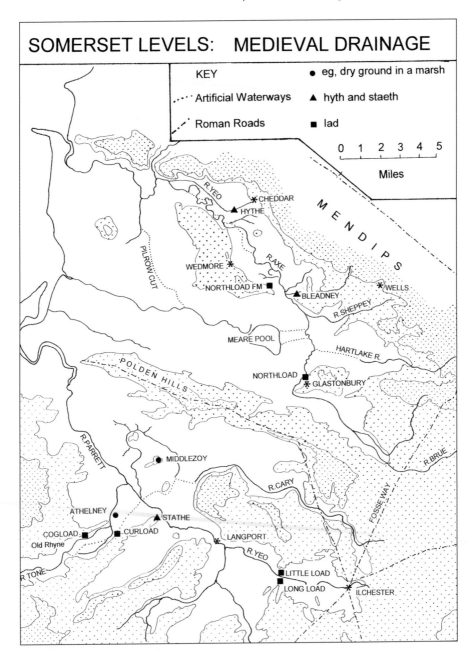

Fig. 17. The Somerset levels: medieval drainage. Compare Figs. 43 and 49.

grid reference SU 43 11 on the banks of the Itchen (990–2, in S 944).[31] There are two further examples on the Thames: *Lundentunes hyth* (743, in S 98)

[31] C. K. Currie, 'Saxon Charters and Landscape Evolution in the South-Central Hampshire Basin', *Proc. Hants. Field Club and Archaeol. Soc.* 50 (1995), 113.

which probably refers to the Anglo-Saxon trading settlement of *Lundenwic*,[32] and *Wealas Huthe* just downstream of Hythe and Glanty near Staines; the last three appear in late Anglo-Saxon bounds attached to an early Chertsey charter (S 1165).

There are two clusters inland. The first is on the margin of the Fens in Yaxley and Farcet near Peterborough; here *Dichyðe*, *Suðhythe*, *Færresheafde Hythe*, and *Norðhythe* (all 956, in S 595) lay beside the old course of the Nene, now Pigs Water, between Yaxley and Farcet;[33] this group is a continuation north-westwards of the series of hythes which ring the Fen/upland margin in the east and south and are discussed above. The second inland cluster is on the eastern edge of the Somerset Levels, where the rivers which emerge from the uplands to the east reach the once marshy Levels. *Cleuan Hithe* in Wootton (Somerset) (946, in S 509) is thought by Grundy to be in the vicinity of Redlake House on the Whitelake River.[34] Four hythes are mentioned in the Butleigh charter: *Hoctanyth*, *Selfith*, *Welesith*, and *Yithe*, all of them thought to lie on a watercourse draining into the River Brue between ST 512 357 and ST 533 346 (801, in S 270a) (Fig. 54). The site of *Middel Hithe* in Podimore (Somerset) (966, in S 743) is not known but may be connected with the River Yeo.

The *hȳðs*, except for Huyton in Lancashire, were landing places in sheltered coastal inlets or on riverbanks. Although some are apparently ill positioned today, in Anglo-Saxon times they were beside navigable water. Some would have had quite extensive hinterlands, especially if they were within easy reach of a Roman or other old road. Others served limited areas such as parts of monastic estates. The sites were usually chosen with care, firm ground such as river terrace gravels being preferred. Whether the boats were pulled up on the bank, or tied floating alongside it, cannot be determined from the place names. However, some of the names, especially those along the Thames, indicate the nature of the cargoes carried (often domestic animals), and others describe the ground or the construction of the hythe. Many of the smaller hythes serving more limited areas have no qualifying element; they are simply the local landing place. There is a notable absence of *hȳð* place names on the rivers Severn (Warwicks.), Avon, and Wye, nor do they occur in Northumberland and Durham.

Staithe, OE *stæð*, is another term meaning landing place, but there are several problems associated with its use in place names. First, its meaning changed from 'riverbank', the usual sense in the OE charter-boundaries, to 'landing place', a meaning well established by later medieval times. It is not always possible to decide which meaning was intended. Secondly, it is not always possible to distinguish it from its cognate, ON *stoð*, and although, in the context of this study, that might not matter very much, a further difficulty arises in distinguishing ON *stoð*, a 'landing place' from ON *staðr*, a 'place'.

---

[32] A. D. Mills, *The Oxford Dictionary of London Place-Names* (Oxford, 2001), 113.

[33] C. R. Hart, *The Early Charters of Eastern England* (Leicester, 1966), 164.

[34] G. B. Grundy, *The Saxon Charters and Field Names of Somerset* (Taunton, 1935), 94–8.

In fact some place names in north-west England might contain any one of the three elements OE *stæð*, ON *stǫð*, ON *staðr*. However, for a few places there seems little uncertainty about the meaning (Fig. 19).

Stathe (ST 374 290, *Stathe* 1233) is beside the Parrett at the end of a long tongue of upland stretching into the marshy Somerset Levels. A route along the ridge links Stathe with the district of Curry (Fig. 17). This area had no Scandinavian settlement, and so the name will be the OE term *stæð*. Its situation is ideal for a landing place on the interface between land and river transport or for exploiting the marshland.

Stafford (SJ 922 233, *Staefford* 913, *Stadford* DB) is from OE *stæð*. This term is used to distinguish this ford from other fords and 'ford by a landing place' is much more specific than 'ford by a riverbank'—as all fords are! A road from *Pennocrucium* (Water Eaton) runs north to cross the River Sow at Stafford. The Sow is navigable for small craft and only about 5 miles from the larger River Trent. 'Landing place' would be an appropriate interpretation of *stæð* here.

Statham (Cheshire) (SJ 670 877, *stathum* 1284–5) is the dative plural of OE *stæth*. It lies on the marshy banks of the Mersey, about 3 1/2 miles east of the point where Margary 70 crosses the river at Wilderspool. There seems little likelihood that the two are connected when they are this far apart.

Birstwith (West Riding) (SE 239 595, *Beristade* DB) is on a rise beside the Nidd, 1 1/2 miles west of Hampsthwaithe where the Roman road from Ilkley (Margary 720b) crosses the river. The Nidd is said to be navigable for small craft,[35] but near Birstwith there are shallows and white water at times, and so this would scarcely be a satisfactory site for a staithe. Etymologists disagree about the origins of the name; for instance Ekwall and Gelling believe it to derive from *byrg-stæð* 'landing place of the fort', while Smith, Mills, and Watts believe it to be from ON *byjar-staðr* (farm built on the site of a lost farm). Although the locational setting is appropriate for a staithe, Birstwith is not a certain example in view of the difficulties of navigating the Nidd.

Hubbersty (Lancs.) (SD 483 546, *Hobyrstath* 1236)[36] is a name no longer in use, and the site of Hubbersty has been affected by the building of a railway, the A6, the Lancashire Canal, and the canalization of the River Condor. The site would appear to have been on the Condor 2 1/2 miles from the coast, and close to the supposed Roman road (Margary 70d) from Ribchester. This would be a useful landing place. The generic could be either the OE or the ON for staithe.

Staithes (NZ 780 183), a tiny harbour on the Yorkshire coast, occurs as *Setonstathes* in 1415. Seaton was a DB estate three-quarters of a mile inland, but is now very much shrunken. Staithes suffers from erosion, which has reduced the size of the harbour and the area suitable for building. It is a late-recorded, atypical staithe site.

---

[35] L. A. Edwards, *The Inland Waterways of Great Britain* (St Ives, 1985) is useful for assessing the navigability of today's waterways.

[36] E. Ekwall, *The Place-Names of Lancashire* (Manchester, 1922).

There are two places where the ON plural, *stǫðvar* occurs: Burton Stather (SE 865 185: Burton appears in DB, and Stather has been appended by 1201) and Flixborough Stather (SE 862 143, Flixborough appears in DB, Stather was appended later).[37] Both Burton and Flixborough are on a bluff overlooking the lower reaches of the Trent, and their staithes are half to three-quarters of a mile away on the riverbanks, now embanked to prevent flooding of the alluvial ground (Fig. 14).

Toxteth (SJ 383 882, *Stochestede* DB, *Tokestath* 1212) is now engulfed by Liverpool. It lay beside the Mersey far from the Roman road system. The first element of the name is ON Toki, and so the second element is more likely to be ON *stǫð* or *staðr* than OE *stæð*. Its estuary-side site would be appropriate to a staithe.

Todderstaffe (SD 368 367, *Taldrestath* 1332)[38] stands on a rise overlooking a formerly ill-drained valley to the east with a straight drainage ditch called Main Dyke running through it to the River Wyre. To the west, a quarter to half a mile away, is a Roman road (Margary 703) leading to Poulton le Fylde. The meaning of the qualifier is not known; the generic is probably OE *stæð*, but could be ON *stǫð* or *staðr*. It would be acceptable as a staithe if Main Dyke or its forerunner had been dug large enough to take boat traffic before the name was coined.

The remaining possible examples of staithe all have more or less inappropriate sites for such a function. Birstath Bryning (SD 401 300, *Birstaf Brinn(ing)* 1201, *Brunigg* 1252)[39] is now 1½ miles from the Ribble estuary. Although it might possibly have been accessible by boat before the marshes were reclaimed, it seems more likely that the name derives from ON *byjarstaðr*, a 'farmstead'. Croxteth (SJ 402 963, *Crocstad* 1257) could be made up of ON *krokr* or, less likely, OE *croc*, a 'bend', with ON *stǫð*, *staðr* or, less likely, OE *stæð*. Its situation near a pronounced bend in the little River Alt and far from the Roman road system makes it a doubtful example of a staithe. It is 3 to 4 miles from the inappropriately named Huyton (above). Equally inappropriately named are Bickerstaffe and Brimstage. Bickerstaffe (Lancs.) (SD446 043, *Bikerstad* 1190) is OE **bicere* 'bee-keeper' with OE *stæð* or ON *stǫð* or *staðr*. Although an OE qualifying element is more likely to be combined with an OE generic, the situation of Bickerstaffe tells against this interpretation because the settlement is on a ridge of higher, drier land a considerable distance from any navigable water, although the ridge itself is a useful route. Topographically, then, Bickerstaffe is more likely to be a *staðr* than a staithe.[40] Brimstage (Cheshire) (SJ 305 828, *Brunstath* 1260) is apparently made up of an OE personal name + OE *stæð*. The Roman road from Chester, Margary 670, is heading straight

[37] K. Cameron, *A Dictionary of Lincolnshire Place-Names* (Nottingham, 1998).
[38] Ekwall, *The Place-Names of Lancashire*.   [39] Ibid.
[40] M. C. Higham, 'The Problems of the Bee Keepers', *Journal of the English Place-Name Society*, 34 (2001–2), 25–6.

for it, but it is 2 1/2 miles from the nearest navigable water and so it cannot be an example of a landing place.

The uncertainties surrounding *staeð* make it a less useful indicator of river traffic than *hȳð* or *port*. Only Stafford, Toxteth, and Birstwith occur as early as DB; only the first two are on navigable rivers and likely to be staithes. Birstwith is doubtful but on the other hand is usefully near the Roman road network. Apart from Stathe in Somerset the examples are found in northern England, more particularly in Lancashire where *hȳð* is rare or absent.

The two place-name terms *lād* and *gelād* have sometimes been confused, but recent studies have established that they were clearly distinguished from each other in Old English place-name formation.[41] The second term, *gelād*, refers to a difficult river crossing, usually to a place where a road running alongside a major river is liable to be swamped at times of flood. The term *lād* means either a road or track, or an artificial waterway or canal.[42] The latter is the more frequent, and the one that concerns us here (Fig. 19). It was often used in Fenland of the ditches constructed by the abbeys for drainage or as canals, but it is not common in settlement names. Whaplode (Lincs.) (TF 323 240, *Copelade* DB, *Quappelada* 1170) is the only place of importance, and in Anglo-Saxon times it would have been much nearer the coast. Possibly the *lād* was a cutting through about 2 miles of saltmarsh linking the settlement to the shore of the Wash (Fig. 15).

Many of the artificial waterways of the Fens have in the past been attributed to the Romans, but more recent work[43] suggests that some of these may be medieval in origin. Among the latter are Bottisham, Swaffham, and Reach Lodes, which are of concern here because the hamlet of Lode is situated at the head of the straight part of Bottisham Lode, which must have been in existence by 1154–89, the earliest extant record for Lode. Horslode Fen Farm (TL 402 829, *Hornigslade* 1240) and Holwoods Farm (*c.* TL 380 804, *Hollode* 1240) in Chatteris, and Crollode's Farm (TL 378 822, *aqua de Grauelode* 1286) take their names from watercourses, most probably from the Roman-built Car Dyke, and were not originally ancient settlement names but imply recognition of the man-made nature of Car Dyke.

The only other area where *lād* occurs as a settlement name is in the Somerset Levels, (Fig. 17; see also Rippon and Hollinrake, below). Unlike the Fens there are no known Roman canals here. Charters of the seventh to ninth centuries only give the barest hints of any draining of the Levels but in the late Anglo-Saxon period recolonization was occurring along the coastal strip of blown sand and on the margins of higher already-occupied ground. The draining of the heart of the Somerset Levels seems only to have begun in earnest in the late

---

[41] M. Gelling, *Place-Names in the Landscape* (London, 1984), 73; Gelling and Cole, *The Landscape of Place-Names*, 81.

[42] See Blair, below pp. 270–1, 278, for this term on the upper Thames.

[43] D. Hall, *The Fenland Project 10: Cambridgeshire Survey, the Isle of Ely and Wisbech*, EAA 79 (1996), 112.

twelfth century.[44] It is in the twelfth and thirteenth centuries that *lād* settlement names appear in the records. Cogload Farm and Curload are near the Tone between Taunton and its confluence with the Parrett, and by a marshy area still very prone to flooding. The Tone has been both diverted and straightened at various times; its old course is represented by the Old Rhyne along the northern fringe of the marshes, passing between Lyng and Athelney to join the Parrett just beyond West Yeo Farm. The greater part of its course is followed by parish boundaries. A southern arm of the Tone passed to the south of Athelney to join the Parrett near the present confluence. It is not known when the western part of the Tone from Knapp to Curland was rerouted. The eastern part was diverted in 1374–5.[45] Cogload (ST 306 275) appears in the cartulary of Buckland Priory as *Coglode* by 1269.[46] It is not possible to tell whether its lode went into the Old Rhyne or on to the present course of the Tone, nor whether it pre-dated the foundation of Buckland Priory or was cut by their monks. Curload (ST 389 273)[47] appears in the cartulary of Athelney Abbey in 1155–9, when one Gilbert of Corilade is mentioned. It cannot refer to the straightened Tone between Curload and Athelney, since the record pre-dates the cutting of the new channel. It is likely to have been a smaller waterway connected with the large Curry estate. It is noteworthy that Stathe is at the end of this long finger of upland which borders the Tone on its south side

Long Load (ST 465 233, *la Lade* 1285) and Little Load (ST 467 239, evidenced by 1364) are on either side of the Yeo between Langport and Ilchester, at a point where higher ground approaches the river on both sides. A short waterway would easily link the Yeo to firm ground and a route south to Martock and the Fosse Way.

The other two *lād* settlement names are north of the Polden Hills by the River Axe. Northload (ST 49 39, *Northlode* c.1180)[48] linked the old course of the Axe to firm ground in Glastonbury, while Northload in Theale (ST 470 463, *Northelode* 1308)[49] would enable boats to reach the Isle of Wedmore from the Axe.

In these examples a *lād* is a waterway dug through ill-drained land to link a river to nearby rising ground, allowing a more efficient movement of goods and people. Some of these watercourses may date back to Roman times, others are more likely to be associated with the activities of religious houses. They seldom gave rise to settlement names.

The term *ēa-tūn* (Modern English Eaton, Eton, Yeaton, Ayton) (Fig. 18), means 'river settlement'. Formally it is a name that could have been given to

---

[44] Rippon, *The Severn Estuary*, 178.

[45] M. Williams, *The Draining of the Somerset Levels* (Cambridge, 1970), 17, 22–3.

[46] F. W. Weaver (ed.), *Cartulary of Buckland Priory: Somerset Record Society*, 25 (1909), 36.

[47] E. H. Bates (ed.), *Cartularies of Muchelney and Athelney Abbeys: Somerset Record Society*, 14 (1899), 164.

[48] A. Watkin (ed.), *The Great Cartulary of Glastonbury*, ii, Somerset Rec. Soc. 63 (Taunton, 1952), 251.

[49] *HMC Wells MSS*, i. 219.

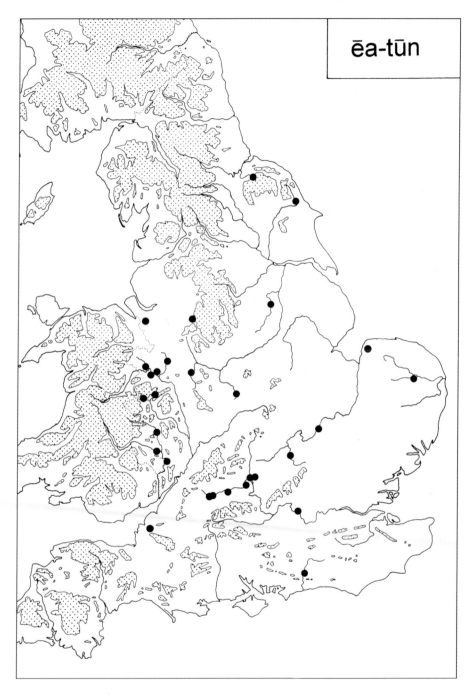

Fig. 18. *ēa-tūn* in place names.

many hundreds of riverside settlements, but the fact that there are only just over thirty suggests that an *ēa-tūn* had some special function relating to the river. The two most plausible suggestions are that this was to provide a ferry service or else to keep the river open for navigation. Their distribution, with so many on the upper reaches of major rivers or their tributaries, favours the latter suggestion, as small rivers are easily blocked by fallen trees, growth of rushes and reeds, shoals, and collapsing banks, and would need frequent maintenance. The lower reaches of major rivers like the Thames would not suffer from these problems to the same extent, but it would be the lower, less fordable reaches where a ferry service would be most needed, and that is not where the *ēa-tūns* are found.

The Thames has four *ēa-tūns* on its banks above Oxford and upstream of the *hȳðs* (see also Blair, below p. 261). At Eaton (SP 448 033, *Eatune* 811 c.1200, *Edtune* DB) the banks slump into the river, and it has to be dredged for today's river traffic. Eaton Hastings (SU 262 985, *Etone* DB) is about 3 1/2 miles downstream of Lechlade, the present head of navigation, where the river is liable to be choked by club rush (*Schoenoplectus lacustris*); it would be important to keep the river navigable as far as Lechlade because it was at the end of a salt route from Droitwich (above, p. 37), and therefore a trans-shipment point.

Castle Eaton (SU 146 960, *Ettone* DB) and Water Eaton (SU 126 938, *Etone* DB) lie between Lechlade and Cricklade. In the 1980s, despite considerable difficulties owing to the growth of club rush and Norfolk reed, and the presence of sand and gravel banks, a rowing boat was taken on up to Cricklade; the river was too narrow for rowing, but punting was quite effective.[50] Goods could have been moved in puntlike craft on the upper reaches of many rivers. Access to Cricklade would have provided another trans-shipment point onto the road system, here the Cirencester to Silchester Roman road (Margary 41b).

Woodeaton (SP 535 119, *Etone* DB) and Water Eaton (SP 515 122, *Eatun* 864 (eleventh century), DB) are situated on opposite banks of the Cherwell just below the confluence with the River Ray at Islip, itself the crossing place of an old route between London and Worcester (cf. Blair, below pp. 263–4).

In Norfolk, Eaton in Sedgefield (TL 695 363) first occurs in the records in the mid thirteenth century.[51] It is on the Heacham River, a small stream which cannot be navigable for far and yet if kept open for just 3 or 4 miles would provide a useful link between the Wash and the Icknield and Peddars Ways.

Eaton's Farm (Sussex) (TQ 187 162, *Etune* DB) is just upstream of Heath Barn and the Roman road Margary 40 (see under *hȳð* above), in what was the Adur estuary. Its presence suggests that silting was already a problem at the time of the Conquest, a serious matter in what was evidently a busy haven.

Although the Wye and Severn and their tributaries have no settlements named *hȳð* along them, they do have *ēa-tūns*—three or four in the Wye basin

---

[50] K. Taplin, 'Beyond the Limit', *The Countryman*, 97/5 (1992), 65–70.
[51] B. Dodwell (ed), *The Charters of Norwich Cathedral Priory*, ii, Pipe Roll Society ns 46 (London, 1978–80), 6–9.

and six in the Severn/Teme basin—the place names suggesting that they too were used for navigation. In the Wye valley the places are:

Eaton Tregoz (SO 605 277, *Edtune* DB),[52] on the Wye.

Eaton Bishop (SO 442 391, *Etune* DB),[53] on the Wye, 1 mile from Margary 630.

Eaton (SO 509 583, *Et(t)on(e)* DB),[54] on the Lugg, 1 mile from Margary 613.

Eaton Hennor (SO 525 588, *Iatton* 1243),[55] 1 mile from the Lugg on Margary 613.

These are above the limestone gorges of the lower Wye, in the gentler country-side around Hereford, and all are on stretches of river within easy reach of a Roman road.

The six in the Severn/Teme basin are less well placed: only Eaton Mascot is close to a Roman road and Eaton Constantine probably is, while Eaton and Eaton under Heywood are close to old tracks. The other two have no obvious connection with the road system. Together they indicate that the upper Severn and its tributaries were used for navigation. The sand and gravel banks which develop near Eaton Mascot, and rushes and reeds which grow at Eaton upon Tern, would need clearing regularly. The *ēa-tūn*s in the Severn basin are:

Eaton Constantine (SJ 599 064, *Etune* DB), on the Severn.

Eaton Mascott (SJ 538 059, *Etune* DB), on the Cound Brook, 1 mile from Margary 6b.

Eaton upon Tern (SJ 654 225, *Eton'* 1255–6), on the Tern.

Yeaton (SJ 433 194, *Aitone* DB), on the Perry.

Eaton in Lydbury north (SO 375 895, *Eton'* 1291–2), on the Onny.

Eaton under Heywood (SO 500 900, *Eton* 1227), on the Eaton Brook.

Other *ēa-tūn*s on small rivers are:

Eton (Lane) (Somerset) (ST 381 615, *Eton* 1325),[56] a lost settlement on the River Banwell, which served a small area locally.

Water Eaton (Staffs) (SJ 903 110, *Eatun*, 940, DB), on the Penk and close to Watling Street.

Nuneaton (Warwicks.) (SP 362 918, *Etone* DB), on the Anker, also close to Watling Street.

East and West Ayton (Yorks.) (SE 991 850,and SE 987 850, both *Atun(e)* in DB), near the edge of the North York Moors were on the upper Derwent, a useful means of moving around in the marshy Vale of Pickering.

---

[52] B. Coplestone-Crow, *Herefordshire Place-Names*, BAR 214 (Oxford, 1989), 86.
[53] Ibid. 80. [54] Ibid. 124. [55] Ibid. [56] *HMC Wells MSS*, i. 95.

Great and Little Ayton (Yorks.) (NZ 557 108 and NZ 570 102, *Atun(a)* DB and *Atun* DB) are on so small and rocky-floored a stream, the Leven, that it would not be navigable under today's conditions.

This leaves about nine *ēa-tūn*s on wider rivers which need much less attention to keep them navigable. About two-thirds are associated with crossing places, mostly fords as noted in the place names, but even, perhaps, some bridges.

Water Eaton (Bucks.) (SP 880 330, *Etone* DB) is 1 mile south of the crossing of Watling Street, Margary 1e, over the Ouzel at Fenny Stratford. It might have been responsible for maintaining the ford so that it was usable by those on foot without causing an obstruction to boats, or perhaps for keeping the arches of subsequent bridges free of debris brought down in time of flood. In 1347 there is a reference to earlier bridges over the Ouzel.[57] Similarly Eaton (Notts.) (SK 710 780, *Etune* DB) is on the Idle, 2 miles upstream of East Retford where one of the routes followed by the Great North Road crossed the river. A bridge had been built by 1385, when there is a reference to pontage.[58] At Eaton Hall (Cheshire) (SJ 415 610, *Eaton c.*1050, DB), the Roman road Margary 6a, from Chester to Wroxeter, crosses the Dee at Aldford a mile upstream. Near Norwich, Eaton (TG 203 061, *Ettune* DB) is on the opposite bank of the Yare to Cringleford, which had a bridge prior to 1539 when it was broken down by floods.[59] Eaton Bishop (Herefordshire) (above) is 1$\frac{1}{2}$ miles from the crossing of Margary 630 over the Wye as it goes from Kenchester to Abergavenny. Eaton Socon (Beds.) (TL 170 589, *Etone* DB) is on the Great Ouse a mile upstream of St Neots where there was a bridge by 1254, superseding a ford.[60] Eton (Bucks.) (SU 965 776, *Ettone* DB) is across the Thames from Windsor. There was a bridge at Windsor prior to 1172 when tolls were being levied. An account in 1793 describes the old timber bridge, and the impediments to water-borne traffic during times of flood by the 'piles, projections and rubbish on the shores at both ends of it'.[61] The weight of flotsam and the force of the floodwater could bring down a bridge if it was not cleared of debris from time to time. The other two *ēa-tūn*s by broad rivers are Eaton Constantine and Eaton Tregoz (above).

The *ēa-tūn*s, then, are most probably connected with keeping the narrower reaches of rivers open for navigation, and for maintaining fords in a fit state to allow both road and river traffic to pass. At a later date, when bridges had replaced fords, they could have been involved in keeping the structures free of debris brought down by floods.

The foregoing discussion is summarized on Fig. 19, where the larger versions of the symbols denote place names recorded in or before Domesday Book. Smaller versions denote place names recorded after 1086 but usually before 1400.

---

[57] E. Jervoise, *The Ancient Bridges of Mid and Eastern England* (London, 1932), 86.
[58] Ibid. 35.   [59] Ibid. 117.   [60] Ibid. 95–7.
[61] F. S. Thacker, *The Thames Highway: A History of the Locks and Weirs* (London, 1920); repr. New York, 1968), 357.

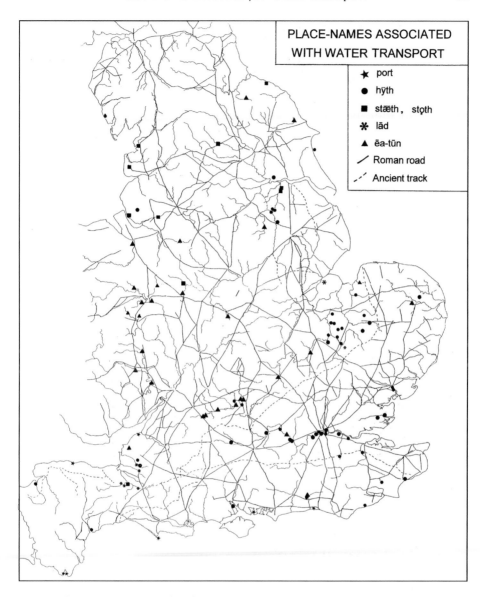

Fig. 19. Place names associated with water transport.

The series of *port* names, well spaced along the south coast and Bristol Channel, are all recorded by 1086; some of the coastal *hȳðs* of southern England are also recorded by then. The Thames was evidently a busy waterway, with numerous *hȳðs* and *ēa-tūns* evidenced by 1086. The DB *ēa-tūns* on the Wye and upper Severn and its tributaries suggest that these rivers too were well used, although they are without any of the landing place terms.

The post-DB *hȳð* names are most frequent in the marshlands where artificial waterways occur. The *hȳðs* are clearly associated with some of them—for instance Stockwith and Walkerith are by Bykers Dyke—and others in the Fens are by the artificial waterways or serve islands such as Ely; they are used similarly in parts of the Somerset Levels. The same link with marshlands occurs with *lād*, which, though often used as a name for a man-made ditch, rarely occurs as a settlement name: only Whaplode is as early as DB. Examples of *stæð* and *stǫð* are difficult, if not impossible, to determine from available spellings, but the few reasonably certain ones are concentrated in northern England.

There are some notable gaps in the distribution. The chalklands, where the occasional rivers are seasonal in flow, have few *hȳð*s for obvious reasons. Large areas of the midlands are devoid of all but *ēa-tūn*s; those that do occur are very close to Watling Street (Margary 1), suggesting an interface between road and river transport at these points. Indeed, many of the port, *hȳð*, and *stæð/stǫð* settlements have links with the Roman road network or ancient long-distance tracks. The use of the waterways must be considered in conjunction with overland routes.

Place names cannot give a complete picture of medieval water routes, but they do emphasize the use of some of them, both coastal and inland, and may be a valuable adjunct to other types of evidence.

4

Hythes, Small Ports, and Other Landing Places in Later Medieval England

MARK GARDINER

Archaeologists and historians alike have paid considerable attention to the larger coastal and riverine ports. Such places, which included Southampton and Sandwich adjoining the English Channel, and King's Lynn and Boston on the Wash, often have significant archaeological remains and were places of considerable economic importance in the later medieval period. The more numerous smaller urban and rural landing places, by contrast, have been very largely ignored. They rarely had the facilities of the large ports. They were unlikely to have warehouses and generally did not have cranes for loading and unloading. They may have had no waterfront structures; boats were often simply drawn up on the mud at the edge of a river or beached on the shore by the ebbing sea-tide. A rural or small urban landing place was simply a location with access to a road or track at which goods could be transferred to and from the land. There were numerous potential sites which met this simple criterion. Individually, these places were of limited economic significance; locally, they may have been of considerable value; collectively, they played an important role in the movement of goods and people in later medieval England.

No single term covers the variety of places at which a boat may be moored or beached. The term 'port' is used here to refer to a settlement, the economy of which was substantially based upon river- or seaborne traffic. A 'hythe' is used to describe a site adjoining a river without a significant settlement at which boats might be pulled up or moored. 'Landing place' is a generic term covering all such sites and any others at which boats might be grounded, pulled up, or moored. The term 'haven' is used for any protected anchorage or mooring on the coast without prejudice to whether it allowed the transfer of goods and people to land. But havens, unless they were also landing places, lie outside the sites discussed here. The problems of terminology and definition reflect the

I am grateful for support for travelling expenses from Queen's University Belfast. I also wish to thank Prof. Bruce Campbell and Ken Bartley for providing access to the inquisitions *post mortem* database. The maps were prepared by Libby Mulqueeney.

diversity of minor landing places. These sites included not only places at which goods were loaded and unloaded, but also fishing ports and even mooring points for ferries. Frequently, a landing place served a number of functions. The present study therefore not only considers landing places for loading and unloading cargo, but also touches briefly upon fishing bases and shipyards, and makes reference to ferry sites.

There are clearly problems not only in the terminology to describe landing places, but also in distinguishing the larger of the minor sites from major ports. It is not possible to draw a sharp boundary between small and large ports on the basis of the size of ships which used them: large ships might occasionally put into small ports. Volume of trade passing through a port is difficult to estimate and often can only be determined impressionistically. It is, however, salutary to consider those few sources which do detail the economic activity at small ports. For example, surviving particulars of the Port of Lynn, a customs jurisdiction which covered the north coast of Norfolk as far east as Yarmouth, show that 15 per cent of all imports in that area came into Holkham, Blakeney, and Burnham.[1] We may assume that by the thirteenth century the larger ports invariably would be found at or close to major urban centres, and smaller ports would lie at lesser settlements, or places for which water-borne trade was not a significant part of the economy. But it is not clear how true this was in earlier centuries; a hierarchy of ports and settlements appears to emerge only towards the end of the eleventh century.[2]

Functions of Minor Landing Places

One way of approaching the multiplicity of minor landing places is to consider the function which they served. It is not suggested that the list of types of landing place below is a strict taxonomy. These sites might simultaneously perform two or more functions, or, as they grew in size, might change from one role to another. The categorization does, however, provide a way of approaching the variety of landing places in medieval England.

Ports at Small Towns, Markets, and Fairs

Many landing places were located at small towns and markets, because these were sites at which goods could be gathered together, having been brought by water from upstream or over land, for onward shipment in larger vessels. London was supplied with corn and other goods by a network of ports on the middle and upper Thames and its tributaries, as well as ports in the estuary and beyond. Henley owed its importance to its position at the highest point on the river which could be regularly reached by boat, and the markets for corn

---

[1] T. H. Lloyd, *Alien Merchants in England in the High Middle Ages* (Brighton, 1982), 48.
[2] M. F. Gardiner, 'Shipping and Trade between England and the Continent during the Eleventh Century', *Anglo-Norman Studies*, 22 (2000), 84–7.

and wood developed there as a consequence.[3] Similar small ports for the supply of London developed at Ware on the River Lea, Faversham on the Thames estuary, Maidstone on the Medway, and Weybridge on the River Wey.[4] There was a comparable network of small ports supplying the harbour at King's Lynn. Lynn, unlike London, was important not so much as an urban centre in which goods were consumed, but as a bulking or collection point at which goods were accumulated before being shipped in larger vessels up the coast or overseas.[5] The network of small ports supplying King's Lynn stretched into the Fens and up the rivers Ouse and Nene into the east midlands. These waterways allowed, for example, corn purchased for the king in 1302 at Cambridge, Reach, St Ives, and Yaxley to be shipped downriver in boats to Lynn.[6]

The importance of water transport to market settlements in the area is apparent from the chorus of complaints which followed from any changes or obstructions in the waterways. When a channel or lode near Whittlesey Mere was blocked in 1342, the men of Walton, Sawtry, and Conington in Huntingdonshire were prevented from taking corn, peat, and other goods downriver to the sea. A similar complaint was made by the men of nearby Holme and Yaxley who had shipped goods to King's Lynn along the Nene. They said that, since it had been stopped up in about 1301, they were forced to take their goods, which included corn, timber, wool, sedge, rushes, turf, stone, meat, and grain, a longer route by *Oldwelnee* (Well Stream) and Littleport. As a consequence, they claimed that the market of Holme, which had been visited by merchants from Lynn, had now declined.[7] Water transport was so vital to the prosperity of the region that some markets in the region, such as that at Swaton (Lincs.), were linked by canals.[8] Elsewhere, markets developed at hythes alongside the waterways. The port at Reach (Cambs.) became the site of a major annual fair at which clunch (chalk), iron, and wood were traded. A number of basins and wharves may still be recognized there.[9]

The association of small ports and markets is found elsewhere. Britnell noted that one-third of all the markets established in Essex in the period 1200–74 were located on the coast and in most cases at small ports. This contrasts

---

[3] Cf. Langdon and Blair, below pp. 128–9, 256–7.

[4] B. M. S. Campbell, J. A. Galloway, D. Keene, and M. Murphy, *A Medieval Capital and its Grain Supply: Agrarian Production and Distribution in the London Region c. 1300* ([London], 1993), 24, 59–63, 194–7; *Ancient Deeds*, i. A1086; W. Walton, 'Accompts of the Manor of Savoy', *Archaeologia*, 24 (1832), 304, 313.

[5] M. Kowaleski, *Local Markets and Regional Trade in Medieval Exeter* (Cambridge, 1995), 232.

[6] D. M. Owen (ed.), *The Making of King's Lynn: A Documentary Survey*, Records of Social and Economic History, NS 9 (London, 1984), 257.

[7] W. H. Hart (ed.), *Chartulariam Monasterii de Ramesia*, Rolls Ser. 79 (London, 1884), i. 174–81; iii. 121–57, esp. 141. Darby provides a useful map of the changes in the watercourses: H. C. Darby, *The Medieval Fenland* (Cambridge, 1940; repr. Newton Abbot, 1974), 97.

[8] *Curia Regis Rolls*, xvi. 490–1.

[9] Royal Commission on the Historical Monuments of England, *An Inventory of Historical Monuments in the County of Cambridgeshire*, ii: *North-East Cambridgeshire* (London, 1972), pp. lxv–lxvi, 85–6, 89; E. Carus-Wilson, 'The Medieval Trade of the Ports of the Wash', *MA* 6–7 (1962–3), 182–201.

with the much smaller proportion of earlier date and seems to suggest the rising importance of water-borne trade.[10] The proportion of coastal markets is probably higher in Essex than elsewhere because the indented shore provides many suitable locations for landing ships, and much of the trade in the county was destined for London. However, ports along the Thames estuary and the east coast supplied not only the capital, but also the Continent. Ships from Barking and Rainham sailed to Calais, Normandy, and Holland. The pattern on the south bank of the Thames estuary was probably similar: there were markets and ports at Greenwich, Gravesend, and Northfleet.[11] Even very minor landing sites might develop into small, informal trading places. Dyer has suggested that grain may have been traded at the hamlet and landing place of Haw near Tirley on the River Severn.[12] The possibility that markets might develop at other landing places clearly concerned the burgesses of a number of well-established towns who sought by various means to restrict unloading and trading to their boroughs.[13]

The smaller landing places, therefore, formed interim stages along the dendritic path which fed goods into the larger ports. This is most clearly illustrated from the purveyors' accounts, which identify the movement of goods in detail.[14] For example, grain was moved to Kingston upon Hull in 1345–6 from various places on the Ouse and Hull river systems, including from Boroughbridge on the River Ure and Wansford on the River Hull using small boats (*batellae*) and ships (*naves*). At Hull it was loaded on a ship for transport by sea. In a similar manner, flour was brought to Shoreham (Sussex) for transport by sea to Newcastle in 1319. Some of the flour reached Shoreham overland from Lewes and Tarring, and the remainder by sea along the coast from the minor hythes at West Wittering and Sidlesham.[15]

The dendritic system which linked major and minor ports did not only operate in one direction. Goods imported into England arrived at the major centres and were distributed, either by land or water, to lower-order centres. However, the pattern of distribution was not simply a mirror-image of the bulking of goods. Imported items from aboard were often high-value items

---

[10] R. H. Britnell, 'Essex Markets before 1350', *Essex Archaeology and History*, 13 (1981), 18–19.

[11] C. Johnson and H. A. Cronne (eds.), *Regesta Regum Anglo-Normannorum 1066–1154*, ii (Oxford, 1956), no. 730; *VCH Essex*, v. 239; vii. 136; viii. 45. Ships from Gravesend are implied by the licence granted to Roger de Gravesende in 1230 and are specifically mentioned in 1298. Ships from Northfleet are recorded in 1295 (*Cal. Pat. R. 1225–32*, 344; *1292–1301*, 134, 337). Gravesend was granted a charter for a market in 1366 (*Cal. Charter R. 1341–1417*, 194); Northfleet was a market by prescription.

[12] M. W. Beresford, *New Towns of the Middle Ages: Town Plantation in England, Wales and Gascony* (London, 1967), 513–14; C. C. Dyer, 'The Hidden Trade of the Middle Ages: Evidence from the West Midlands', *JHG* 18 (1992), 150.

[13] R. H. Britnell, 'English Markets and Royal Administration before 1200', *EcHR* 2nd ser. 31 (1978), 194–6; S. H. Rigby, *Medieval Grimsby: Growth and Decline* (Hull, 1993), 31–2.

[14] See also Langdon below.

[15] R. A. Pelham, 'Fourteenth-Century England', in H. C. Darby (ed.), *An Historical Geography of England before A.D. 1800* (Cambridge, 1936), 264, fig. 47; R. A. Pelham, 'Studies in the Historical Geography of Medieval Sussex', *SxAC*, 72 (1931), 169–70.

and were rarely marketed in small towns. Their sale was confined to the larger ports or other major towns where they were sought out by consumers. For example, gentry households near the north Somerset coast did not buy imports at the local, minor ports of Minehead or Watchet, but travelled to Exeter. A wider range of goods could be obtained there and could be bought more economically by purchasing larger quantities. The difference between the pattern of distribution of cheaper local items and higher-cost imports is well illustrated by comparing the places of purchase of foreign and English millstones. The former were generally sold in the major ports, but the cheaper English stones were available more widely in many towns, both coastal and inland.[16]

Landing Places at Estate Centres

Both lay and religious estates often preferred to 'buy' or transfer goods from manors within the estate rather than purchase them in the open market. The practice is sometimes known as 'costless consumption', a term which might be accurate insofar as no money changed hands, although manorial accounts include the value of grain, stock, and other items transferred to other parts of the estate. The movement of goods over considerable distances might not have been economically viable in other contexts, but carriage within an estate could be managed by means of the labour services given by villeins or performed by demesne employees.[17] The tenants of the east Kent manor of Northbourne (near Deal, Kent) had the work service of bringing timber by ship from the demesne woodlands in the Weald near Tenterden. The timber was presumably loaded at one of the small ports on the River Rother close to the woods in which it was cut, then taken by ship around the east Kent coast to an unidentified place called *Greistonehende*, and completed the final part of the journey overland to Northbourne.[18] The tenants of each hide of land at the St Paul's Cathedral manor of Kirby with Horlock in Essex carried $4\frac{1}{2}$ seams from the lord's barn to an unspecified, but presumably local, port.[19] The labour services owed by villeins in Fenland commonly included moving goods by water. Crowland Abbey tenants on the manors of Oakington, Cottenham, and Dry Drayton gave the payments of *menyngpeni* and *schiphire* which evidently replaced the tasks of crewing and providing vessels. Those at Morborne in Huntingtonshire

[16] Kowaleski, *Local Markets*, 269; C. C. Dyer, 'The Consumer and the Market in the Later Middle Ages', EcHR 2nd ser. 42 (1989), 308–12; D. L. Farmer, 'Millstones for Medieval Manors', *Agricultural History Review*, 40 (1992), 101.

[17] D. Postles, 'Customary Carrying Services', *Journal of Transport History*, 3rd ser. 5/2, (1985), 1–15.

[18] R. H. Hilton, *A Medieval Society: The West Midlands at the End of the Thirteenth Century* (Cambridge, 1966), 177–8; G. J. Turner and H. E. Salter (eds.), *The Register of St Augustine's Abbey Canterbury Commonly Called the Black Book*, British Academy Records of the Social and Economic History of England and Wales 2 (London, 1915), i. 78–9.

[19] W. H. Hale (ed.), *The Domesday of St. Paul's of the Year MCCXXII*, Camden Soc. 69 (London, 1858), 47.

paid *rowyngsilver*. Though these tasks had been commuted by the middle of the thirteenth century, corn and malt was still delivered by water to the abbey from Cottenham in the early fourteenth century, and the demesne lease of the same manor dated 1430 included the delivery of malt by boat to Crowland, when there was sufficient water, and by cart otherwise.[20]

Estate centres had landing places to receive the incoming produce or dispatch it to the market. An excavated dock or wharf at Waltham Abbey (Essex) close to the River Lea was sited to provide access to barns within the grange adjoining the abbey. Goods could be collected here for the use of the household or shipped downstream to London. Other manors in the vicinity of London possessed their own boats for a similar purpose, including Chingford which lay south of Waltham Abbey on the Lea, Laleham (Middx.) on the Thames, Thorncroft (Surrey) on the River Mole, and perhaps Faversham where the townsmen had to provide and pay a steersman.[21] Boats were used on the flat land around Caister Castle (Norfolk), both for the carriage of building materials during the castle's construction, and for the movement of goods from the nearby port of Great Yarmouth to the moat and into the castle itself.[22] Moated houses, such as Caister Castle, were often situated in low-lying spots which were suitable for the construction of docks. The moat at East Haddlesey (W. Yorks.) was enlarged to take boats and was linked to the River Aire. Similarly, excavations at the moated site of Stretham (Sussex) held by the Bishop of Chichester have uncovered a waterfront on the River Adur.[23]

Landing Places at Points of Production

Minor ports were particularly important for the movement of items of low value and high weight, such as stone, ore, coal, and wood. Transport charges could amount to a significant proportion of the purchasers' costs, and these goods could only be marketed economically over long distances if they could be loaded on to boats or ships near their place of production. Salzman estimated that, so great were the costs of transport by land, they exceeded the cost of the building stone if the journey was greater than 12 miles.[24] Direct transport from producer to consumer reduced the cost which might otherwise have formed an uneconomic proportion of the cost of the goods. Building stone was commonly

[20] F. M. Page, *The Estates of Crowland Abbey: A Study in Manorial Organisation* (Cambridge, 1934).

[21] P. J. Huggins, 'Monastic Grange and Outer Close Excavations, Waltham Abbey, Essex, 1970–1972', *Essex Archaeology and History*, 4 (1972), 81–8; A. Bennett, 'Archaeology in Essex 1998', *Essex Archaeology and History*, 30 (1999), 210–31; Campbell et al., *A Medieval Capital*, 59; *Placitorum Abbreviatio, Richard I–Edward II* (London, 1811), 140.

[22] H. D. Barnes and W. D. Simpson, 'Caister Castle', *Antiquaries Journal*, 32 (1952) 36, 38; H. D. Barnes and W. D. Simpson, 'The Building Accounts of Caister Castle A.D. 1432–1435', *Norfolk Archaeology*, 30 (1949–52), 178–88.

[23] H. E. J. le Patourel, *The Moated Sites of Yorkshire*, Society for Medieval Archaeology Monograph 5 (London, 1973), 23; unpublished excavation by A. Barr-Hamilton.

[24] L. F. Salzman, *Building in England down to 1540: A Documentary History* (Oxford, 1952), 119.

purchased at the quarry and then transported directly to the construction site, either by the quarry operator or by a contractor; it was rarely sent to merchants who sold it on to builders. Firewood was sometimes obtained in the same manner, direct from the coppice woodland at which it was cut.[25]

Two types of landing places can be identified for the purpose of shipping heavy goods: those which were located at or very close to the place of production, and loading wharves which were some distance away on the nearest navigable waterway and may therefore have served more than one production centre. Comparatively few landing places fall in the first of these categories. Sea-salt was one of the few commodities which had to be produced close to tidal waters and might be loaded directly into ships. However, the exposures of salt-laden silt necessary for its manufacture were often found in shallow waters which were not suitable for larger vessels. The port of *Hirnflete* near Holbeach in the Wash was visited by ships for salt in the thirteenth century, and at the end of the previous century ships were moored at Scrane.[26] Some stone quarries also lay very close to or adjoining the sea or a river. The stone quarries at Beer in Devon and those at Fairlight in Sussex were close to the sea.[27] Where a quarry was not situated close to a waterway, one might be constructed. Bishops Dyke may have been built to facilitate the movement of stone from the quarries at Huddleston (W. Yorks.) to the quay at Cawood on the River Ouse.[28] Another canal was constructed a few miles to the south initially to carry stone from the Monk Fryston quarries to Selby Abbey and by the fourteenth century was used to take wood from the abbey's lands through which it passed.[29]

The distance over which a commodity was carried was related to the scarcity of the resource and its value. As far as possible, low-value stone was brought from close to the building site, but some materials were not available in the immediate vicinity. Chalk was used in London for footings, rubble infill, and, most of all, for lime for mortar. The chalk Downs in Kent provided a convenient source. It was cut from quarries at Greenhithe and Northfleet on the Thames, and near the River Medway. The chalk might be burnt for lime

[25] D. Knoop and G. P. Jones, 'The English Medieval Quarry', EcHR 1st ser. 9 (1938–9), 26–9; C. C. Dyer, 'Trade, Towns and the Church: Ecclesiastical Consumers and the Urban Economy of the West Midlands, 1290–1540', in T. Slater and G. Rosser (eds.), *The Church in the Medieval Town* (Aldershot, 1998), 65–6.

[26] N. Nielson (ed.), *A Terrier of Fleet, Lincolnshire, from a Manuscript in the British Museum*, British Academy Records of the Social and Economic History of England and Wales 4 (London, 1920), 83.

[27] N. J. G. Pounds, ' Buildings, Building Stones and Building Accounts in South-West England', in D. Parsons (ed.), *Stone: Quarrying and Building in England AD 43–1525* (Chichester, 1990), 234; L. F. Salzman, *English Industries of the Middle Ages* (Oxford, 1923), 87.

[28] J. S. Miller and E. A. Gee, 'Bishops Dyke and Huddleston Quarry', *YAJ* 55 (1983), 167–8. Parsons has raised doubts about the evidence: D. Parsons, 'Stone', in J. Blair and N. Ramsay (eds.), *English Medieval Industries: Craftsmen, Techniques, Products* (London, 1991), 22 n. 78.

[29] S. Moorhouse, 'Medieval Yorkshire: A Rural Landscape for the Future', in T. G. Manby, S. Moorhouse, and P. Ottaway (eds.), *The Archaeology of Yorkshire: An Assessment at the Beginning of the 21st Century* (Leeds, 2003), 194.

near the quarry, or loaded straight into a boat and burnt elsewhere. Much of the chalk for London was taken to Limehouse, a name derived from the oasts or kilns in which the lime was heated.[30] There were even quarries as far east as Sarre on the Wantsum Channel, but whether these supplied the London market, or only that in Kent, is not known.[31] Roofing slate had a greater value in relationship to its weight and was distributed over a wider area. It was shipped in the later medieval period from quarries adjoining the Kingsbridge estuary and from Beesands Quarry near Torcross on the coast in Devon. The area of distribution extended along the south coast as far as east Kent and even London.[32]

Since transport costs could form a considerable part of the cost of a product, much of the initial processing of materials took place near the point of production to reduce the volume. The degree to which stone was shaped at the quarry varied. Stone was generally shipped in blocks, but mouldings might be prepared at the quarry, presumably from supplied templates.[33] Slate was split and cut to size at the quarry, although the holes for pegs were made on site.[34] Metal ores were generally refined close to the point of extraction because of the cost of transport of low-value materials. However, this was not always the case. In more remote areas such as Wales, unprocessed iron ore was moved overland and by water to the furnaces at which it was converted into blooms.[35]

Loading Wharves

Loading wharves were found some distance from the point of production, but on the nearest suitable waterway and often at rural locations. A series of landing places have been identified on the tributaries of the Humber. These allowed stones to be shipped to many places in eastern and southern England. Stone

---

[30] Salzman, *Building in England*, 128–9; V. Harding and L. Wright (eds.), *London Bridge: Selected Accounts and Rentals, 1381–1538*, London Record Soc. 31 (London, 1995), nos. 262, 323, 469; P. Marsden, *Ships of the Port of London: Twelfth to Seventeenth Centuries AD*, English Heritage Archaeological Rep. 5 (London, 1996), 105–6, 212–13; R. H. Britnell, 'Rochester Bridge 1381–1530', in N. Yates and J. M. Gibson (eds.), *Traffic and Politics: The Construction and Management of Rochester Bridge AD 43–1993* (Woodbridge, 1994), 64–8; C. Phillpotts, 'Landscape into Townscape: An Historical and Archaeological Investigation of the Limehouse Area, East London', *Landscape History*, 21 (1999), 65.

[31] *Ancient Deeds*, vi. C5364.

[32] B. Tyson, 'Transportation and the Supply of Construction Materials: An Aspect of Traditional Building Management', *Vernacular Architecture*, 29 (1998), 69; E. M. Jope and G. C. Dunning, 'The Use of Blue Slates for Roofing in Medieval England', *Antiquaries Journal*, 34 (1954), 209–17; J. W. Murray, 'The Origin of Some Medieval Roofing Slates from Sussex', *SxAC* 103 (1965), 79–82.

[33] D. Knoop and G. P. Jones, *The Medieval Mason: An Economic History of English Stone Building* (New York 1933), 76, 78. Large blocks of limestone were found in Whittlesey Mere when it was drained. These were evidently being carried by water to a building site: D. Hall, *The Fenland Project 6: The South-Western Cambridgeshire Fenlands*, EAA 56 (Cambridge, 1992), 32.

[34] E. W. Holden, 'Slate Roofing in Medieval Sussex: A Reappraisal', *SxAC* 127 (1989), 79.

[35] N. Nayling, *The Magor Pill Medieval Wreck*, CBA Research Rep. 115 (York, 1998); T. P. Young and G. R. Thomas, 'Provenancing Iron Ore from the Bristol Channel Orefield: The Cargo of the Medieval Magor Pill Boat', in A. M. Pollard, *Geoarchaeology: Exploration, Environments, Resources*, Geological Society Special Publication 165 (London, 1999), 103–21.

for York Minster was cut at quarries at Thevesdale and carried to Cawood where it was loaded on to ships. Other stone from the quarry at Bramham was loaded at Tadcaster. There was a further quay on the River Wharfe at *Ketilbarnbrigg*. The Thevesdale and Huddleston quarries also supplied stone for King's College, Cambridge, to which it was carried almost all the way by water after loading on to ships at the River Wharfe. Stone for Eton College from Huddleston was taken to Cawood and shipped in a similar way to the Thames.[36] Other stone was used for specialist purposes. Purbeck marble, for example, was shipped from the quays at Ower, Swanage, and Ware around Poole harbour and distributed throughout England. A glimpse of the coastal distribution is given in an entry in the Close Rolls recording a ship arrested in 1230 at Portsmouth carrying a cargo of marble intended for Waltham Abbey.[37]

Building stone was only one of the products moved by water on a large scale. In some areas, where markets lacked a landing place in the vicinity, they might establish one on a more distant waterway to provide access to shipping. The land-locked town of Bodmin had a small quay on the Fowey estuary at Bodmin Pill, one of a number of landing places which have been recently identified on the inlet.[38] Fuel, both coal and wood, was also carried on ships over considerable distances. The main medieval source of sea-coal was around Newcastle, from whence it was shipped to Scotland and the east and south-east England, as well as continental Europe, but it was also loaded at North Shields and on the south of the Tyne.[39] Firewood was shipped over shorter distances. London was supplied from numerous small landing places on the middle Thames. The Weald, which was mostly too remote for the London market, exported wood to the Continent from numerous small quays along the River Rother on the Kent–Sussex border. These are discussed further below.[40]

Landing Places of Local Function

Water transport was widely used in marshlands where there was a network of minor waterways allowing the movement of boats to many places. Travel overland in marshlands could be difficult because of the large number of water-filled ditches and limited number of bridges. Water transport might be

[36] J. Raine (ed.), *The Fabric Rolls of York Minster*, Surtees Soc. 35 (Durham, 1858), 6, 34, 49; E. Gee, 'Stone from the Medieval Limestone Quarries of South Yorkshire', in A. Detsicas (ed.), *Collectanea Historica: Essays in Memory of Stuart Rigold* (Maidstone, 1981), 247–55; R. A. Brown, H. M. Colvin, and A. J. Taylor, *The History of the King's Works*, i (London, 1963), 274–5, 281–2.

[37] G.D. Drury, 'The Use of Purbeck Marble in Mediaeval Times', *Proceedings of the Dorset Natural History and Archaeological Society*, 70 (1948), 77; J. Blair, 'Purbeck Marble', in J. Blair and N. Ramsay (eds.), *English Medieval Industries: Craftsmen, Techniques, Products* (London, 1991), 43; *Cal. Close R. 1227–31*, 198.

[38] C. Parkes, *Fowey Estuary Historic Audit* (Truro, 2000), 14.

[39] J. B. Blake, 'The Medieval Coal Trade of North-East England: Some Fourteenth-Century Evidence', *Northern History*, 2 (1967), 24–5.

[40] J. A. Galloway, D. Keen, and M. Murphy, 'Fuelling the City: Production and Distribution of Firewood and Fuel in London's Region, 1290–1400', *EcHR* 2nd ser. 49 (1996), 465.

easier, more direct, and often faster. As a consequence, goods were routinely transported by water. Firewood was cut from the alder groves in the Somerset Levels and then moved by boat. In the Pevensey Levels (Sussex), hay was carried from Pevensey Castle to the dairy to feed cattle and to an area of sheep pasture called *Ylond*. Sand was shipped from the coast to the castle for building, and timber and straw taken for the same purpose to the sheephouse. In the Fenland, even sheep were sent by water, confined in crates if necessary, to the abbey at Crowland.[41] Local transport of goods also occurred in some river valleys. Turf cut in the Norfolk Broads at Hoveton was carried by cart to a boat and then taken by water a few miles down the River Yare and along the River Thurne to Ludham, where it was unloaded and again put in a cart to take to the manor house.[42] Travellers also went by boat. The Bishop of Ely moved between his estates in the Fens by water, and in Sussex the mayor of Rye and his councillors travelled by boat up the Rother estuary from the town to Small Hythe in Kent in 1489–90.[43]

Such local movement was only viable where there were numerous landing sites. Generally, water-borne transport was not suited to short distances, because the costs of loading and unloading goods into a cart to complete the journey made it uneconomic. Where a watercourse lay close to the destination, or could be constructed to allow boats to be brought there, the costs of unloading might be no more than those incurred if the goods had been brought over land. Some lodes or artificial channels were joined to other watercourses at both ends, while others had one blind end and terminated at a hythe. It has been suggested that some of the blind lodes may have been initially dug to facilitate the shipment of building materials for a major construction project, but once constructed they would have been used more widely for other goods.[44] The basin near to a place called the Waits in Cottenham (Cambs.) is mentioned in an account of 1454–6 and a port is recorded at Swavesey in *c*.1200, which was evidently linked by a lode to the nearby Ouse.[45] Lakenheath on the fen-edge had a wharf from which corn was transported to Ely, King's Lynn, and Bury St Edmunds.[46] The nearby village of Isleham was evidently connected by a lode to the River Lark. Although it has been recently described as an entrepôt, there

[41] M. Williams, *The Draining of the Somerset Levels* (Cambridge, 1970), 30; TNA, PRO SC6/1027/17, m. 3; L. F. Salzmann, 'Documents Relating to Pevensey Castle', *SxAC* 49 (1906), 12; TNA, PRO SC 6/1027/21, m. 1; F. M. Page, 'Bidentes Hoylandie', *Economic History*, 1 (1926–9), 603–13.

[42] J. N. Lambert, J. N. Jennings, C. T. Smith, C. Green, and J. N. Hutchinson, *The Making of the Broads: A Reconsideration of their Origin in the Light of New Evidence*, RGS Research Series 3 (London, 1960), 84.

[43] E. Miller, *The Abbey and Bishopric of Ely: The Social History of an Ecclesiastical Estate* (Cambridge, 1951), 78 n. 2. East Sussex Record Office, RYE 60/3, fo. 68.

[44] Cf. J. R. Ravensdale, *Liable to Floods: Village Landscape on the Edge of the Fens, AD 450–1850* (Cambridge, 1974), 25.

[45] Ibid.; *VCH Cambridge and the Isle of Ely*, ix. 390–1.

[46] M. Bailey, *A Marginal Economy? East Anglian Breckland in the Later Middle Ages* (Cambridge, 1989), 155.

is no record of a market or fair in the village, and it is more likely that the port served only the immediate area.[47]

Silvester has provided a preliminary survey of hythes in the Fens and Russett has considered similar sites in the Somerset Levels, but it seems probable that they may have identified only the more important landing places.[48] Sites for loading hay, reeds, or stock may have been simply a convenient stretch of riverbank against which a boat could be moored. Some more substantial landing sites were constructed in the Fens, often at the end or alongside the artificial canals or lodes.[49] The full network of Fenland lodes and location of landing places remains to be plotted, and the map here, which extends Silvester's work, is certainly not complete (Fig. 20). The date of the construction of the Fenland lodes remains uncertain, but there is little to support the view that they are Roman in origin.[50] Some may be as early as the tenth century, a date suggested by Hall for Cnut's Dyke in Ramsey.[51] *Cotingelade* near the Isle of Ely is mentioned in accounts of William I's siege of Hereward, and the same sources record that ships from the area were used to transport timber and stone from that place to a landing site at Aldreth, suggesting that the practice of the moving goods by water was already established.[52]

Shipbuilding Sites

It is hardly surprising that most ships were built at the major ports, even though materials had been transported over a considerable distance to the shipyard. A small number of shipyards developed in rural locations where timber might be readily obtained. The most important of the rural yards were possibly those on the River Rother at Small Hythe and Reading Street in Kent. Both places appear to have been small ports by the early fourteenth century, but access was improved in 1332 when the river was diverted to the north of the Isle of Oxney to run past them. By 1402 Small Hythe was established as a shipbuilding centre. These two sites and also Newhythe, which lay on the River Medway to the north of Maidstone, became established as shipbuilding centres because

---

[47] S. Oosthuizen, 'Isleham: A Medieval Inland Port', *Landscape History*, 15 (1993), 29–35.

[48] R. J. Silvester, ' "The Addition of More-or-Less Undifferentiated Dots to a Distribution Map?" The Fenland Project in Retrospect', in J. Gardiner, *Flatlands and Wetlands: Current Themes in East Anglian Archaeology*, EAA 50 (Cambridge, 1993), 34–37; V. E. J. Russett, 'Hythes and Bows: Aspects of River Transport in Somerset', in G. L. Good, R. H. Jones, and M. W. Ponsford (eds.), *Waterfront Archaeology: Proceedings of the Third International Conference on Waterfront Archaeology*, CBA Research Rep. 74 (London, 1991), 60–6.

[49] Hart (ed.), *Chartularium Monasterii de Ramesia*, i. 346. Bailey, *A Marginal Economy?*, 152–4.

[50] D. Hall, *The Fenland Project 10: Cambridgeshire Suyrvey, the Isle of Ely and Wisbech*, EAA 79 (Cambridge, 1996), 112; G. Fowler, 'The Extinct Waterways of the Fens', *Geographical Journal*, 83 (1934), 30–9.

[51] D. Hall and J. M. Coles, *Fenland Survey: An Essay in Landscape and Persistence*, English Heritage Archaeological Rep. 1 (London, 1994), 137; Hall, *The Fenland Project 6*, 42.

[52] T. D. Hardy and C. T. Martin (eds.) *Gesta Herwardi Incliti Exulis et Militis*, Rolls Series 91 (London, 1888), i. 388–90. See E. O. Blake (ed.), *Liber Eliensis*, Camden Soc., 3rd ser. 92 (London, 1962), pp. lvii, 3 n. 1 for discussion of the location of *Alreheðe* and *Cotinglade*.

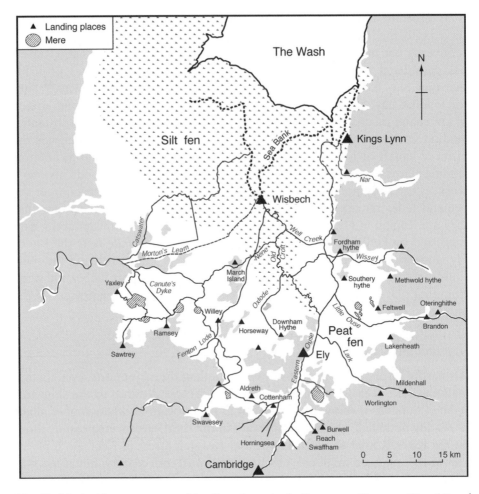

Fig. 20. Navigable waterways and landing places in the Fens area. Compare Figs. 15 and 37.

they had ready access to abundant supplies of timber from the Weald and the North Downs. There was a similar pattern in Yorkshire, where shipyards were to be found not only at the major centres of York and Hull, but in many other smaller places along the Don, Aire, Wharfe, and Ouse rivers. Here too, the availability of timber compensated for their rural locations away from any major commercial centre.[53]

[53] *Cal. Pat. R. 1324–27*, 215; J. Eddison, 'Developments in the Lower Rother Valleys up to 1600', *Archaeologia Cantiana*, 102 (1985), 95–110; Centre for Kentish Studies, NR/FAc2, fo. 55v; P. S. Bellamy and G. Milne, 'An Archaeological Evaluation of the Medieval Shipyard Facilities at Small Hythe', *Archaeologia Cantiana*, 123 (2003), 353–82; I. Friel, *The Good Ship: Ships, Shipbuilding and Technology in England 1200–1520* (London, 1995), 52–3; Moorhouse, 'Medieval Yorkshire: A Rural Landscape', 197.

Ferry Points

Bridges allowed many of the major roads in later medieval England to cross rivers, but ferries were used on the lower reaches and over estuaries. Ferries are recorded over the Thames estuary from numerous points between Aveley and Tilbury and across the Humber estuaries between Whitgift and Paulfleet.[54] Other ferries operated higher up the river where it was crossed by routes too minor to justify the construction of a bridge, such as those across the Thames upstream at Datchet, Hambleden, Caversham, Streatley, Shillingford, Clifton Hampden, Sandford, and Nuneham Courtenay.[55] Ferries might also serve as a means of crossing if a bridge was temporarily unavailable. The Medway at Rochester (Kent) was provided with a ferry between 1381 and 1391 after the Roman bridge had collapsed and before the new bridge was complete. Some ferries were seasonal, such as that over the River Trent at Stoke Bardolph (Notts.), which operated only in the winter. Presumably at other times, when the water level was lower, the river was forded.[56]

The right to operate a ferry was a common manorial monopoly and the income is often noted in manorial account rolls. The monopoly was often farmed for an annual payment rather than operated directly by the lord, which does not allow the economics of ferrying or the details of their operation to be readily studied. Alternatively, the ferry might be operated by an unfree tenant as part of his or her labour services. The ferry of Pevensey (Sussex), which probably carried travellers eastwards towards Bexhill, was farmed for about £4 in the late thirteenth century. That income was considerably greater than the sums spent annually to maintain the ferry boat. A new boat bought in 1283–4 cost only 40s.[57] The potential for profit is confirmed by the figures for the ferry across the Tamar estuary in Cornwall which was let for a farm of £8 10s. in 1301 and later for £10. New ferry boats cost 49s. 4d. in 1298 and 66s. 8d. in 1351.[58]

Victualling Ports

Most of the sites discussed above lay on inland waterways or estuaries. The final two categories of landing place generally lay on the coast. Victualling

---

[54] D. F. Harrison, 'Bridges and Economic Development, 1300–1800', *EcHR* 2nd ser. 45 (1992), 245. Ferries are recorded over the Thames estuary at Aveley, Grays Thurrock, West Thurrock, and Tilbury in Essex and Erith, Swanscombe and Gravesend in Kent (TNA, PRO C134/21, C135/43, C134/8, C135/26, C135/23, C134/37). For the Tilbury–Higham ferry, see *PWML* ii. 306, 312; TNA, PRO C135/40.

[55] TNA, PRO C135/32 (Datchet), C135/56 (Hambledon), C133/128 (Caversham), Harrison, 'Bridges and Economic Development', 245.

[56] Britnell, 'Rochester Bridge, 1381–1530', 43–6; TNA, PRO C133/115.

[57] TNA, PRO SC6/1027/17.

[58] J. Hatcher, *Rural Economy and Society in the Duchy of Cornwall 1300–1500* (Cambridge, 1970), 192–3. For other Cornish ferries, see C. Henderson, *Essays in Cornish History* (Oxford, 1935), 163–7.

ports were situated on the shore adjoining major maritime routes. Mariners adopted a cautious approach to long voyages. They tended to follow routes parallel to the coast and preferred to ride out poor weather in ports and havens rather than risk being caught at sea in storms or adverse winds. Bad weather could prolong a voyage and revictualling at ports along the route was often necessary. The ship the *Margaret Cely* left London for Bordeaux for a cargo of wine in September 1487 and did not return until February the following year, although the same voyage could be completed in three or four weeks in favourable conditions. There were ports of call to which ships could put in along the length of the Channel, but they were particularly important on the south Devon and Cornish coasts where they provided a landfall before crossing to Brittany, where the port of Saint-Mathieu near Finistère served a similar function.[59] These landing places offered opportunities for trade. The Cornish ports of Fowey, Falmouth, and Mousehole were visited on occasion by ships travelling to and from the north coast of Spain, and the inhabitants there sold fish and some poor-quality cloth in return for imported goods. A ship arrested in 1230 in the Cornish port of Helford, and another from the Ponthieu district of north-east France carrying garlic and onions held in the port of Branksea Island (Dorset), may have been sheltering or revictualling. Similarly, eleven Venetian galleys loaded with wool took shelter in the port of Penryn (Cornwall) in 1324.[60] Evidently it was common practice in the late fourteenth century for ships, seeking shelter in Plymouth Sound, to land on St Nicholas Island (now Drake's Island) near the mouth of the estuary. They were revictualled either at the island or in Plymouth itself.[61]

Inland waterways are also likely to have had points for sailors to stop overnight and to purchase food, but few of these are recorded. One exception is the building which was constructed on the lode leading from Sawtry Abbey to Whittlesey Mere. It provided accommodation for the sailors of the barges carrying stone for the construction of the abbey.[62]

Fishing Bases

Landing places for fishermen served two roles. The first was as a permanent base from which fishermen might operate. They commonly had buildings for tackle to be stored and places for boats to be drawn up. The second type were places at which fishermen operating some distance from home might land, dry their nets, and preserve their catch, generally by drying it. The home base of local fishermen might also serve as a landing place for those working away from

[59] M. K. James, *Studies in the Medieval Wine Trade*, ed. E. M. Veale (Oxford, 1971), 119–24, 170; Hatcher, *Rural Economy and Society*, 35.

[60] W. R. Childs, *Anglo-Castilian Trade in the Later Middle Ages* (Manchester, 1978), 179; *Cal. Close R. 1227–31*, 301; *Cal. Close R. 1234–37*, 38; Lloyd, *Alien Merchants in England in the High Middle Ages*, 200.

[61] *Cal. Close R. 1396–99*, 33.   [62] Hat (ed.), *Chartularium Monasterii de Ramesia*, i. 166.

home. Fox has shown that fishing villages—permanent settlements situated on the coastline, whose main function was fishing—are very rarely found before the later fifteenth century. Fishing vessels operated either from ports which had a mercantile function or from bases which were occupied solely during the season.[63] The pursuit of the herring shoals, which swam down the east coast, reaching Scarborough in August and September, and the Straits of Dover in November and December, often required fishing vessels to travel considerable distances.[64] In the late eleventh and early twelfth centuries ships were travelling from the east of Sussex and the south-west of Kent as far as Yarmouth where they established the right to land and dry their nets. In the late thirteenth century ships from Hove (Sussex) kept nets permanently at Yarmouth for use during the herring season.[65] Accommodation was established at fishing bases in coastal settlements, such as that at Dungeness (Kent), for both local fishermen and those coming from a greater distance. Land obtained in *c.*1240 by Beaulieu Abbey at Gillan in St Anthony in Meneage and Porthoustock (Cornwall) was for drying fish, and probably to provide temporary accommodation for their fishermen operating far from home.[66]

Fishing vessels might be launched from numerous places along the more gently sloping lengths of coastline. They were able to land on poorly protected coasts so long as they could be hauled up onto the beach out of the reach of the sea, sometimes using capstans, as at Hythe (Kent).[67] A list of places drawn up in 1385 charged with the payment of a levy on fish landed in Kent and Sussex includes most settlements near the coast between Selsey Bill and the Isle of Sheppey.[68] Even quite small settlements supported substantial numbers of boats. The best recorded are perhaps those of Thorpe and Sizewell in Suffolk, where boats operated off the beach to fish in the coastal waters. The masters of the ships attended an unusual court held by the Abbot of Leiston known as *hethewarmoot*. The number of masters recorded in the annual rolls therefore allows minimum numbers of vessels to be calculated. An average of 10.7 masters were recorded in the 1380s from Thorpe and 8.5 from Sizewell. That rose to 13.8 in the former place in 1430s and 11.9 in the latter, and continued to increase until the 1470s, the last decade for which there is surviving evidence,

[63] H. S. A. Fox, *The Evolution of the Fishing Village: Landscape and Society along the South Devon Coast, 1086–1550* (Leicester, 2001), 145–69.

[64] B. Waites, 'The Medieval Ports and Trade of North-Eastern Yorkshire', *Mariners Mirror*, 63(1977), 146; A. L. F. Dulley, 'The Early History of the Rye Fishing Industry', *SxAC* 107 (1969), 44.

[65] K. M. E. Murray, *The Constitutional History of the Cinque Ports* (Manchester, 1935), 18–19; Gardiner, 'Shipping and Trade', 77–93 71–93; W. D. Peckham (ed.), *Thirteen Custumals of the Sussex Manors of the Bishop of Chichester*, Sussex Record Soc. 31 (Lewes, 1925), 84.

[66] H. S. A. Fox, 'Cellar Settlements along the South Devon Coastline', in H. S. A. Fox (ed.), *Seasonal Settlement* (Leicester, 1996), 61–9; M. F. Gardiner, 'A Seasonal Fishermen's Settlement at Dungeness, Kent', *Medieval Settlement Research Group Annual Report*, 11 (1996), 18–20.

[67] A. J. F. Dulley, 'Four Kent Towns at the End of the Middle Ages', *Archaeologia Cantiana*, 81 (1966), 105.

[68] *Cal. Patent R. 1381–85*, 588: S. F. Hockey (ed.), *The Beaulieu Cartulary*, Southampton Records Series 17 (Southampton, 1974), 215–17.

when there were 18 and 14.2 respectively. The list drawn up in 1336/7 of fishing vessels at Holkham and Wells-next-the-Sea (Norfolk) suggests that they were of a similar size.[69]

Two Regional Examples

The diversity of activity in small ports can best be illustrated by two local studies. The first attempts to identify the use made of one area of marshland and the adjoining rivers to indicate the role played by water-borne transport. The second uses the unusually complete record of ships berthing at the port of Scarborough in 1321–2 to consider the origin of vessels working the North Sea fishing grounds and trading with the town. No length of English river or coastline can be said to be representative of the whole, and the examples chosen are intended to be indicative. The Rother estuary and adjoining marshland in east Sussex has been chosen for no better reason than familiarity with the diverse records which may mention riverine and maritime activity. Small ports are poorly documented and a complete picture has to be constructed from many, often incidental references. Manorial documents may not survive and the use of the landing place may have been too minor to have been mentioned in the records of the Chancery or Exchequer. The paucity of records makes the identification of changes in the use of sites very difficult. Changes in the coastline and the silting of rivers and harbours will have brought some landing places to prominence, just as others were precipitated into decline.

The morphology of the estuary of the River Rother near the Kent–Sussex border changed dramatically during the late thirteenth and early fourteenth century when the shingle barrier beach to the south of Rye was eroded and then swept away. Its removal created a broad arm of the sea which extended inland as far north as Appledore (Fig. 21). The coastal erosion exposed the important port of Old Winchelsea to the full force of the sea and led to its eventual disappearance. Its replacement, the town of New Winchelsea, was founded on the upland at *Iham*, although the payment of a rents resolute of herrings implies that there was an earlier fishing port on the site.[70] The coastal changes improved access to the River Rother from the sea and established it as an important artery for the export of wood and timber from the east end of the Weald. The greater tidal flow up into the Rother valley led to flooding some

---

[69] Suffolk Record Office (Ipswich branch), HD 371/1, 2; HD 1032/87, 88. The figures calculated here are slightly different from those given by M. Bailey, 'Coastal Fishing off South East Suffolk in the Century after the Black Death', *Proceedings of the Suffolk Institute of Archaeology*, 37 (1990), 102–14. For details of the figures see M. F. Gardiner, 'The Exploitation of Sea-Mammals in Medieval England: Bones and their Social Context', *Arch J* 154(1997), fig. 1. W. Hassall and J. Beauroy (eds.), *Lordship and Landscape in Norfolk 1250–1350: The Early Records of Holkham*, Records of Social and Economic History, NS 20 (London, 1997), 451–2.

[70] J. Eddison, 'Catastrophic Changes: A Multidisciplinary Study of the Evolution of the Barrier Beaches of Rye Bay', in J. Eddison, M. Gardiner, and A. Long (eds.), *Romney Marsh: Environmental Change and Human Occupation* (Oxford, 1998), 68–70; W. M. Homan, 'The Founding of New Winchelsea', *SxAC* 88 (1949), 26.

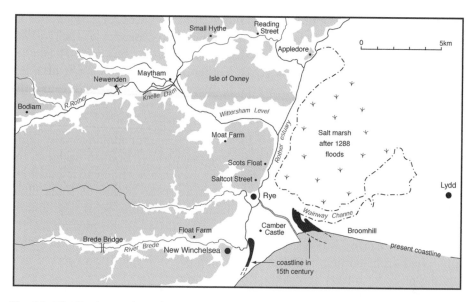

Fig. 21. The Rother and Brede rivers, and the west side of Walland Marsh, showing the hythes. Compare Fig. 12.

distance inland, where drainage of the river water was impeded at high tide. A sea wall called the Knelle Dam was constructed in 1332 across the valley to prevent the ingress of sea water from the south of the Isle of Oxney. Whatever its original intention, it had the effect of diverting the River Rother to the north of the Isle.[71]

The upper limit of ship navigation on the Rother was determined by the presence of a bridge at Newenden. In 1482 the town of Rye sent a rowing boat upriver as far as that village to seek cogs and arrest shipping.[72] Ships from France and Flanders in 1354 were said to sail up the Rother to ports at Reading Street, Small Hythe, Maytham, Knelle Dam, Newenden, and Bodiam to load with firewood. The last of these places is above Newenden and it is unlikely, in fact, to have been accessible to ships; when timber was transported in 1367 for work on Calais harbour, it had to be brought down the river on shouts from Bodiam bridge to be loaded at Knelle Dam and an unidentified place, *Waterwassh*, near Rye.[73]

Among the ships transporting the timber across the Channel were those from Reading Street. Reading Street and Small Hythe were both landing places and ship building centres on the Rother. The two were already of such importance as seagoing ports before the diversion of the Rother to the north of the Isle of

[71] Eddison, 'Developments in the Lower Rother Valleys', 99–103. For sea walls constructed on a number of Sussex rivers, see M. F. Gardiner, 'The Geography and Peasant Rural Economy of the Eastern Sussex High Weald, 1300–1420', *SxAC* 134 (1996), 125–6.

[72] The bridge at Newenden was derelict for some time in the mid-14th century, when it was replaced by a ferry: *Cal. Pat. R. 1364–67*, 150. East Sussex Record Office, RYE 60/3, fo. 23ʳ.

[73] *Cal. Pat. R. 1354–58*, 70, 578–9; TNA, PRO E101/178/8.

Oxney that agents of the king had enquired at both places for sailors. One of the ships based at Small Hythe in the mid-fourteenth century was the *Gabriel*, which was of sufficient size that it sailed to Gascony to buy wine.[74] Smaller vessels were able to pass up the River Rother as far as Salehurst, which lies upstream from Bodiam, and some even further: a millstone brought round the coast from Bulverhythe (near Hastings) was taken by boat along the Rother to be landed at Bivelham which was upriver from Salehurst.[75]

The ports of the Rother were of local importance for moving many commodities. Battle Abbey established a landing place above Bodiam bridge in the early twelfth century, probably to transport hay cut in the nearby meadow. Further downstream nearer the port of Rye were the hythes of Scots Float and Saltcot Street. Scots Float took its name from Sir John Scott who greatly enlarged it in 1480 to serve as a loading place for firewood destined for Calais. Saltcot Street nearby was a small fishing port, recorded from 1385 when duty was levied on fishermen in Playden, in which parish it lay. It also served as a landing place for a ferry across the Rother estuary.[76]

The Camber at the southern end of the Rother was the enclosed anchorage and haven. The name is cognate with the French *chambre*, implying that it was an embayment. The first reference to the name is in a document of 1330. It was later used by the Venetians as a point for assembling their fleets for sailing back to the republic and is depicted on a chart of 1436 as a large protected inlet. It was certainly in use as an anchorage by 1400 when ships lying there were arrested. There was a landing place for fishing boats at Broomhill on the south side of the Camber. Broomhill even contributed to Romney's Cinque Port obligations of supplying ships for occasional service for the king.[77] A long creek on the east side of the Camber, known as the Wainway Channel, allowed boats to bring timber to Lydd Bridge, an unidentified landing place which served the town of Lydd. In 1336 the timbers of a barn were carried by water from the Battle Abbey grange at *Fother* in Pett Level (south-west of New Winchelsea) to Broomhill, from whence they were then taken by land to the abbey's manor of Denge and re-erected. The River Brede, a tributary of the Rother, was accessible as far as the *Damme* at Float Farm, where there was a landing place used for the loading of firewood and occasionally for carrying beer and wine from Winchelsea. In the 1440s, the course and character of the river were changed in major works almost certainly intended to promote a greater tidal flow and so scour the port of Winchelsea. The works extended the

---

[74] *Cal. Pat. R. 1324–27*, 215; *Cal. Pat. R. 1364–67*, 16; TNA, PRO E101/178/8.

[75] *Cal. Pat. R. 1348–50*, 80; TNA, PRO SC6/1148/13.

[76] E. Searle (ed.), *The Chronicle of Battle Abbey* (Oxford, 1980), 256–7; East Sussex Record Office, NOR 15/117; *Cal. Pat. R. 1381–85*, 588; M. Biddle, H. M. Colvin, and J. Summerson 'The Defences in Detail', in H. M. Colvin (ed.), *History of the King's Works, iv: 1485–1660* (London, 1982), 423; *VCH Sussex*, ix. 151.

[77] British Library, Add. Ch. 18623; *Cal. State Papers Venice*, i, nos. 120, 126, 209, and frontispiece map; British Library, Add. Roll 16432. East Sussex Record Office, RYE 99/5; Murray, *Constitutional History of the Cinque Ports*, 240.

limit of navigation for ships to Brede Bridge, though the quay at Float Farm continued to be in use.[78]

The rivers and creeks which converged to the south of Rye therefore provided a network of routes which linked the settlements around the Rother. Goods were routinely moved by water. For example, a shout carried goods from Lydd Bridge to Bodiam in 1377–8, and barley was moved from the Wainway to Brede Bridge in 1492.[79] The burgesses of Rye found it convenient to take a boat upriver to Small Hythe when they were visiting that port or travelling to Tenterden. Building stone for the construction of Camber Castle was taken from the quarries by the river at Saltcot Street to the building site. In the same way stone for the manor house at Mote (Moat Farm) in Iden was shipped along the coast from Eastbourne and then upriver to John Mayne's 'gutter' or drainage channel in the Wittersham Level. Indeed, the Mote manor accounts show that many bulky goods were routinely moved by water, including grain sent to the mill at Appledore.[80]

These places on or near the Rother estuary are indicative of the diversity of types of landing places which developed on a river and in wetlands. The second example illustrates the numerous landing places for ships found along the east coast of England and in the Humber estuary. In 1318 a grant was given for five years for the collection of lastage for the repair of the quay at Scarborough. A surviving account of February 1321–February 1322 lists the vessels putting in to the town, recording the shipowners and their port of origin week by week. Scarborough was a port of moderate size from which corn and some wool was shipped, but its greatest significance was as a fishing centre. From early January to mid-June there was a small but significant number of ships arriving at the port. These came from both major and minor east coast ports and the larger continental ports, including Calais, Harfleur, and Iser. Ships from more distant ports, both in England and abroad, often arrived in pairs or threes, suggesting that they sailed in small convoys for mutual assistance. In mid June there was a significant increase in the number of ships arriving, particularly from the Continent, and there was intense activity in the port until early August, a period which coincided with the herring and cod season. From mid-September very few vessels put in at the port.[81]

---

[78] TNA, PRO SC 6/889/14; M. F. Gardiner, 'Medieval Farming and Flooding in the Brede Valley', in J. Eddison (ed.), *Romney Marsh: The Debatable Ground* (Oxford, 1995), 127–37; East Sussex Record Office, ACC 7024 (Udimore accounts); Henry E. Huntington Library, BA Stewards' Accounts 1489 (consulted from microfilm: East Sussex Record Office, XA3/20).

[79] TNA, PRO SC6/889/22; SC6/889/25; SC6/890/1; Henry E. Huntington Library, BA Stewards' Accounts 1489 (consulted from East Sussex Record Office, XA3/20).

[80] East Sussex Record Office, RYE 60/3, fo. 68; Biddle et al., 'The Defences in Detail', 423; East Sussex Record Office, NOR 15/103, /106.

[81] *Cal. Pat. R, 1317–21*, 318; TNA, PRO E122/134/3. The Scarborough account is discussed in considerably greater detail in Waites, 'The Medieval Ports', 137–49. The fishing season at Scarborough was said in 1348 to last from midsummer to Michaelmas.

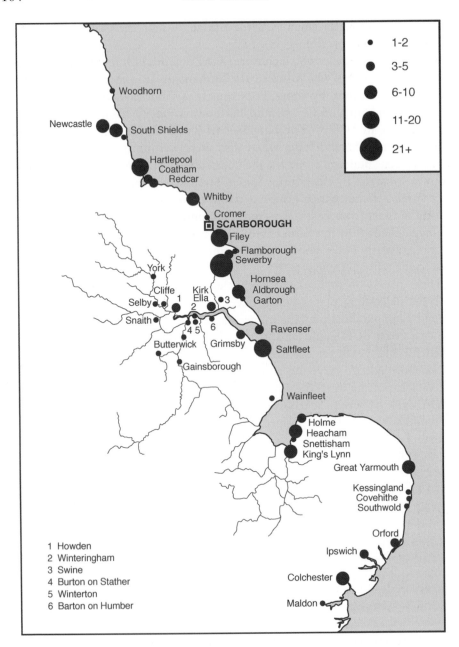

Fig. 22. The home ports of ships landing at Scarborough, February 1321–February 1322.

The present interest is not so much with the activity at Scarborough as with the ports of origin of the ships. Analysis of the places of origin and the dates of arrival allows patterns of shipping to be identified. A significant number of boats from minor ports in East Anglia, including Heacham, Holme, Kessingland, and Southwold, travelled up the coast to Scarborough (Fig. 22).

They almost all arrived between mid-June and mid-July, and were evidently there for the herring fishing. By contrast, vessels from small ports along the Humber and its tributaries, including Burton on Stather, Butterwick, Faxfleet, Winterton, and Howden, visited Scarborough over a longer period during the year, suggesting that fishing was not the main reason for their voyages. Ships from these ports are more likely to have been participating in local trade. The places of origin of the ships landing at Scarborough does not provide a complete record of the landing places on the east coast. Landing sites in Lincolnshire, in particular, are under-represented, possibly because fishing vessels could operate from their home base without putting into Scarborough. Ships from East Anglia were a greater distance from home and a larger number of places of origin are represented, although the account does not provide a full record of sites used by fishing vessels even for that region.[82]

Landing Places in Perspective

The discussion above has taken an entirely different view of river transport from that presented in a recent series of articles on river navigation in England.[83] The other work considered, in particular, the 'inter-urban routes of regional and national importance'.[84] Amongst the records used by participants in the debate were the purveyance accounts which record the large-scale movement of goods for the army and the king's household. These cannot be used to stand for all water-borne cargoes, for they were a record only of a specific type of shipment: they were concerned with the rapid movement of large quantities of goods, generally over considerable distances. Most of the goods were directed into a small number of nodal points from which they could be loaded onto ships. The needs of the king's purveyors were not necessarily those of others moving goods, and the economics of moving smaller quantities shorter distances and at less speed were different. Thus, the River Axe in Somerset, although not identified as a route for water transport by Langdon on the evidence of the purveyors' accounts, was used by the cargo boats of the Abbot of Glastonbury in the thirteenth century.[85] Equally, boats on the River Rother could travel well past Small Hythe, considered by Langdon to be the limit of shipping. The heads of navigation which he identified were rarely the absolute limits of water-borne transport, though in some places the presence of a weir or low

---

[82] S. Pawley, 'Maritime Trade and Fishing in the Middle Ages', in S. Bennett and N. Bennett (eds.), *An Historical Atlas of Lincolnshire* (Hull, 1993), 56–7. The major fishing ports in Norfolk and Suffolk are listed in 1318 as Burnham, Thornham, Blakeney, Cromer, King's Lynn, and Great Yarmouth, Kyrkele, Dunwich, Orford, Gosford, Orwell, and Ipswich (*Cal. Pat. R. 1317–21*, 107–8).

[83] J. F. Edwards and B. P. Hindle, 'The Transportation System of Medieval England and Wales', *JHG* 17 (1991), 123–34; J. Langdon, 'Inland Water Transport in Medieval England', *JHG* 19 (1993), 1–11; E. Jones, 'River Navigation in Medieval England', *JHG* 26 (2000), 60–75; J. Langdon, 'Inland Water Transport in Medieval England: The View from the Mills; A Response to Jones', *JHG* 26 (2000), 75–82.

[84] Edwards and Hindle, 'Transportation System', 125; Jones, 'River Navigation', 69.

[85] Langdon, 'Inland Water Transport', 4–5; Williams, *The Draining of the Somerset Levels*, 65.

bridge might create a physical impediment which was difficult for larger vessels to pass. The limit of navigation was defined solely by the size of vessel, and small boats might carry goods much higher up a river than larger ships.

Water transport operated at different levels. There is a problem that in focusing on the movement of larger vessels we may lose sight of the greater number of journeys made by smaller boats and of their cumulative economic significance. It has been argued that decisions about whether to use water transport were made largely on economic grounds and were related to the distance travelled, the relationship of the value of the goods to the cost of carriage, and the proximity of landing places to the origin and destination of the journey. One further factor was also important: goods were more likely to be moved by water if the estate or the individual held a boat or had access to one. In wetlands or on coastal sites, where boats were common, goods and people may have moved more frequently by water because of the ease of transport.

The number of journeys made by water was also dependent upon the varying condition of rivers. Seasonal variation in water level and the general condition of the waterway would have been particularly important on smaller rivers. Jones has argued that the number of navigable rivers declined between the late thirteenth and fifteenth centuries, though the reasons for this change remain uncertain. It is not entirely clear whether the decline was due to physical causes—a failure to maintain waterways or an increase in the number of physical obstructions—or to economic reasons.[86] Indeed, the two may be connected, for there would have been little benefit in maintaining river navigations if the traffic was much reduced. One of the clearest statements of the problem of minor waterways is the presentment in 1360 of the River Ant (*Smalee*) in Norfolk. It was said that the river had fallen out of use for the carriage of goods at the time of the Black Death (1348–9) and since then it had been choked by weeds. Local factors may have been important here. Peat was increasingly difficult to cut from the Broads which were flooded in the first half of the fourteenth century. The peat had been transported by water to Norwich, but the supply seems to have decreased in the last quarter of the fourteenth century, if the evidence for the usage of fuel in the cathedral priory is typical. Peat is likely to have been one of the major cargoes on the River Ant.[87]

The evidence above for water transport from minor landing places has been drawn mainly from the thirteenth and fourteenth centuries. It remains to be established whether these places played a similar role in earlier centuries. The issue may be examined by considering the types of journeys made to

[86] Jones, 'River Navigation', 69–72; Langdon, 'Inland Water Transport in Medieval England: The View from the Mills', 79–80. Cf. Blair, above pp. 5, 12.

[87] *PWML* i. 88–90. I am grateful to John Langdon for drawing my attention to this reference. Lambert et al., *The Making of the Broads*, 99–102. The relevant length of the river and the adjoining Broads are illustrated in T. Williamson, *The Norfolk Boards: A Landscape History* (Manchester, 1997), fig. 25.

and from minor landing places in the eleventh and twelfth centuries. These may be loosely grouped under five headings. The first was the distribution of goods by water to trading centres. It hardly needs to be said that there were fewer trading places in England in 1100 than 200 years later when the level of commercial activity had greatly increased. But it is also arguable that many early markets were not necessarily well situated to take advantage of water-borne transport. Their trade was largely local: specialized production of commodities was poorly developed, and it was possible for international ports to operate without a network of towns from which goods for exports might be drawn or to which imports might be distributed.[88] However, the attempts in the twelfth century to restrict the landing of goods from ships to established markets suggest that trade was occurring at new landing places.[89] It is likely that the coastal markets which Britnell observed in thirteenth-century Essex did not originate *de novo* as seigniorial creations with their grants of charters, but at an earlier date as informal trading places on the banks of creeks. The relatively well-documented rise of King's Lynn may be only one example, albeit an extraordinary one, of the development of a port. It began as a trading site on the marshes at the edge of the Wash where traders landed to purchase wool and salt, and was already a flourishing market when a church was established there *c.*1096.

Ports were not merely towns by the sea, and could not be called into existence at the stroke of a seigniorial quill; the landing place frequently preceded the foundation of the town.[90] The development of Brighton (Sussex) provides a good example of this. It was a fishing base by the late eleventh century when the manor paid a render of 4,000 herrings. No burgesses are mentioned in Domesday Book, and the regular form of the town, bounded by West, North, and East Streets, suggests that it was a later planned foundation. The parish church lay outside the town, on the hill to the north-west, where it may have served as a landmark for ships at sea. It is significant that it was dedicated to the patron saint of sailors, St Nicholas. The town itself was served by the chapel of St Bartholomew and the establishment of a dependent chapel is a common feature of new town development.[91] It seems likely that the town was founded beside the sea below the earlier parish church to take advantage of existing fishing and possibly trading activity. A similar pattern of development also took place on inland waterways and has been suggested for the town

---

[88] R. H. Britnell, *The Commercialisation of English Society, 1000–1500* (Manchester, 1996), 9. Britnell contrasts the small number of recorded markets in the east of England adjoining the North Sea ports with the greater number in western England.

[89] Britnell, 'English Markets and Royal Administration before 1200'.

[90] Britnell, 'Essex Markets before 1350', 18–19; D. M. Owen, 'Bishop's Lynn: The First Century of a New Town', *Anglo-Norman Studies*, 2 (1979), 146–8; Gardiner, 'Shipping and Trade', 84–5, 92–3.

[91] DB i. 26ᵛ; TNA, PRO SC6/HEN VII/1474, m. 2, an account of 1497–8 mentions the street names; British Library, Cotton MS Aug. I.i.18 for mid-16th-century plan of the town. [W. Budgen], 'A Lewes Priory Charter', *Sussex Notes and Queries*, 2/8 (1929), 252; Beresford, *New Towns of the Middle Ages*, 169–75 for the ecclesiastical dependency of new towns.

of Bawtry (W. Yorks.) which lay on the River Idle, a tributary of the Trent. The church, similarly dedicated to St Nicholas, stands at the edge of the town close to the river. Its location suggests that it preceded the establishment of the town at the beginning of the thirteenth century, so that here, too, water-borne commerce seems to have been established before the commercial boom of the thirteenth century.[92]

The second type of water-borne journey which may be identified was for the distribution of goods within an estate. The twelfth-century or earlier documentary evidence for the practice is limited, although the service of transporting corn to the coast given by villeins on the St Paul's Abbey manor of Kirby with Horlock recorded in 1222 has already been noted. A rather different type of distribution within an estate is indicated by the movement of produce across the Channel. Monasteries in northern France were certainly shipping food across the Channel from their lands in England in the late eleventh century and perhaps earlier.[93] There is more abundant evidence for the third journey type, the direct shipment of goods from producers to consumers. The demand for stone produced by the upsurge in building work in the late eleventh and early twelfth century was met by opening new quarries and shipping stone over considerable distances. New quays were established as close as possible to the building site and, where necessary, canals were constructed to assist in the movement of the numerous cargoes.[94] The landing places established during the construction phase of monastic buildings continued to be of value even after stone ceased to be required because they served just as well for the movement of other goods.

The fourth type of journey was those made for purely local purposes. It is apparent from the use of Fenland boats in the campaign to repress Hereward the Wake that the practice of water transport was already established in that region. The tenth-century Graveney Boat contained fragments of continental lava quern stones, suggesting that it had sailed across the Channel. However, its shallow draught suggests that it was suitable for use in small creeks and may have been used for coastal or local movement and, indeed, it was finally abandoned at a landing place in the north Kent marshes.[95] Other local journeys may have been made in even smaller boats; scientific dating methods have shown that some logboats belong to the medieval period and may have been made as late as the fourteenth century. Boats such as these would have

---

[92] K. Collis, 'Documentary Evidence', in J. A. Dunkley and C. G. Cumberpatch, *Excavations at 16–22 Church Street, Bawtry*, BAR British Ser., 248 (Oxford, 1996), 184–7; Beresford, *New Towns of the Middle Ages*, 522–3.

[93] Gardiner, 'Shipping and Trade', 78–9.

[94] For a possible quay at Lewes Priory evidently associated with the landing of stone, see M. Lyne, *Lewes Priory: Excavations by Richard Lewis 1969–82* (Lewes, 1997), 43. Johnson and Cronne (eds.), *Regesta Regum Anglo-Normannorum*, ii, no. 1843.

[95] V. Fenwick, 'Structural Evidence for the Origin of the Ship and Comparison with Other Early N. European Ship-Finds', in V. Fenwick, '*The Graveney Boat*, BAR British Ser. 53(Oxford, 1978), 254–5; E. McKee, 'Reconstructing the Hull', ibid. 285–8.

been suitable for journeys on small waterways and rivers, where the majority have been found.[96]

The final type of water-borne journey was made in connection with fishing expeditions. Domesday Book has a very limited record of fishing activity in eleventh-century England, and yet it is clear that it was widely practised along the English coast. The use of fishing bases, both at home and when operating away from home, was very probably established by the end of the eleventh century and before. A fishing base seems to have been established on the shingle waste of Dungeness (Kent) by the 1050s, and the later town of Great Yarmouth developed from a site where fishermen, some operating far from their base, dried their nets and gutted their catches.[97] The evidence, slight though it is, appears to suggest that minor landing places of many types were used in the eleventh and twelfth centuries, much as they were at a later date.

The discussion above has sought to show that water transport was widely used by medieval communities in certain areas of England. Investment in boats and, where necessary, in waterfront structures was modest and largely unrecorded. The boats themselves survive only in small numbers and cast only a little light on the scale of water-borne movement. Equally few landing sites have been identified and have been investigated by archaeologists.[98] Indeed, little evidence can be expected of the numerous places at which ships were drawn on the seashore or riverbank. However, cumulatively the evidence of physical remains and written sources seems to suggest that small-scale water transport was certainly more frequent than the transport of goods and people between the major ports in great ships; indeed, it is possible that the total volume of goods moved in small craft might have exceeded that carried in larger ships. Small ports and landing places deserve greater attention from both historians and archaeologists than they have received.

---

[96] S. McGrail and R. Switsur, 'Medieval Logboats', *MA* 23 (1979), 229–31; S. McGrail, 'A Medieval Logboat from the R. Calder at Stanley Ferry, Wakefield, Yorkshire', *MA* 25 (1981), 160–4; S. McGrail, *Logboats of England and Wales*, BAR British Ser. 51 (Oxford, 1978), 330–1.

[97] Gardiner, 'A Seasonal Fishermen's Settlement'; Murray, *History of the Cinque Ports*, 18–19.

[98] For a few examples of minor landing places which have been reported, see M. Johnson, 'A Medieval Harbour at Flamborough', *YAJ* 60 (1988), 105–11; M. Johnson, 'The Medieval and Post-Medieval Port of Filey', *YAJ* 70 (1998), 73–84; M. G. Fulford, S. Rippon, J. R. L. Allen, and J. Hillam, 'The Medieval Quay at Woolaston Grange, Glos.', *Transactions of the Bristol and Gloucestershire Archaeological Society*, 110 (1992), 101–27; J. R. L. Allen, 'A Possible Medieval Trade in Iron Ores in the Severn Estuary', *MA* 40 (1996), 226–30.

5

The Efficiency of Inland Water Transport in Medieval England

JOHN LANGDON

The inland waterway system of medieval England, particularly just before the advent of the Black Death, has become the subject of considerable interest and debate in the past decade. Not only has the extent of England's navigable waterways in the period been disputed, but also how it interacted with the medieval English economy as a whole. Nor has it been a narrow discussion of largely antiquarian interest, since the emergence of some very large-scale studies into the regional variation of agricultural productivity has made the issue of transport—especially water transport—a critical feature in trying to explain rather more complicated patterns of medieval land exploitation than we would have assumed twenty years ago.[1] Over the past twenty years or so, we have found out much more about the extent of the inland waterway system,[2] its evolution over time,[3] its complexity in terms of the hydrological, environmental, and topographical conditions under which it had to operate,[4] its interaction with other economic interests,[5] its cost benefit relative to road

I would like to thank Robert Peberdy in particular for his careful reading of an earlier draft of this chapter and his sage comments on many of the issues involved, as indicated on many occasions in the notes. Mark Gardiner and Derek Keene have also contributed much useful information. I am also grateful to the Social Sciences and Humanities Research Council of Canada for funding the research from which the article derives and to Margot Mortensen for assisting in the collection of data.

[1] Most notably in B. M. S. Campbell, J. A. Galloway, D. Keene and Murphy, *A Medieval Capital and its Grain Supply: Agrarian Production and Distribution in the London Region c.1300* ([London], 1993), and Bruce M. S. Campbell, *English Seigniorial Agriculture 1250–1450* (Cambridge, 2000).

[2] J. F. Edwards and B. P. Hindle, 'The Transportation System of Medieval England and Wales', *JHG* 17 (1991) 123–34; J. Langdon, 'Inland Water Transport in Medieval England', *JHG* 19 (1993), 1–11; J. F. Edwards and B. P. Hindle, 'Comment: Inland Water Transportation in Medieval England', *JHG* 19 (1993), 12–14.

[3] E. Jones, 'River Navigation in Medieval England', *JHG* 26 (2000), 60–75; J. Langdon, 'Inland Water Transport in Medieval England: The View from the Mills: A Response to Jones', *JHG* 26 (2000), 75–82.

[4] R. Peberdy, 'Navigation on the River Thames between London and Oxford in the Late Middle Ages: A Reconsideration', *Oxoniensia*, 61 (1996), 311–40.

[5] Langdon, 'Inland Water Transport in Medieval England: The View from the Mills'.

transport,[6] and its regional and national impact.[7] Although the research that has been undertaken has sometimes tended to pull in different directions,[8] the importance of inland water transport in helping to explain the development of various economic patterns in medieval England has been broadly accepted.[9]

It is the purpose of this chapter to elaborate on many of these issues by looking at the so-called 'efficiency'of the inland water transport system, including that taken around the mouths of rivers, that is, voyages on estuaries and from the mouth of one river system to another through short coastal hops. 'Efficiency' is a very slippery concept. In our own times it tends to describe how much in the way of resources—time, money, etc.—is needed to get a particular job done and whether it could be accomplished with less. Economists often look at this from the point of price structures, where efficiency is defined by the lowest price at which a product can be delivered to the customer, and it is true that transport services in the middle ages were sufficiently commoditized to allow examination from such a perspective.[10] But there were many less measurable factors that impinged upon its ability to command frequent and effective usage: reliability; the capacity to handle both large and small amounts of traffic; the degree of professionalization; and the degree to which inland water transport had access to essential materials, labour, and capital. We are, in fact, only at the beginning of trying to elucidate these issues, and it is the purpose of this chapter to make a modest beginning at trying to figure out how we should assess this particular transport activity in terms of its effectiveness in meeting the basic needs of the economy of the time. In any event, I do want to go beyond a simple price-structure approach and to look at inland water transport from the point of view of those things, such as hydrological limitations, which would undercut its price advantage against land transport in particular. I plan to do this through an examination of vessel types and sizes used on inland waterways or for short coastal voyages. The variations in vessel types and sizes can tell us much about the characteristics of inland and coastal water transport and, to some extent, about hydrological conditions underpinning them.

A particularly fine source for supplying information about vessels and their sizes is the accounts for purveyance mostly entered in the E101 class (sheriffs' miscellaneous accounts) at the Public Record Office. Although there are certainly many other sources giving similar information about vessel types and sizes,[11] the purveyance accounts provide a sufficiently concentrated core of material to make possible some systematic analysis. The documentation created by the process of securing provisions from the countryside became very sizeable from

---

[6] J. Masschaele, 'Transport Costs in Medieval England', *EcHR* 46 (1993), 266–79.

[7] See, for example, Campbell et al., *Medieval Capital*, 60–3.

[8] Most notably over the question of the extent of the inland water transport system. Edwards and Hindle, 'Transportation System'; Langdon, 'Inland Water Transport'; Edwards and Hindle, 'Comment'; Jones, 'River Navigation'.

[9] Perhaps most obviously evident in Campbell, et al., *A Medieval Capital, passim*.

[10] Masschaele, 'Transport Costs'; Campbell et al., *A Medieval Capital*, 61–2.

[11] E.g., Edwards and Hindle, 'Transportation System'.

the 1290s to the 1350s, after which it began to taper off, as the Hundred Years War reached its first major break with the Treaty of Brétigny in 1360 and as the unpopularity of the system drove English monarchs to seek other options.[12] For about sixty years it provides a concentrated spotlight which provides materials for an in-depth assessment of inland water transport.[13] In this study it was decided to stop at 1348, since the advent of the plague dramatically altered the economic conditions under which transport functioned, but even with this slightly narrower focus there is a very substantial amount of information with which to work.

Finally, a few words should be said about the limitations of the purveyance as evidence of the inland water transport system in medieval England. As Fig. 23 suggests and as has been demonstrated by a number of recent publications,[14] it is not an absolutely comprehensive coverage of the entire inland water transport system, but mostly of its key commercial arteries. Lesser courses or isolated stretches of river, used for limited local water transport but not connecting all the way to the coast, were clearly not considered as serious transport options by the various sheriffs and other officials in charge of purveyance.[15] Nor even on water routes along which purveyance was conducted does it capture all the subtleties of such systems. In particular, small landing places in between or even upstream of the more important inland water ports may have been ignored by the purveyors (see Gardiner, above p. 105). Nevertheless, the purveyance accounts do provide extremely rich information about many aspects of the inland (and coastal) water transport systems, especially, as we shall see in this chapter, about the types, dimensions, and loading capacities of the boats themselves, the logic of trans-shipment points, and the potential for satisfying regional and national carrying requirements.

The Distribution of Vessel Types for River, Estuary, and Short Coastal Travel

As a first step, all references to vessels with loads were abstracted from purveyance accounts as early as 1294 up to 1348, resulting in 288 cases of a vessel or vessels (of less than full seagoing ship size) making a trip down, or sometimes up, rivers, through estuaries, or along the coast. As I have indicated in an earlier work,[16] purveyance was not spread evenly over the country, but largely limited to its eastern and southern parts, where the various provisions required were more easily obtained and more easily transferred, especially by river, to various ports for shipment to the Continent or to Scotland. When

[12] M. Prestwich, *Armies and Warfare in the Middle Ages: The English Experience* (New Haven, 1996), ch. 10.

[13] For example, see Langdon, 'Inland Water Transport', esp. 2–7, for an assessment of the extent of inland water transport, as reflected in purveyance accounts.

[14] For example, see Jones, 'River Navigation' for a thoughtful consideration of the evidence from purveyance accounts and other sources.

[15] Langdon, 'Inland Water Transport', 6–7.   [16] Ibid.

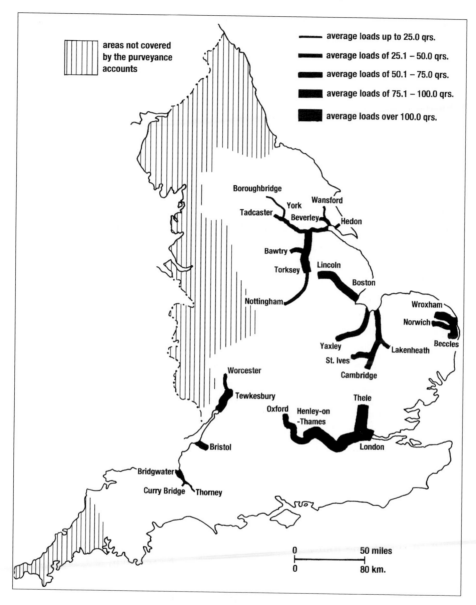

Fig. 23. Average loads carried on rivers in England, 1294–1348.

invasion was threatened, as in the latter years of Edward II, provisions were also sent to various castles within the kingdom. The goods conveyed ranged from grains for human consumption (wheat, barley, and malt), flour (in barrels), fodder for horses (oats, peas, beans, hay), meat (often in barrels, but also as sides of bacon), fish, cheeses, salt, horseshoes, and, finally, wooden hurdles and canvas for facilitating the storage of these goods on board or for making temporary horse stalls within the ship. For the 288 cases indicated above,

Table 1. Types of boats used for river, estuary, and short coastal travel

River systems, estuaries, etc.	Number of times types of boats mentioned							
	Batella	'Shout'	'Keel'	'Catch'	'Crayer'	'Barge'	*Navicula*	Others
Ure/Ouse/ Wharfe	7	—	—	—	—	—	2	—
Humber	10	—	—	5	—	—	7	2 *naves*
Trent	12	—	—	—	—	—	—	—
Witham	5	—	—	—	—	—	—	—
Great Ouse/Cam/ Nene, etc.	52	—	2	—	—	3	—	—
Yare/Bure/ Waveney	24	—	—	—	—	—	—	—
Thames	56	25	—	—	1	1	2	1 'rysbote'
Parrett/Tone	9	—	—	—	—	—	—	—
Severn	32	—	—	—	—	—	—	1 *farcosta*
Coastal	26	—	—	—	1	1	1	—
Total	233	25	2	5	2	5	12	4
%	80.9	8.7	0.7	1.7	0.7	1.7	4.2	1.1

Note: In order as listed, the Ure, Ouse, and Wharfe rivers accommodated water traffic from Boroughbridge, Tadcaster, and York through to Kingston upon Hull on the Humber. The Humber traffic indicated here includes that on the Hull from Beverley and Wansford to Kingston upon Hull, plus travel across the Humber from northern Lincolnshire. The Trent traffic includes all that from Nottingham north to Kingston upon Hull. The Witham traffic is that from Lincoln to Boston. The Great Ouse, Little Ouse, Cam, Nene, and Wissey traffic includes all the Fenland water transport to King's Lynn. The Yare, Bure, and Waveney rivers carried goods from Norwich, Wroxham, and Beccles to Yarmouth. The Thames here includes traffic on the river itself, and in its estuary, plus traffic on the Lea and Medway. The Parrett and Tone includes traffic through the Somerset Levels from Thorney, Langport, and Curry Bridge to Bridgwater. The Severn includes traffic from Worcester downstream and also traffic in the estuary from Bridgwater and other places. The coastal traffic here includes short trips along the southern and eastern coasts from Exeter round to King's Lynn.

Table 1 shows the distribution of the various vessel types across the river systems, estuaries, and coastal stretches that they served.

By far the most common term used for inland water vessels was the simple *batella* (boat), being found in over 80 per cent of the voyages sampled. What exactly the *batella* was is difficult to categorize. As we shall see, it tended to include boats of many different sizes, large and small. It presumably was a boat with a keel, ranging from perhaps a rowboat to larger keeled boats that could be fitted with sails.[17] The latter often occurred when these *batellae* were used for coastal travel or for estuaries like the Humber or the mouths of the Severn or Thames, as in the case of two Essex *batellae* for which sails were hired to take 277 quarters of beans and oats from Colchester through the Blackwater estuary to Maldon in 1338.[18] It is likely that sails were also used for river

[17] Masschaele, 'Transport Costs', 271–2, suggests that *batellae* may have been barges, but the use of these boats for coastal travel in particular suggests that they had more pronounced keels.

[18] TNA: PRO E358/3, m. 8. See also the reference to an old boat with mast sold for 21s. 8d. on the Duchy of Lancaster manor of Embleton (Northumberland) in 1367–8. In this same account a new

travel, particularly for those going downstream.[19] The mode of propulsion for upstream travel is unfortunately not indicated in any of the cases where it occurred. For upstream travel on lower sections of major rivers such as the Thames, Trent, and Severn, it was possibly sail, but rowing, poling, or hauling from shore by way of ropes might otherwise have been the preferred method.[20]

Well behind the *batellae* but still found in significant numbers were those vessels termed 'shouts' (*shoutae* in the Latin). In the purveyance accounts they were found just under 9 per cent of the time, and were limited strictly to the Thames river system. Other sources show that shouts could be found elsewhere in England,[21] but it does appear that they were a particular feature of Thames navigation. Shouts seemingly took their name from the Dutch *schuit* or *schuyt*, a flat-bottomed vessel, built somewhat like a barge but pointed at both ends.[22] They were ideal for rivers prone to shallowness and were particularly necessary for negotiating the 'flashes'[23] of mill-weirs which occasionally blocked the Thames, starting at Maidenhead, some 10 miles downstream from Henley. Peberdy has identified twenty-five possible mills between Maidenhead and

---

batella was built for £6 13s., which may well have been the boat leased out (in the 'Farm' section) for 70s. per year to John of Quarneby that same year: TNA, PRO DL29 354/5837. Masschaele, 'Transport Costs', 272, also cites a *batella* being built for more than £20, including £2 10s. 'for unspecified tackle and a sail'.

[19] D. G. Wilson, *The Thames: Record of a Working Waterway* (London, 1987), 30–1, provides two later illustrations—one from the 17th century and the other from the early 18th—of barges using sails in the downstream direction. See also P. Marsden, *Ships of the Port of London: Twelfth to Seventeenth Centuries AD*, English Heritage Archaeological Rep. 5 (London, 1996), 92–3, for medieval shouts using sails on the Thames.

[20] Because of the lack of the Priory of Worcester's own sailors (*nautae*), a number were hired to haul a (ship) (*navis*) up the Severn from Longdon (near Tewkesbury) to Worcester (as recorded in the 1351–2 cellarer's account): James M. Wilson and Cosmo Gordon (eds.), *Early Compotus Rolls of the Priory of Worcester*, Worcestershire Historical Society (Oxford, 1908), 50. I am grateful to an anonymous referee of this volume for this reference. Wilson, *The Thames*, also provides a number of later illustrations from the 16th to the 19th centuries of boats and barges being poled or hauled up the Thames (e.g. 32, 33, 49). F. S. Thacker, *The Thames Highway: A History of the Inland Navigation* (London, 1914; repr. New York 1968), 147–8, gives a 1793 example of teams from three to twelve horses being required to haul a 70-ton barge up the Thames from Stadbury (in Shepperton) to Lechlade, depending upon the force of the current against the boat. I am grateful to Robert Peberdy for this reference. Finally, the Luttrell Psalter shows a somewhat eccentric depiction of a boat being presumably towed upstream by two men pulling on a cable; they are not being helped by the people in the boats, who seem to be rowing against the towers: British Library MS Add. 42130, fo. 160. The scene is reproduced in colour in Janet Backhouse, *Medieval Rural Life in the Luttrell Psalter* (Toronto, 2000), 53.

[21] A shout (*scouta*) was hired to carry timber along the south coast from Rye to Dover in 1295, as recorded in a building account of a tower windmill at Dover Castle: TNA, PRO E101/462/14, m. 3. For other 14th- and 15th-century examples of shouts found outside the Thames system see Gardiner above, p. 103; J. F. Willard, 'Inland Transportation in England during the Fourteenth Century', *Speculum*, 1 (1926), 372; D. Burwash, *English Merchant Shipping 1460–1540* (Toronto, 1947; repr. Newton Abbot, 1969), 139–40. I am grateful to Mark Gardiner for directing me to the last two references.

[22] As in the case of the 'Blackfriars 3' boat excavated in 1970: Marsden, *Ships of the Port of London*, 55–104.

[23] Flashes or flash-locks were removable pieces of board on the top of a mill-weir or dam, which allowed a temporary 'flash' or flow of water over the weir, which boats could 'shoot' if going downstream or, as indicated below, be winched over if going upstream: see esp. M. J. T. Lewis, W. N. Slatcher, and P. N. Jarvis, 'Flashlocks on English Waterways', *Industrial Archaeology*, 6 (1969), 209–53. See above, pp. 9–10 and Fig. 3.

Oxford,[24] many of which seem to have been equipped with a cable and winch for hauling the shouts up over the mill-weir flash.[25] Indeed, the Thames system shows a striking co-operation between boat-owners wishing to use the river and mill-owners wishing to harness water power for their own uses, accommodated by the use of the flat-bottomed shout.[26] This arrangement was necessitated by the increasing movement of milling from lesser to larger watercourses from the eleventh to the thirteenth century.[27] This was especially the case on the Thames below Oxford, where a sharper drop in gradient encouraged the construction of watermills.[28] But such a refined system seems to have been limited to the Thames. In comparison, all other river systems seem to have relied upon the ill-defined *batella*, and there was seemingly much less in the way of accommodation between boat- and mill-owners, where inland river navigation often tended to stop sharply at the first mill-weir encountered.[29]

A third reasonably well-used term for vessels was the *navicula*, found in over 4 per cent of cases. It was most frequently used on the Trent, and it was here, too, that the term *navis* was occasionally used for larger boats on the river. It would appear that the *navicula* occupied a position similar to and possibly synonymous with the *batella*, a supposition supported by the fact that both types of vessels seemingly carried loads of about the same magnitude (Table 2). A 1457 will from York, however, equates a *navicula* to a shout,[30] so that it is difficult to be categorical about what the term meant.

Other types of boats were found less frequently. Barges (*bargeae*) were occasionally found (just under 2 per cent of cases) and may have been similar to shouts, although over time (i.e., fourteenth to sixteenth centuries) they may have become more narrowly identified as larger flat-bottomed boats than shouts had been.[31] Much less mentioned (a little over 1 per cent of cases) were vessels called 'keels' (*kelae*). These again may have been more often lumped with *batellae* in the documents, but when loads are given for them (see below) they seem to have been in the upper range, size-wise, of *batellae*. In this sense,

---

[24] Peberdy, 'Navigation', 318.   [25] Ibid. 313, 325–6.

[26] For more on this compromise, see Lewis, Slatcher, and Jarvis, 'Flashlocks', 211–12. On other rivers, a similar accommodation between fishermen and boat-owners was made where the nets of fishermen were only allowed to jut half-way into the stream, thus allowing boats to get by—an arrangement which was not always observed, according to cases presented to the royal courts: *PWML*, ii, pp. xxv–xxvi, 300–6.

[27] Langdon, 'Inland Water Transport in Medieval England: The View from the Mills', 76–7.

[28] M. Prior, *Fisher Row: Fishermen, Bargemen and Canal Boatmen in Oxford, 1500–1900* (Oxford, 1982), 107–8. Wilson, *The Thames*, 31–2, indicates that the rate of fall in the River Thames for the 13-mile stretch immediately downstream from Oxford is about twice that of the equivalent stretches of the river (in terms of distance) immediately above and below.

[29] As in the case the mill at Over Colwick (near Nottingham) on the Trent and the mill at Hemingford Grey (near Huntingdon) on the Great Ouse: Langdon, 'Inland Water Transport', 4–5; see also Jones, 'River Navigation', 67–8.

[30] 'Lego Thomae Dauson apprenticio meo. j. naviculam vocatam le Showte' (in will of John Rodes of York, fishmonger): J. Raine (ed.), *Testamenta Eboracensia*, ii, Surtees Soc. 30 (London 1855), 209. I am grateful to Mark Gardiner for directing me to this reference.

[31] Peberdy, 'Navigation', 330–2.

they may possibly have been similar to barges.[32] 'Catches'[33] (*cacchae*) and 'crayers' (*craierae*) were mentioned over 2 per cent of the time altogether and seem in the context of the purveyance accounts to have functioned as lighters for taking goods from shore to the big ships in harbours.[34] Presumably they were oared vessels of reasonable size (see below) which could speedily transport these goods back and forth. Finally, there were occasional terms which indicate quite specialized types of boats, such as the two *batellae* 'called rysbotes' used to carry provisions from Manningtree (in Essex) to Ipswich in 1340.[35] These were probably 'rushboats', more normally used to carry rushes to London or other urban centres.[36] Similarly, two 'farcosts' (*farcostae*) were used to ferry 72 quarters of wheat and 30 quarters of oats from Bridgwater to Bristol in 1345.[37] These were probably fitted with sails to make the trip across the Severn estuary, and may have been usually a little larger than a *batella*, since, uniquely, the crew size for these vessels was also indicated at six men per *farcosta* compared to four men per *batella* (given in the same document).[38] Finally, in two cases *naves* were used on the Trent. These were not full-sized seagoing ships, but large boats, possibly for use in taking trans-shipped goods onto the lower stretches of river, or for when water was sufficient to take these ships all the way to the ultimate head of navigation. The former was indicated in 1346 when 16 tuns of flour, probably containing around 104 quarters, were trans-shipped from 4 *batellae*, which had brought the flour downstream from Newark, onto a *navis* for transport to Kingston upon Hull.[39] The necessity of trans-shipment onto more seaworthy ships for venturing onto the Humber was a common feature of inland water transport both on the Trent and the Yorkshire Ouse.[40] In the second case (also for 1346), however, two *naves* carried 29 tuns of flour (about 190 quarters) from 'Colwykeshore' (Over Colwick near Nottingham, the effective head of navigation for the Trent[41]) all the way to Kingston upon Hull.

---

[32] *Oxford English Dictionary*, 2nd edn., viii. 367; L. C. Wright, 'Technical Vocabulary to do with Life on the River Thames in London *c*. AD 1270–1500' (University of Oxford D.Phil. thesis, 1988), 154, 155 (under 'Keel' and 'Lighter').

[33] Possibly more accurately spelled as 'ketch', which Wright ('Technical Vocabulary', 155) gives as a small fishing vessel, although the *Oxford English Dictionary* (ii. 973) has 'catch' as a term signifying a two-masted vessel of 100 to 150 tons.

[34] Although Burwash, *English Merchant Shipping*, 122–3, characterizes them as small boats used for cross-Channel traffic.

[35] TNA, PRO E101/556/21.   [36] Wright, 'Technical Vocabulary', 158.

[37] TNA, PRO E101/585/27. Farcosts were sometimes associated with the carriage of stone: Wright, 'Technical Vocabulary', 151; see also Marsden's description of 'Blackfriars 4', possibly a 15th-century farcost involved in the carriage of ragstone from the Medway to London: Marsden, *Ships of the Port of London*, 105–6.

[38] Burwash (*English Merchant Shipping*, 124) feels that farcosts were 'small cargo vessels plying the cross-channel trade', perhaps capable of carrying around 40 tons. This would be a description not entirely out of place with the two farcosts making the Bridgwater to Bristol run in 1345, although, at a combined 102 quarters of grain, each of these vessels was probably carrying less than 10 tons (see Table 4).

[39] TNA, PRO E358/2, m. 32.   [40] See also Langdon, 'Inland Water Transport', 7.
[41] Ibid. 4.

The Carrying Capacity of Vessels

The vessels outlined above could carry all sorts of loads, as, for example, the eight *batellae* that carried 12 empty barrels or tuns (*dolea*), 36 sides of bacon, 9 beef carcasses, 40 sheep carcasses, 603 horseshoes, and 860 horse nails from St Ives in Huntingdonshire to (King's) Lynn (Norfolk) in 1338.[42] For mixed loads like this it is difficult to judge either the weight or volume, but with loads solely restricted to grain, an estimate of volume and weight is easier to make. For example, the two *farcostae* mentioned above, together carrying 72 quarters of wheat and 30 quarters of oats, can be considered as averaging 51 quarters each. It has been assumed here that the quarter was the same volume whether the grain or good was wheat, malt, oats, beans, or peas (the most common crops taken by the purveyors). This is a debatable assumption, but it does at least provide a rough indication of load volume that can be used for comparative purposes. Also, loads including or consisting solely of tuns (*dolea*) of flour were also added to the sample, since a rough measurement of the volume of these barrels can be derived from cases where the purveyed grain was milled. A small sample of such instances gives figures ranging from 5.5 to 7.5 quarters of flour per tun.[43] I have assumed an average of 6.5 quarters per tun and have applied that to all references to tuns of flour. When pipes (*pipae*) of flour were mentioned, it was assumed, as is the tradition, that these were half the size of a tun.[44]

Taking all this into consideration, one can look at these load sizes from a number of perspectives. We shall start from the standpoint of vessel type, as shown in Table 2, where data concerning vessel loads were abstracted from 108 separate voyages. The most commonly found term where the load in quarters could be measured was the *batella*, where eighty-four voyages of *batella* or *batellae* carrying loads measurable in quarters were recorded. Perhaps the most striking feature of these eighty-four cases was their variation in capacity, ranging from very small vessels carrying barely 10 quarters to very large vessels carrying in excess of 200 quarters. This reinforces the impression of *batella* being a very generalized term, encompassing a great variety in boat size and possibly design. The mean and median values for these eighty-four cases, however, both fell within the 40–50 quarter mark.[45] As mentioned above, nine

---

[42] TNA, PRO E101/552/17.

[43] The cases, five in all, of grain milled for campaigns and put into tuns (*dolea*), gave the following figures in quarters of flour per tun (years in brackets): 7.50 (1298); 6.63 (1322–3); 6.87 (1333); 5.53 (1346); 5.96 (1347). Sources: TNA, PRO E101/592/3; 585/12; 559/9; 592/8; E358/2, m. 25ᵛ.

[44] In two cases goods other than crops or flour were included, because they were a small proportion of the total. The first involved 15 quarters of salt carried with over 400 quarters of wheat, malt, and peas in eight *batellae* from St Ives and Cambridge to King's Lynn in 1337 (TNA, PRO E358/2, m. 13). Here each quarter of salt was considered as being equivalent in weight and volume to a quarter of grain. Similarly each of the 19 tuns of meat carried along with 30 tuns of flour and 526 quarters of wheat in four shouts from Henley-on-Thames to London in 1347 was considered as equal in weight and volume to a tun of flour (TNA, PRO E358/4, m. 26; see also below).

[45] It should be said here that the mean and median figures were those of the 84 voyages, *not* those of all the vessels involved. For example, the 10 *batellae* that carried 494 quarters and 7 bushels of malt

Table 2. Load sizes in quarters by vessel type

Vessel type	No. of voyages	No. of quarters per vessel		
		Mean	Median	Range
Batella	84	50.4	41.5	10.0–235.5
'Shout'	7	121.6	105.0	60.0–211.1
'Keel'	1	65.2	65.2	65.2
'Catch'	3	48.1	50.0	19.2–75.0
'Barge'	1	129.7	129.7	129.7
Navicula	9	53.3	52.0	20.0–84.2
Navis	2	99.2	99.2	94.3–104.0
Farcosta	1	51.0	51.0	51.0

cases of *naviculae* yielded similar mean and median loads, but with a much smaller range in load size. The seven cases of shouts for which loads were measurable in quarters also produced a more closely grouped set of figures, but with a much higher mean and median, again suggesting that the shout was a rather more specialized vessel suitable for large commercial traffic. The other vehicle types yielded only a few load examples, from which it is difficult to make much in the way of meaningful comments. The one measurable case of barges being used (to carry grain from London to ships moored in the Thames in 1346[46]) yielded a figure comparable to those for shouts, which is plausible given that the two were probably very similar in type and dimensions. The two cases of *naves* equated to the upper range of *batellae*, again indicating vessels that could function both as river and seagoing craft. In the one case of 'keels' that occurred with loads on the Cam and Great Ouse, they seem equivalent with the other *batellae* plying the same routes.[47] Catches and *farcostae* fell generally in the mid-*batella* range in terms of load and seemingly functioned as boats that could both penetrate rivers and sail on the open sea. The three cases of catches, in particular, indicated a wide variety of uses, five being used to carry 96 quarters of wheat (at an average 19.2 quarters per catch) along the River Hull from Wansford to Kingston upon Hull in 1338, while in the same account twenty catches carried 1,500 quarters of wheat, malt, oats, peas, and

from Cambridge to King's Lynn in 1333 were calculated to carry 49.5 quarters each (i.e. $\frac{494.9}{10}$): TNA, PRO E101/552/14. In calculating the mean and median *batella* figures for Table 2, however, the 49.5 figure for this voyage was only counted once (that is, not ten times for each of the vessels involved). This was done so as not to distort the figures in Table 2 in favour of the large flotillas. It should also be said here that single vessel loads were considered as full loads. In some cases they might have been part loads only, although the load similarity of the single boat loads compared to those found for multiple vessel journeys in a particular area suggest that it was rare to send a boat only partially loaded (for example, nine separate journeys on the Parrett and Tone rivers in 1344, three for single boats and six for vessel groups up to twelve, yielded mean loads per voyage from 11.9 to 19.7 quarters per vessel, the single vessel voyages being more or less in the middle of the range: TNA, PRO E101/585/13).

[46] TNA, PRO E358/2, m. 22ᵛ.

[47] TNA, PRO E101/552/26 (1347); cf. E101/552/14 (1333), 16 (1328), 50 (1347); E358/2, m. 13 (1337); E358/3, m. 7 (1324–5).

Table 3. Load sizes in quarters for various rivers, estuaries, and coastal travel

River systems, estuaries, etc.	No. of voyages	No. of quarters per vessel		
		Mean	Median	Range
Ure/Ouse/Wharfe	3	28.1	30.0	14.3–40.0
Humber	7	52.8	50.0	19.2–87.8
Trent	13	60.8	66.2	20.0–104.0
Witham	1	78.0	78.0	78.0
Great Ouse/Cam/Nene, etc.	17	55.1	55.2	27.8–99.8
Yare/Bure/Waveney	12	54.2	42.5	10.0–125.0
Thames	20	81.8	60.0	20.0–211.1
Parrett/Tone	9	15.5	15.5	11.9–19.5
Severn	15	45.4	37.2	14.0–99.0
Coastal	11	78.8	47.5	20.0–235.5
Whole country	108	55.2	50.0	10.0–235.5

beans (averaging 75 quarters per catch) from Kingston upon Hull to ships moored in the Humber.[48] Similarly, in 1340, twenty catches carried 1,000 quarters of mixed grains (averaging 50 quarters per catch) from Barton upon Humber (Lincs.) across to ships at Kingston upon Hull.[49]

Such a wide variety of uses and load sizes for all the vessel types indicated above makes it difficult to be precise in attaching specific patterns of employment to particular vessel terminologies; indeed, any type of vessel might have been used for a range of different transport activities. Ignoring the various vessel types, in fact, allows other perspectives to be taken of the data. Arranging the same 108 voyages according to the river, estuary, or coastal areas that they serviced indicates a considerable geographical variation, as shown in Table 3. The Witham and Thames systems clearly could handle vessels with quite large loads, while at the other extreme the Parrett and Tone rivers in the Somerset Levels could clearly only accommodate very small-scale river transport. Only slightly larger was the traffic on the Ure, Ouse, and Wharfe rivers. Most other systems—the Trent; the Fenland water systems of the Great Ouse, Nene, and so on; the East Anglian Yare, Bure, and Waveney; and the Severn—averaged in the 40–60 range. An even more nuanced view, however, can be gained by looking at the specific stretches of rivers. Of the 108 voyages, 75 related to river travel, the other 33 to estuary and coastal trips. The Appendix lists the various load sizes that could be carried by various parts of river systems (as indicated by these 75 voyages) and Fig. 23 shows these in graphical form, with average load size indicated by the thickness of the lines representing the rivers.

One thing that seems clear from this particular breakdown is that many river systems had important 'break points' that separated more small-scale and perhaps more sporadic upstream travel from larger-scale and more professionally organized downstream travel, perhaps best expressed by Peberdy's categorization of medieval river travel in terms of 'primary' and 'secondary'

48 TNA, PRO E101/597/26.   49 TNA, PRO E358/2, m. 11.

navigation.[50] It has long been recognized that Henley-on-Thames provided this function for the River Thames (though see Blair, below pp. 254–7, 285),[51] but the Appendix indicates that this probably occurred on at least three other river systems. The most obvious of these was the Trent. The purveyance accounts suggest possible break points at Dunham on Trent, Torksey, and Gainsborough, along a 10-mile stretch of river that separated small-scale carriage down from Nottingham and Newark-on-Trent to these towns from that carried further north to the Humber.[52] Although the sample sizes for these respective stretches of river are small (seven and five respectively) the differences in loads carried by the two were significantly different, being 48.4 quarters per vessel on average for trips on the upper Trent compared to 80.1 quarters per vessel on the lower Trent (see Appendix 5A). This seemingly could vary markedly, however, perhaps in the same year. As indicated above, two *naves* carrying 29 tuns of flour (about 94.3 quarters apiece) negotiated the entire length of the navigable Trent in 1346, while another 16 tuns of flour in the same year (containing about 104 quarters) were brought down from Newark in four *batellae* to Torksey and then trans-shipped onto a single *navis* for travel down to Kingston.[53] The rationale behind this move is hard to fathom given the ability of the two *naves* above to make the entire trip, but it does show that Torksey in this case did function as a seemingly important trans-shipment point. Indeed, the appearance of Dunham and Gainsborough as competitors to Torksey in this way suggests the gradual decline of Torksey as the pre-eminent trans-shipment point on the Trent. This was possibly due to the silting-up of the Foss Dyke, the western outlet of which was situated at Torksey and which previously had given the town a considerable prominence on the river.[54]

Fairly obvious 'break points' on other rivers are suggested on the Severn for Tewkesbury and Gloucester, where a series of voyages—five in all—going upstream to Worcester averaged out at 26.4 quarters per vessel, while in two cases proceeding downstream from Tewkesbury and Gloucester to Framilode at the start of the Severn estuary averaged a more impressive 41.9 quarters per vessel. Although less clear, York and Acaster also served as break points on the Yorkshire Ouse, since boats coming down from places like Boroughbridge averaged only 22.2 quarters per vessel and had to be transferred to larger boats for the journey on the Humber.[55] All of these indicate that England's

[50] Peberdy, 'Navigation', 333.

[51] It is difficult to know whether Henley was a 'break point' because of the difficulties of navigation above Henley or because the river changed direction sharply towards the south, making road transport to Oxford more competitive with the longer way around the river. I am grateful to Robert Peberdy for making this point.

[52] Robert Peberdy (pers. comm.) has suggested that demand from Lincoln might also have been a reason for these trans-shipment points on the Trent.

[53] TNA, PRO E358/2, m. 32.

[54] M. W. Barley, 'Lincolnshire Rivers in the Middle Ages', *Lincolnshire Architectural and Archaeological Society*, NS 1 (1938), 13–16; Langdon, 'Inland Water Transport', 4, 6, 7.

[55] TNA, PRO E101/597/1, 31; Langdon, 'Inland Water Transport', 7.

river systems were becoming more adapted to the complexities and needs of water-borne trade.

But it was the Thames which had the most obviously attuned system for commercial river travel. Shouts seemingly dominated water transport from Oxford to London, although again load sizes became larger after Henley. The one measurable voyage from Oxford to Henley in 1324–5 averaged 61.4 quarters per vessel. In contrast, the three measurable voyages from Henley to London, two of which also occurred in the same 1324–5 account, averaged 159.5 quarters per vessel, a powerful testimony to the volume of trade that the river could carry from Henley, even with at least five mill-weirs to negotiate on the way to London.[56] Indeed, travel in the Thames estuary was often carried on smaller boats than in the Henley–London stretch, as indicated by a series of voyages along the northern Kent coast, carrying grain from places like Faversham, Maidstone, Gillingham, Cliffe, Gravesend, Erith, and Woolwich—eventually to ships in the Thames estuary or the Swale—during the period 1318–20, all carried on *batellae* with loads ranging from 20.0 to 74.9 quarters per vessel.[57] Altogether, a great number of vehicle types and sizes converged to make the Thames a capacious and flexible transport system.

The Dimensions and Weight-Carrying Capacities of Vessels

How did the number of quarters transported by a vessel translate into actual boat dimensions and the weight that it could carry? For estimating weight, this is most safely done with wheat, for which the growing consensus is that a medieval quarter of the grain weighed 384 lbs avoirdupois.[58] For volume, I have assumed a Winchester bushel of 1.24 cubic feet, which equates to a quarter of 9.92 cubic feet or 0.3674 cubic yards. From these basic premises, Table 4 provides weight and volume conversions for various loads of wheat at 25-quarter intervals. It would be useful here to give a specific example and try to figure out what these figures meant in terms of the actual size of a boat. Thus, in 1327, two *batellae* carried 100 quarters of wheat from Heybridge (near Maldon, Essex) to a ship moored at Stangate Abbey in the Blackwater estuary, a short trip of about 4–5 miles for which the boat-owners were paid a modest halfpenny per quarter.[59] Assuming the two boats carried equal loads this would make 50.0 quarters per vessel. From Table 3 this gives a weight

---

56 TNA, PRO E101/582/11 (1324–5; two cases); E358/4, m. 16 (1345; one case); Peberdy, 'Navigation', 318. Two more cases of shouts carrying oats from London to Westminster during the crisis of 1326–7 averaged 150 and 101.5 quarters per vessel respectively: TNA, PRO E101/556/12.

57 TNA, PRO E101/566/8.

58 H. Hall and F. J. Nicholson (eds.), *Select Tracts and Table Books Relating to English Weights and Measures, 1100–1742*, Camden Miscellany xv (London, 1929), 8; John Langdon. 'Note, 1987', in T. H. Aston (ed.), *Landlords, Peasants and Politics in Medieval England* (Cambridge, 1987), 63–4; Masschaele, 'Transport Costs', 277–8.

59 TNA, PRO E101/556/14. A summary of this voyage, entered later into E358/3, m. 15ᵛ, gave the total load for the two *batellae* at 105 quarters. It has been assumed here that the original entry of 100 quarters in E101/556/14 is the correct one.

of 8.57 imperial tons or 8.71 metric tonnes. In terms of volume, a load of 50 quarters would be $50 \times 0.3674 = 18.37$ cubic yards (or 14.05 cubic metres). To compare these figures with those from an actual medieval boat, it has been estimated from hydrostatic analysis that the shout 'Blackfriars 3', excavated at London and dating from the late fourteenth or early fifteenth century, could carry about 7.5 tons or just over 86 per cent of each of the Heybridge *batellae*.[60] Since 'Blackfriars 3' was 16 yards long, 4.7 yards wide, and was a yard high at the gunwales at its middle point,[61] the Heybridge *batellae* were probably a little longer (20 yards?). Also, given that they were almost certainly not as flat-bottomed as 'Blackfriars 3', they were probably less broad at the beam (about 3 yards?) and deeper in draught (about 2 yards?). If the 18 or so cubic yards implied by the purveyance account for the individual Heybridge to Stangate *batella* loads were laid more or less end to end and with some allowance for the boats being narrower at the ends than at their broadest point, it would seem that a load of 50 quarters would fit reasonably comfortably into a boat of about 20 yards long.

However, there were certainly much larger boats than this. In particular, the shouts that regularly plied the Thames usually exceeded 100 quarters and occasionally 200 quarters. Unfortunately none of the shout-loads recorded in this study consisted of wheat only. The closest instance involved four shouts that carried 526 quarters of wheat, 30 tuns of flour, and 19 tuns of pork and beef from Henley-on-Thames to London in 1345.[62] Assuming the above 6.5 quarter per tun equivalent for the flour and (more questionably) the meat this would add another 318.5 quarters to the wheat, making 844.5 quarters

Table 4. Weight and volume equivalents for medieval quarters of wheat

Number of quarters of wheat	Imperial tons[a]	Metric tonnes[b]	Cubic yards[c]
25	4.29	4.36	9.19
50	8.57	8.71	18.37
75	12.86	13.07	27.56
100	17.14	17.41	36.74
125	21.43	21.77	45.93
150	25.71	26.12	55.11
175	30.00	30.48	64.30
200	34.29	34.84	73.48
225	38.57	39.19	82.67
250	42.86	43.55	91.85

[a] Assumes a quarter of 384 lbs avoirdupois and 2,240 lbs per imperial ton.
[b] One imperial ton equals 1.0160 metric tonnes.
[c] Assumes a Winchester bushel, probably the one used in the early 14th century, of 2148.28 cubic inches (R. D. Connor, *The Weights and Measure of England* (London, 1987), 164), which equals 1.24 cubic feet. At 8 bushels to a quarter, the volume of one quarter would equal 9.92 cubic feet or 0.3674 cubic yards.

[60] Marsden, *Ships of the Port of London*, 98.   [61] Ibid. 55; Peberdy, 'Navigation', 330.
[62] TNA, PRO E358/4, m. 16.

in all or an average 211.1 quarters per shout. Although the density of the flour in particular would be less than that of the wheat, an equivalent 200 quarters of wheat per shout would not seem out of the question here. Again using Table 4, this suggests quite large boats carrying nearly 35 tons. If the 'Blackfriars 3' shout did indeed carry only 7.5 tons, then the shouts coming from Henley in 1345 were considerably bigger vessels, having a capacity four and a half times that of 'Blackfriars 3'. Since it is unlikely that these boats could have been much deeper in draught, in order to be able to make it down or up mill-flashes, then the extra size probably had to be made up through greater breadth or length. Breadth was probably more problematic than length, again for the reason of having to make it through mill-flashes and perhaps occasionally constricted sections of the river. If we allow a maximum of 50 per cent more breadth (say 7 yards?[63]) for the 1345 shouts than 'Blackfriars 3' then the 1345 shouts would have to be around three times as long,[64] that is, around 48 yards in length (since 'Blackfriars 3' was 16 yards long). Even if draught could be increased by another 50 per cent (to, say, a yard and a half) along with breadth, the length of these boats would still have to be twice as long as the 'Blackfriars 3' shout—say 32 yards. Dimensions of this magnitude are equivalent to those for eighteenth-century barges capable of hauling up to 170 tons.[65]

Perhaps greater realism might be achieved by using a more adventurous estimate for the weight that could be carried by 'Blackfriars 3'. Marsden calculated the 7.5-ton figure assuming that the vessel could only be loaded to the point where the freeboard (that is, the distance from the waterline to the top of the gunwale at the mid-point of the boat) was about 55 per cent of the total depth of the vessel; otherwise the boat was in danger of capsizing. Using a minimum 40 per cent freeboard figure, however, as suggested by medieval Icelandic law, Marsden arrived at a cargo load of 12.385 tons,[66] or slightly more than a third of the 1345 shouts that carried their 211.1 quarters (on average) from Henley to London. Assuming again that the depths of 'Blackfriars 3' and the 1345 shouts were the same but that the 1345 shouts were 1.5 times broader in the beam, then each 1345 shout would seemingly have to be about twice the length of 'Blackfriars 3', again around 32 yards. If the depth of the 1345 vessels was also 1.5 times then the length of these boats would be only a third longer than 'Blackfriars 3', that is, about 21 or 22 yards. These are probably more realistic dimensions, such that we should see medieval river vessels as seldom exceeding 30 yards in length and that they routinely loaded their vessels to the same adventurous extent as did their Icelandic cousins.

---

[63] This would probably represent a maximum breadth, since even in the 18th century, flash-locks were generally only about 20 feet wide: Thacker, *Thames Highway*, 121.

[64] That is, if the 1345 shouts could carry four and a half times the 'Blackfriars 3' vessel and both had the same depth and breadth could only be increased by 50% or a factor of 1.5, then the 1345 shouts would have to be $\frac{4.5}{1.5} = 3$ times longer.

[65] Wilson, *The Thames*, 40–2.   [66] Marsden, *Ships of the Port of London*, 96–7.

Given that these figures are based on what amounts to little more than guesses, confirmation (or not) of vessels of these dimensions must wait for further archaeological evidence in particular, but it does again suggest the potentiality of a very impressive volume of trade along the Thames, and that hydrological limitations or the presence of mill-weirs were not a huge problem before the plague, at least as far upstream as Henley. The predilection for vessels of very high capacity for the Henley to London route does, however, tend to support a contention that I have made about the essential one-way (i.e., downstream) nature of that traffic.[67] Getting fully loaded boats up over mill-flashes was bound to be increasingly difficult as vessels got larger.[68] The decision to use the vessel sizes suggested in this study then seems very odd if there was much concern at all to get goods upstream, since this would appear to be better served by using smaller boats. On the contrary, it seems likely that the shouts heading for Henley were taken upstream unloaded (or very lightly loaded) as a matter of course and that their purpose was mainly to facilitate shipment of goods (especially grains) from the interior of England to London and further abroad. The building of very large shouts may have been a way of offsetting losses on return voyages. This is not to deny that some goods did make their way upstream,[69] but clearly this was the secondary purpose of the traffic for these large vessels in particular. Part of this picture, however, might well be qualified by the fact that purveyance required these larger vessels, and that smaller vessels involved in upstream trade just did not make it into these particular records.

Looking at the other extreme, the dimensions of boats on the Parrett and Tone must have been very tiny. Thus in 1324–5 four *batellae* carried 50 quarters of wheat from Curry Bridge down the Tone and Parrett to Bridgwater.[70] The average 12.5 quarters per vessel load here, according to Table 4, would work out to 2.2 tons and 4.7 cubic yards per vessel, suggesting vessels of scarcely more than five yards in length, which seems to have been typical of the Parrett/Tone system.[71] Boat dimensions on other systems can similarly be calculated. Assuming that 50 quarters was a load that could normally be expected for vessels on systems like the Trent, Severn, the Great Ouse, Nene, Yare, and so forth (see Table 3), Table 4 would suggest that these systems all tended to support medium-sized boats somewhere near the size of the 'Blackfriars 3' shout, that is, somewhere in the range of 15–20 yards long. The

---

[67] Langdon, 'Inland Water Transport', 6.

[68] During the 18th and 19th centuries barges capable of carrying up to 200 tons were apparently hauled up over flashes, although the prevalence of boats of this size was a relatively recent phenomenon at the time: Thacker, *Thames Highway*, 118–20, 177.

[69] Wine in particular was a bulky item much sought after in the interior, as indicated by the hogsheads of wine sent to Henley for the Stonors in the 15th century: Wilson, *The Thames*, 29.

[70] TNA, PRO E101/585/13.

[71] Again TNA, PRO E101/585/13 for 1324–5 provides boat capacities for nine of these small-scale voyages (including that from Curry Bridge), ranging from 11.9 to 19.5 quarters per vessel and averaging around 15 quarters per vessel with a total range of 11.9 to 19.5 quarters per vessel: see Appendix.

underlying point behind all of this, though, is the variety that existed in boat size and probably design. No matter what the size of the waterway, a vessel could be produced for it. Failure of waterways to be navigable in medieval England thus rests with other reasons than the inability of boat technologies to adjust, as the Thames experience amply proves.

The 'Efficiency' of the English Water Transport System

How does this translate into assessments of the effectiveness or 'efficiency' of medieval English inland water transport, particularly in regard to what various waterways could carry and when? Edwards and Hindle's view of the medieval English inland water transport was very sweeping, claiming that '[i]n the final instance there were only a few areas which were more than 15 miles from navigable water'.[72] In a later response, I argued that this was too simplistic, and that issues such as, among others, seasonality and the difficulty of carrying profitable loads upstream should be taken into consideration.[73] The analysis of vessel size clearly emphasizes the complexity of the matter. For one thing, there were marked regional differences in the contribution that river transport could make to local economies. It would be foolish, for example, to consider the Parrett/Tone system as equivalent to that of the Thames, where the former at best could only support boat sizes of a tenth or so of the latter. Also, the hydrological limitations of upper stretches of rivers clearly affected the size of boats that could be carried, and required ongoing adjustments, such as the insertion of trans-shipment points, as was evident in the Trent and Thames. Vessel size was clearly important to the issue in enabling—or not—upstream cargoes, especially on the Thames, where a trend to larger vessels, if it happened, may have been a strategy on the part of boat-owners to limit the disability of having profitable goods carriage in the downstream direction only.

Given these seeming complications and often rather uncomfortable realities for transporters, what can we really say about the 'efficiency' of the medieval English inland water transport system? First, its usefulness varied considerably according to region. As Edwards and Hindle indicate from their study,[74] the eastern part of the country was particularly favoured with effective navigable rivers, and the purveyance accounts very much back this up.[75] This is especially the case when the size of the vessels that plied these eastern waterways is taken into consideration. Whatever its peculiarities, the Henley–London leg of the Thames River was a major conduit supplying the city and parts of the estuary, a conduit that—in the other direction from Henley—may have extended in more attenuated form as far west as Oxford or beyond (cf. Blair below). The Trent and Witham similarly could accommodate relatively large

---

[72] Edwards and Hindle, 'Transportation System', 129; see also the map on 130.
[73] Langdon, 'Inland Water Transport', esp. 5–6.
[74] Edwards and Hindle, 'Transportation System,' 128.
[75] Langdon, 'Inland Water Transport', 7–9.

boat sizes, although in the case of the Trent again probably, for the most part, only in its lower section. In the west, only the Severn could compete with these eastern waterways in terms of the size of vessels that could be accommodated by the river. But the relative lack of information for the Severn in the purveyance accounts makes it uncertain whether economic opportunities were sufficient to make the route truly important in the economy of western England.[76]

Second, even with the better sense that this chapter gives about vessel sizes for particular stretches of rivers, it still leaves the great unknown about the number of boats that actually used the river, since the volume of goods travelling down (or up) rivers is equal to the amount that the boats carried times the number of the vessels themselves times the number of trips each could make per year. It is unlikely that we will ever have a very accurate sense of the number of boats at any one time on any stretch of England's navigable rivers in the middle ages, or of the number of trips they would make in a particular period of time. Part of this is because we have little sense of the number of boat-owners or of how professionalized they were. The number of trips made per year by a transporter, for example, would be affected considerably by the fact of whether this work was full-time or part-time.

Any answers to these questions can, for the moment, only be approached indirectly. In terms of purveyance, we do at least have a sense of the amount of provisions that the king wanted. This tended to vary from campaign to campaign and from reign to reign. Certainly, it seems to have reached its maximum during the reign of Edward I, when 26,500 quarters of grain were required in the spring of 1296 and a further 63,200 quarters in November of that year, during a spell in the 1290s when war was being fought in Wales, Scotland, Flanders, and Gascony.[77] Dealing with the November request, 63,200 quarters would seem well within the capabilities of English agricultural production.[78] Was it also within the capabilities of the English transport system, especially in terms of the number of vessels available? If we assume that the entire 63,200 had to be shipped by the inland water transport system (unlikely, since a good portion was sent by cart to the various ports) and that, as in Table 3, the median load per vessel was 50 quarters, then the number of voyages needed to get these goods to port was $63,200/50 = 1,264$. Given the hurry in which purveyance campaigns were conducted, it is probable that boat-owners were only required to serve Crown needs once in a campaign, so that a maximum of 1,264 vessels would be involved in moving grain in late 1296 (although, as we have seen, sometimes grain had to be carried by two different vessels to get it from the upper reaches of a river to the port). Assuming that grain only covered half of the needs of a campaign and that an

---

[76] See my comments in this regard in Langdon, 'Inland Water Transport', 7–8.

[77] Prestwich, *Armies and Warfare*, 251.

[78] 63,200 quarters would be about 1% of the yearly crop production from England as a whole (as calculated from figures from Campbell et al., *A Medieval Capital*, 38–43).

equal number of vessels would be needed for other goods such as meat, cheese, fish, iron goods (including horseshoes), and wooden hurdles, then something like 2,528 vessels would be required overall. Would the Crown be able to count on this number of vessels? If they were actually used more than once during a purveyance campaign, say twice (probably the upper limit), then the number would shrink back to 1,264. It is difficult to say whether the number of vessels on England's inland waterways or around the coast approached numbers like these. Given that purveyance campaigns seldom raised the exact amount of provisions requested,[79] it might be that transport was a problem.

Perhaps the issue can be approached in a more limited regional way and in a time of peace rather than war. Here we might take the stretch along the Thames from Henley to London as a test case. Campbell, Galloway, Keene, and Murphy have recently provided a very useful analysis of London's provisioning requirements, and these can be used as a way of checking the number of boats that would be needed on the Thames waterway from Henley to London, given the size of the shouts that we have estimated from this study. Their estimate is that 'a mean of 165,000 quarters is the best single guide to the general magnitude of London's [annual] grain requirements at this date' and that 'well over half the city's grain supply was brought by water'.[80] If we assume, say, that two-thirds of the grain was in fact brought by water, then this would yield a figure of 110,000 quarters coming to the city annually by water. But much would come from the east, particularly from Essex and Kent, rather than the west. If we say that half of the grain brought by water came along the Henley to London stretch, this would mean 55,000 quarters per year. The number of shouts and other vessels needed would depend upon the number of trips that could be made each year. It would appear that one return voyage from London to Henley per month would be a likely minimum.[81] If we assume a mean load of 120 quarters per shout, as in Table 2, and twelve return voyages per year per shout, then the number of vessels needed to carry 55,000 quarters would be $\frac{55,000}{(120 \times 12)} = 38.2$. If the average 150 quarters per vessel figure for those three cases which actually made the Henley to London voyage in the sample was used, the number of vessels would drop further to 30.7. These shouts may well have had to carry other goods down to London than just grain (for example, wood, iron goods, livestock products), as well as the occasional part load plus goods for other parts beyond London, but nonetheless 50 and certainly 100 vessels would seem plenty to handle the trade. This has implications for

---

[79] Of the 89,700 quarters requested in 1296, only 25,700 were sent on 36 ships to Gascony and Flanders in 1297, making it necessary to buy more grain in the Low Countries: Prestwich, *Armies and Warfare*, 251.

[80] Campbell et al., *A Medieval Capital*, 35.

[81] Wilson, *The Thames*, 29, suggests that the voyage from London to Henley normally took five days. It would seem unlikely that the downstream journey would have taken any longer.

the number of vessels, shouts or otherwise, which would have to be built per year. The boats plying the Thames in the middle ages seem to have had a working life of around seventy-five years.[82] If so, the construction of only one or two new boats per year on average would have been needed to maintain the 'fleet' required to handle the Thames water traffic upstream of London and perhaps the same for that downstream and in the estuary.[83] Exactly how the industry for building boats was structured is uncertain, but evidence from the bridgewardens' accounts for London Bridge in the late fourteenth century indicates that various shipwrights, most probably residing in London, were normally hired by piece work to construct new boats.[84]

As yet, the available evidence does not easily allow us to make the same sort of calculations and conjectures for other river systems, but further research and, above all, further excavation should continue to illuminate the picture. Certainly, we should be cautious in claiming too much for the figures presented in this study. It is clear, particularly from the work of Edwards and Hindle,[85] that there is a considerable amount of information to be added to that of purveyance, which will certainly add yet more detail concerning the capacity of the river system, and could well overturn some of the conclusions made about specific river systems here. We should, however, be very careful about not mixing evidence from various time periods. As pointed out by Jones,[86] river capacities could change rapidly over time, and the one advantage of using the purveyance accounts is that, as a source of evidence, they were restricted to a relatively narrow time frame. In any case, whatever its faults, one thing seems abundantly obvious from the evidence provided by the purveyance accounts. The inland water transport system was already, by the early fourteenth century, exceedingly complex, capable of making adjustments on several levels, from the types of boats used to the establishment of trans-shipment points. It could, if needed, seek out accommodations of considerable sophistication when the needs of local and national economies collided, as they did in the case of the Thames. The exact ways in which these adjustments and accommodations were fashioned are for the most part still a mystery, but that they existed at all is a testimony to the powers of resourcefulness and compromise in medieval society.

---

[82] Marsden, *Ships of the Port of London*, 24–5.

[83] I am grateful to Robert Peberdy for pointing out these implications.

[84] B. Spencer, 'Expenditure on Shipbuilding and Repair by London Bridge, 1382–98', and T. Dyson, 'Three Medieval London Shipbuilders', in Marsden, *Ships of the Port of London*, 209–15. A lay subsidy return for London in 1319 indicates that there was a cluster of up to eight shipwrights in the city's Tower Ward (just upstream from the Tower of London): E. Ekwall, *Two Early London Subsidy Rolls* (Lund, 1951), 346, 348–50. See also the shipwrights resident across the river in Southwark: M. Carlin, *Medieval Southwark* (London, 1996), 148, 280. I am grateful to Derek Keene for directing me to these last two sources.

[85] Edwards and Hindle, 'Transportation System', 124–6.

[86] Jones, 'River Navigation', see esp. 69–72.

Appendix

Table 5. Loads (in Quarters per Vessel) Carried on Various Rivers or River Sections

River (river section)	No. of cases	Mean load (qrs/vessel)	Scale (on Fig. 23)[a]
Ure/Ouse (from Boroughbridge to York)	2	22.2	1
Wharfe/Ouse (from Tadcaster to Kingston upon Hull)	1	40.0	2
Hull (from Wansford to Kingston upon Hull)	1	19.2	1
Hull (from Beverley to Kingston upon Hull)	2	40.0	2
Hedon (from Hedon to Kingston upon Hull)	1	42.0	2
Upper Trent (from Nottingham or Newark to Torksey, Gainsborough, or Kingston upon Hull)	7	48.4	2
Lower Trent (from Dunham, Torksey, or Gainsborough to Kingston upon Hull)	5	80.1	4
Idle (from Bawtry to Kingston upon Hull)	1	52.0	3
Witham (from Lincoln to Boston)	1	78.0	4
Nene (from Yaxley to King's Lynn)	6	55.4	3
Little Ouse (from Lakenheath to King's Lynn)	1	42.0	2
Cam and Great Ouse (from Cambridge or St Ives to King's Lynn)	9	56.7	3
Yare (from Norwich to Yarmouth)	9	50.2	3
Bure (from Wroxham to Yarmouth)	1	76.5	4
Waveney (from Beccles to Yarmouth)	1	100.0	4
Upper Thames (from Oxford to Henley-on-Thames)	1	61.4	3
Lower Thames (from Henley-on-Thames to London)	3	159.5	5
Lower Thames (from London to Westminster)	2	125.8	5
Lea (from Thele to London)	1	153.0	5
Upper Parrett (from Thorney or Langport to Bridgwater)	7	15.5	1
Tone/Upper Parrett (from Curry Bridge to Bridgwater)	2	14.6	1
Lower Parrett (from Bridgwater to Bristol (twice) and from Redland to Bridgwater (once))	3	54.7	3
Upper Severn (from Tewkesbury to Worcester)	5	26.4	2
Lower Severn (from Tewkesbury and Gloucester to Framilode and Bristol)	3	87.7	4

[a] A scale of 1 equals 0–25.0 quarters per vessel; 2 equals 25.1–50.0 quarters per vessel; 3 equals 50.1–75.0 quarters per vessel; 4 equals 75.1–100.0 quarters per vessel; 5 equals over 100.0 quarters per vessel.

PART II

IMPROVED WATERWAYS
AND CANALS

6

Identifying Human Modification of River Channels

ED RHODES

Water flow in river and stream channels is not a simple phenomenon, and its study has developed into the discipline of hydrology. Flowing water is in an ever-changing relationship with the bedrock or unconsolidated sediments through which it flows, and with the sediment load being transported by the moving water. However, some useful generalizations can be made about the pattern and behaviour of river flow, the channels in which flow occurs, and the nature of sediments deposited by flowing water. It is these general principles which may be of interest to a non-technical audience; in particular, the way in which the study of contemporary river channels may help elucidate previous human interaction or interference with these structures.

In this short chapter, I do not intend to provide a full introduction to the principles of hydrology, fluvial geomorphology, or sedimentology, but rather I aim to raise the level of awareness of relevant research directions in a different discipline. I highlight the potential of using some of the ideas suggested below to assist in the identification, either putative or definitive, of past human modification of fluvial channel systems, particularly when combined with other research efforts such as archaeological excavation, place-name investigation, and historical study based on documentary or cartographic evidence.

I explain the limitations of the geomorphological concepts which are introduced and the approaches to determining the degree and timing of past human channel modification. I hope to stimulate the reader into attempting to place historical or prehistoric events into a more complete environmental setting, in which human events are viewed within the wider context of landscapes and landscape evolution, occurring as a result of natural or human events, and in most cases probably a combination of the two.

River Channel Function and Form: How Rivers Work

Two distinct approaches to fluvial geomorphology, that is the study of the landforms associated with rivers and streams, have developed. These are the theoretical and empirical approaches. In the former approach, predictions of the flow of water, and its effect on sediment, are constructed from relatively simple hydrological equations such as the Raleigh equation or Stokes Law.[1] In the empirical approach, studies of patterns of landform distribution, for example the complex interlinked channels of ephemeral fluvial systems in the New South Wales desert, Australia,[2] are performed, in order to draw out general principles relating to the behaviour of river systems. Huge quantities of field data have been collected relating to such features (or variables) as channel gradient, channel discharge (volume of water flow), sediment transport, channel pattern, and many more. Current understanding of the behaviour and morphology of fluvial systems is based on an integration of these two approaches, at different spatial and time scales. The discipline is an active area of research, and many of the key concepts are relatively immature; ongoing and future research has the potential to overthrow or significantly modify some of these ideas.

One of the first and most basic concepts of fluvial geomorphology is that of channel form. Rivers develop their drainage networks over long periods of geological time, superimposed on the backdrop of the geological conditions of the basin, and in response to changing environmental and climatic conditions. In general, three distinct channel types are recognized, termed meandering, braided, and anastamosing (see Fig. 24). The first type typically consists of a single sinuous channel, which flows through a series of tortuous bends which often locally reverse the direction of flow. These systems will probably be familiar to readers, as they are often found in lowland Britain, as well as being typical of tropical rivers. Meandering channels are associated with low channel gradients, generally slow water flow, a predominance of fine-grained (silt and clay) sediment, and a relatively invariant discharge (water flow).

In contrast, braided river systems comprise many sub-parallel channels which show a high degree of interaction, separated by sand and gravel bars. They occur where channel gradients are steep, in areas where flow is variable through time, and in association with the transport of significant loads of coarse-grained sediment (sand, gravel, and boulders). They tend to be unstable, the channel positions changing in response to changes associated with high-flow events such as floods. They are typically found in highland areas, as glacial outflow rivers, and across the surface of large desert alluvial fans.

The third type of channel pattern recognized is termed anastomosing or anabranching. However, this group is much more poorly delineated than either of the two above, and classified into this group are systems with rather different characteristics. Huge ephemeral fluvial systems in deserts in many cases are only activated on a decadal or centennial time scale (possibly longer), and can form an interlinking network of many channels, often terminating

---

[1] D. Knighton, *Fluvial Forms and Processes: A New Perspective* (London, 1998).

[2] G. C. Nansen, R. W. Young, D. M. Price, and B. R. Rust, 'Stratigraphy, Sedimentology and Late Quaternary Chronology of the Channel Country of Western Queensland', in R. F. Warner (ed.), *Fluvial Geomorphology of Australia*. (Sydney, 1988).

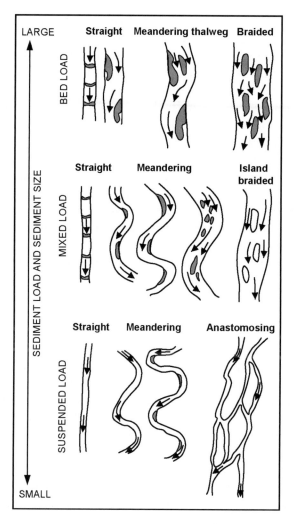

Fig. 24. Classification of channel pattern (redrawn and modified after Knighton, Fluvial Forms). Bars within channels are shown in grey.

within the desert as flow dwindles through transmission loss.[3] In contrast, the geological record provides evidence for very stable, multi-channelled fluvial systems flowing though fine-grained (silt and clay) sediment, where channel positions are retained over considerable time.

This last pattern is of interest, as in some ways it shares characteristics with the rivers of lowland Britain. Multi-strand channel systems, flowing apart and meeting occasionally, with some degree of meandering within individual channels, but with reaches that have a more braided nature with sand and gravel bars, are found throughout lowland Britain. The origin and geological

---

[3] Knighton, *Fluvial Forms*.

evolution of these systems are currently a matter of research interest, and it is possible that the degree of human modification of these systems has been previously underestimated. One possibility is that they represent the adjustment to a lower level of environmental activity within the context of relative climatic stability within the Holocene period (the last 11,500 years, approximately), superimposed on the relict geomorphology of a much higher-energy regime associated with more severe climatic conditions of the Younger Dryas stadial (roughly 13,000 to 11,500 years ago) or even from the Last Glacial Maximum (LGM) of around 20,000 years ago. During these more dramatic and harsh climatic conditions, when the environment of much of lowland Britain was sparsely vegetated arctic tundra, and considerable floods from the spring melt of significant snow accumulation were common, many lowland valleys were occupied by high-energy, braided fluvial systems, with much higher discharges than under contemporary climatic conditions. The remnant features of these systems (e.g. large gravel bars) may still impose a significant influence on the evolution and development of modern lowland river organization.

Summarizing the above section, it is not entirely clear to fluvial geomorphologists exactly what forces and processes have governed the evolution of lowland British river systems. The common pattern of multiple sub-parallel channels, each having a tendency to develop meandering reaches, does not fit simply into a single large-scale classification of channel morphology. This emphasizes the point made above, that rivers are inherently unstable and dynamic. A river may be in equilibrium with its environment, but this does not mean that its channel locations are stationary. In any one drainage system, the characteristics of the channel form may vary from one location to another, and at any one location, different channel types may evolve and replace each other through time. As such, rivers are good examples of complex systems, in which a rich array of different behaviours may emerge from the interactions of interdependent controls and responses. In such complex systems, useful prediction of future response using techniques such as numerical modelling has intrinsic limitations, as does the use of similar techniques to estimate details of rates of change in channel morphology. This latter point is of particular interest, as it may in some cases hold the key for estimating the timing of human intervention in channel morphology, and is discussed further below. Rivers tend to behave as if they have an unconscious desire to achieve what may be termed a minimum energy configuration. It is not easy to predict the effects of this in detail, but they usually include, amongst other features, a smooth downstream profile.

Processes that Control River Channel Form

Some of the reasons that rivers have a tendency to adopt specific channel forms have been introduced above. Specifically, these include geology (rock types, or geological structures such as fault zones) and geomorphology (relief, slope angles, drainage network density) as invariant or only slowly changing factors; soils and vegetation as more rapidly varying factors (varying over the centennial to millennial scale); and climatic variation, which operates on all time scales including the most rapid (i.e. hours or minutes). Another important concept is the 'flashiness' of the hydrological regime to climatic variation, that is how rapidly variations are reflected in changes in discharge, and in a similar manner, the variability of the climatic regime itself. Clearly this list includes a large number of factors, of which many are interdependent, such as the dependence of vegetation status on climate and soils, or of the geomorphology (steepness of slopes, drainage density, etc.) on long-term climatic trends. Local and short-lived factors can clearly play a role in this highly complex web of control and response. However, the fact that geographically related valleys contain rivers with many similar characteristics to each other demonstrates that these details do not control all aspects, and regional trends are significant.

Perhaps more significant for the issue under consideration in this chapter are the processes that are operating on a more local scale of metres to hundreds of metres. It is on this scale that we may be able to distinguish features of channel morphology which may suggest or indicate human intervention. The processes that control channel morphology and the response of the fluvial system to changes in any of the controlling variables are intimately linked to sediments, and the interaction between water and sediment (and in some cases vegetation also plays a significant role).

In lowland Britain, river valleys tend to be characterized by a relatively flat surface (the floodplain) cut by one or more distinct channels where water flow is concentrated except during floods. Geological investigations often reveal a significant thickness (often 5 m or greater) of sediments underlying the active channel(s) and floodplain, deposited by flowing water, and separating the flow from bedrock. On either side of many valleys remnants of similar sediments are preserved at higher altitudes. These 'terrace' deposits represent the valley fill of a former fluvial system flowing at a higher altitude, before erosion to the present level took place. These fragments are of some significant interest: as the sediments provide evidence for the type of rivers flowing previously, they may contain archaeological or biological records and may often be dated using radiocarbon[4] or optically stimulated luminescence dating.[5] Sediments found down a river tend to be rather similar, with a gradual increase in clasts (pieces such as sand grains, boulders) from the local geological formations as each is traversed by the river. This sediment mixing has a tendency to reduce sharp variation in channel behaviour as different rock types are encountered.

---

[4] See as an example P. L. Gibbard, G. R. Coope, A. R. Hall, R. C. Preece, and J. E. Robinson, 'Middle Devensian Deposits beneath the "Upper Floodplain" Terrace of the River Thames at Kempton Park, Sunbury, Surrey, England', *Proceedings of the Geologists' Association*, 93 (1982), 275–89.

[5] See as an example N. K. Perkins and E. J. Rhodes, 'Optical Dating of Fluvial Sediments from Tattershall, UK', *Quaternary Geochronology (Quaternary Science Reviews)*, 13 (1994), 517–20.

At any one time within a single channel, water will be both eroding and depositing sediment. The locus of erosion will tend to vary in response to flow conditions, the nature of the bank and bed, and the load already being transported. Deposition may occur nearby (e.g. within a few metres or tens of metres), or further downstream. Finer material (usually silt and clay only) may be transported out to sea, where eventual deposition will incorporate it into marine deposits. Clearly fluvial sediment transport will always convey material downstream, and there will be a net gradual erosion within the river catchment. However, at any one location, deposition may dominate, depending on the subtle interplay of these finely balanced factors.

Transport of material (predominantly the inorganic products of bedrock weathering and erosion, but organic material such as tree trunks is also carried) occurs in three ways: as dissolved material (e.g. dissolved carbonate in a limestone area); as suspended loads such as clay and silt, kept in suspension by the turbid action of the flowing water; or as coarser material either rolled along the bed (bedload), or "bounced" along it by a process known as saltation. Boulders, pebbles, and sand grains may spend significant periods within constructional bedforms such as ripples or dunes within the channel between active periods of transport during the higher-flow regimes.

For material to be transported by water, the water must have sufficient carrying power to entrain it. The carrying power of flowing water rises dramatically with velocity, which is why surprisingly large quantities of coarse material can be moved by fast-flowing flood waters. Sediment load entrainment is perhaps best represented by the Hjulström diagram, which relates the velocity of water flow to its ability to erode, entrain, and transport particles of different grain size (Fig. 25).[6] These processes take place when velocity thresholds are crossed: when the water velocity drops (e.g. water entering a lake or body of slower-flowing water), the sediment load will be deposited. Note that the erosion of silt and clay requires more energy (greater velocity of flow) than does that of sand. This is because of the cohesive forces acting between the grains of silt and clay, in other words, its "stickiness".

The morphology of a channel can only be maintained if the delicate balance between erosion and deposition within that reach is maintained. The introduction of excess material from upstream may lead to deposition within the channel, and possibly an associated shallowing of the channel, while a reduction in sediment load from upstream (e.g. by the construction of a dam or weir) may lead to increased erosion, bank collapse, channel widening, and other processes. Even where no change in sediment supply or erosion occurs, gradual migration of the channel can occur. This is a normal property of meandering channels, as witnessed by the frequent fully or partially cut-off meander loops observed on floodplains.

[6] F. Hjulström, 'Studies of the Morphological Activity of Rivers as Illustrated by the River Fyris', *Bulletin of the Geological Institute of the University of Uppsala*, 25 (1935), 221–527.

The reason why meander migration occurs, at least in a general sense, is the key to the understanding of the causes of channel form change. This is that the water velocity at different parts of a channel that is turning through a bend is not constant. In other words, water will speed up and slow down as it negotiates a bend or meander loop. The way in which the water changes in velocity is not entirely simple, and beyond the scope of this discussion, but relates to flowlines within the water body and the frictional forces within the water and with the bank. The outermost part of a bend tends to have significantly higher water velocities than does the inner part, as well as being deeper. The faster-moving water causes erosion of the bank material as the threshold to entrainment and transport are crossed by the faster water (Fig. 25). Hence, erosion occurs at the outside of the bend, while at the more slowly moving inner part, the water tends to deposit its load, leading to the build-up of a bar.

If the above point is appreciated, it can be seen that only in very special circumstances will there be no tendency to erode or deposit sediment within a channel. These special circumstances would require that even the highest velocities achieved within a channel were lower than the threshold required to erode the bank material, or that at all points within the channel the velocity was constant. Any disturbance to the flow (bank irregularity, bend, bridge pier, submerged tree trunk) will cause flow irregularities and an associated variation in flow velocity. In a straight channel section, a temporary small localized flow disturbance may lead to bank erosion, which then persists as an independent flow disturbance. Once such a disturbance has been created, it is not likely that the flow regime of the channel will do anything but exaggerate it, as the greater the magnitude of the disturbance, the greater the difference in velocities and the greater the erosion. In this manner, even a very minor disturbance can eventually lead to the development of a meander loop owing to the positive feedback system which pertains.

The key concepts here are the erodibility of the bank material, that is how easy to erode the material is, and the erosivity of the water, that is how easily

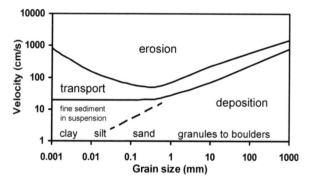

Fig. 25. Hjulström diagram, showing erosion and deposition criteria defined in terms of grain size and flow velocity. Zones for erosion, deposition, and transport are shown.

it can erode the bank. The erodibility of the bank can be reduced significantly in a typical lowland British context where the bank is constructed (naturally) dominantly of silt, clay, and loam (a mixture of silt, clay, and sand), and by vegetation such as tree roots, grass, and other plants. The action of animals such as cattle drinking or crossing, or of small mammals burrowing into the bank material, can lead to a significant reduction in erodibility. In other words, the river valley ecosystem, and any management strategies applied to that ecosystem or parts of it, can be important determinants of the bank stability and hence the nature of the channel.

The erosivity of water flow is enhanced at greater velocities. Consequently, a significant proportion of the erosive action of rivers takes place during flood events. Floods of higher magnitude take place less often, and it may be these rare events (one per century, or even less often) that are responsible for a great deal of the detail of the channel morphologies we observe today.

In summary, river channels are in a continual state of erosion and deposition at different locations within the channel. The location of channels can migrate across the floodplain, and similar processes can lead to the development of meander loops in formerly straight channels. It is difficult to provide theoretical estimates of the rate of these processes, as details such as bank stability are important factors, and much change can occur during a small number of high-magnitude flood events.

Human Channel Modification

Much of the relevant empirical research undertaken on small temperate rivers has been performed in the UK.[7] Sometimes there is an implicit assumption that the river channels are in their "natural" state, with the exception of obvious major human intervention such as the diversion of flow into a mill-race. However, the concept of "natural" is a very unclear one, in a context where a high degree of the sediment transported by British rivers represents soil erosion related to ploughing (since the advent of farming in the neolithic), and down channels which have been the focus of much economic activity including transport, fishing, and water power. Of course, for some studies, the underlying causes of river behaviour may not be directly relevant (e.g. flood prediction). However, we should be very cautious of characterizing the pattern of channels in any British river as "natural".

Some studies chart changes that have occurred over relatively recent periods when reasonable records of any intervention may be available.[8] Fig. 26 shows

---

[7] For examples see R. H. Johnson and J. Paynter, 'The Development of a Cut-off on the River Irk at Chadderton, Lancashire', *Geography*, 52 (1967), 41–9, and R. I. Ferguson, 'River Meanders: Regular or Random?', in N. Wrigley (ed.), *Statistical Applications in the Spatial Sciences*, (London, 1979).

[8] For examples see J. M. Hooke and C. E. Redmond, 'Causes and Nature of River Planform Change', in P. Billi, R. D. Hey, C. R. Thorne, and P. Tacconi (eds.), *Dynamics of Gravel-Bed Rivers*, (Chichester, 1992).

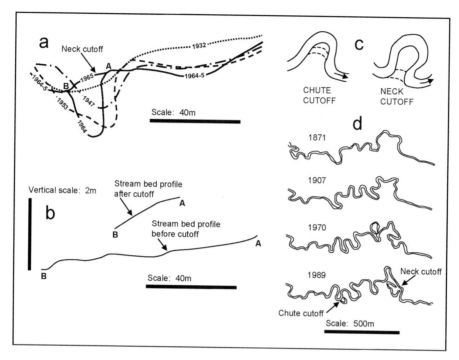

Fig. 26. Evolution of channel position, and cut-off development (redrawn and modified after Knighton, *Fluvial Forms*). (a) River Irk, Lancashire (after Johnson and Paynter, 'Development of a Cut-off' op. cit. note 7), showing changes prior to the cut-off. (b) The same, showing change in stream-bed profile before and after the cut-off. (c) Types of cut-off. (d) Sequence of changes on the River Bollin, Cheshire (after Hooke and Redmond, 'Causes and Nature').

the results from studies[9] which have monitored the development of channel and meander form and profile for two British rivers in the north-west of England, the Irk and the Bollin. In particular the study by Hooke and Redmond of the Bollin (Fig. 26d) provides an excellent series of snapshots in the evolution of meanders over a period of approximately 120 years. For this small, slow-flowing river, with heavily vegetated clay-rich banks, significant reorganization has clearly taken place over this time period.

Bearing in mind the conclusion of the last section, it may be possible to comment in specific or simply in general terms as to the degree of direct human intervention in a given river channel, based on the identification of features which are discordant with the state of other parts of the same river, or similar-scale nearby rivers.

I first consider the reasons why people should have made the effort to modify river channels in the past. Table 5 lists some of the main reasons for human

---

[9] These are Johnson and Paynter, 'The Development of a Cut-off', and R. I. Ferguson, 'River Meanders: Regular or Random?', in N. Wrigley (ed.), *Statistical Applications in the Spatial Sciences*, (London, 1979).

Table 5. Reasons for modifying a water-course

Mill construction
River crossing
Bridge
Ford
Ferry
Improve navigation
Increase depth
Stabilize banks
Increase width
Reduce sinuosity
Cut new channel
Fishing
Reduce erosion
Structural reinforcement
Reduce loss of farmland
Flood protection
Defence
Provide water supply
Provide drainage
Modify boundary
Modify flow rate
Mitigating the effects of channel modification
Aesthetic reasons

intervention. This may be a useful first step in considering rivers in a historical context, as the locations of many of these activities may be relatively well defined, such as river crossings and mills.

Examples of Suggested River Channel Modification

Illustrated above (Fig. 26d) was the development of meander loops and cut-offs for a small river, the Bollin. This is a tributary of the larger River Mersey which, to the south of Manchester, formed the pre-1974 Lancashire–Cheshire boundary. The main channel is represented by a series of wide sinuous meanders, and cuts through what was marshy floodplain before drainage for agricultural and recreational use. Large levees have been constructed as flood defences. The administrative map of the region shows the former county boundary, also following a series of wide sinuous meanders, with the occasional redirection. However, the boundary does not ubiquitously follow the present river course, but at some points follows its own route across the floodplain in a series of sinuous curves (Fig. 27). Even northern doggedness seems an untenable explanation for this misfit, and it appears more likely that the positions of some of the meanders have migrated, probably without significant intentional human intervention, since the boundary was established, perhaps in pre-Norman times. These observations highlight the fact that many possible responses to modification can take place over centennial time scales, and our ability to recognize

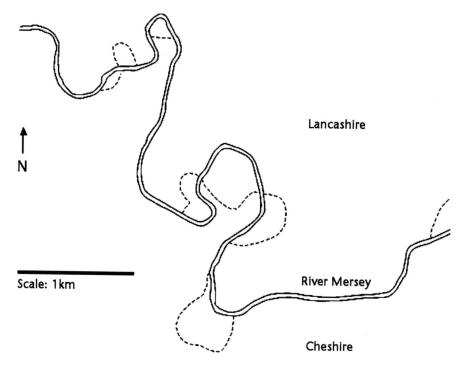

Fig. 27. The River Mersey south of Didsbury, formerly Lancashire, showing the Lancashire–Cheshire county boundary where it did not follow the modern river course. The scale is approximate.

modification which was engineered on the millennial time scale may be limited in many contexts.

It seems likely that the potential malign effects of modifying existing river channels may not have been obvious to former generations. Indeed, recent engineering work sometimes appears to have been planned in the absence of a clear understanding of the consequences. The time scale of response to modification was in many cases probably long enough not to alter the planning of engineering works significantly, as the builder was unlikely to be directly responsible for future repairs or upkeep.

Where channels were straightened, possibly to improve flow, but more likely to improve navigation ease and reduce distance, the gradient of the resulting channel is inevitably steeper than the reach it replaces (Fig 26b). As a result, water velocity is likely to be increased through the channel, possibly leading to erosion and widening, while silting might occur just downstream, where the water slows again. Unexpected consequences such as increased flood risk may have required the construction of further modifications. Changes in the flow regime can lead to silting or narrowing of what were formerly relatively major channels.

The identification of human channel modification is primarily a matter of finding discordant geomorphological features or situations, that is to say, situations which are not likely to arise by natural agency alone, as they are unstable. Three examples of this are often very clear, namely straight channel sections, stepped downstream profiles, and channels retained behind banks or levees at a height above that of the floodplain surface. Note that all of these conditions (straight channels, stepped profiles, and raised levees) can and do occur naturally, but it is the scale of these features which is important in the identification of human intervention.

The first of these is probably the easiest to identify, most successfully using maps. Aerial photography is also possible as a tool, but is probably generally less clear. One risk with maps is the possibility that the mapmaker will draw a water-course 'as it ought to be', following implicit expectations rather than faithfully recording anomalies such as variable channel width, secondary channels etc.

County Ditch in Yeoveny, Staines, now Surrey, takes water to the Thames from the complex of channels that comprise the River Colne. This ditch formed the boundary between Middlesex and Buckinghamshire before nineteenth-century political reorganization. The boundary (which is now a minor admin-istrative boundary) undergoes a series of tight turns, several of which appear to form a rather loose alignment with the more northerly straighter section. This is shown with a broken line in Fig. 28. This part of the ditch has been

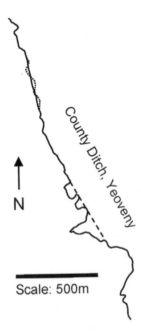

Fig. 28. County Ditch, Yeoveny, Staines (Surrey). This stream was formerly the boundary between Middlesex and Berkshire. The close dotted line in the upper part of the figure represents the administrative boundary; the dashed line in the lower part illustrates the inferred former stream course before the meanders were established. Large parts of this stream have now been removed by gravel extraction and motorway construction.

entirely removed by gravel extraction and subsequent construction of the M25 motorway. However, further north the ditch has been straightened between two gravel pits, while the boundary gently oscillates from side to side, reflecting the former course of the ditch shown as a dashed line. Interestingly, the 6-inch map of 1881 shows a pattern of meanders and channel location which is indistinguishable from those shown on the modern map.

Examples abound of straight channel sections, or sections that appear to have been straightened in the past, and a few have been highlighted here. The area around Oxford has been used for most of the examples presented here, but they are common on many British rivers. Two very pronounced straight reaches are the Great Brook, south of Bampton (Oxon.) (Fig. 29a: see also Fig. 64, and Blair, below pp. 272–8), and the New River Ray skirting Otmoor (Oxon.). The name of the latter helps elucidate its anthropogenic origin. The 1884 OS 6-inch map of Otmoor shows a very similar pattern of channels to those seen on the modern 1:25000 map, though there are some minor changes both in channel location and, more significantly, the apparent relative importance of the (now minor) channel flowing south-west from the start of the New River Ray. This channel forms a series of large open bends, and the pattern of field boundaries shown on the 1884 map hints at these bends having developed within a channel which was formerly artificially straightened into two reaches heading approximately south-west across Otmoor. A further minor channel appears on the modern and 1884 maps, initially heading south-east from the start of the New River Ray, but forming a series of large regular meanders around 500 m across until it too becomes canalized into a series of straight reaches which take it south then west to join the other two channels.

North of the town of Thame (Oxon.) the River Thame is met by a tributary flowing from the east, shown in Fig. 29b. Over a distance of around 1,000 m, the present main channel of this tributary runs in a pronouncedly straight course, with a number of small sinuous bends superimposed. Field ditches link this channel to a smaller channel which flows to the south, then crosses to flow to the north, meeting the Thame upstream of this main channel. Where this smaller channel is north of the present main channel, it is followed by the county boundary between Oxfordshire and Buckinghamshire. It seems likely that the now smaller channel may at one time have been the main watercourse, and that a new straight channel was dug. This may be related to Scotsgrove Mill, located at the east end of this channel on Fig. 29b. The small bends have presumably developed since the channel was constructed. These bends appear very similar in 1885, suggesting significant antiquity to this construction. At that time, the straight channel did not reach the Thame as shown in Fig. 29b, but veered south beside the Aylesbury Road (not shown) to Thame Mill.

Pronounced changes in the sinuosity of a channel may indicate a modification of one of the contrasting sections. While in most cases this is more likely to

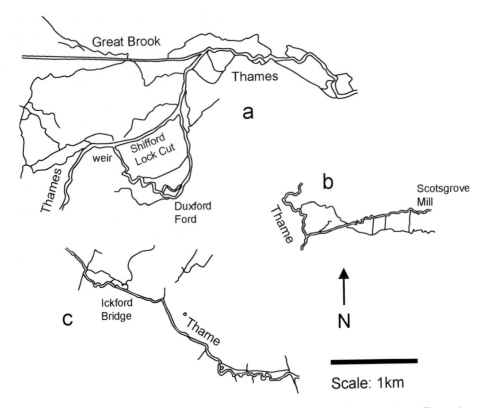

Fig. 29. Watercourses in three locations in Oxfordshire and Buckinghamshire, illustrating a variety of situations: see text for details. (a) The junction of the perfectly straight Great Brook with the Thames south-east of Bampton (Oxon.). (b) The junction of the River Thame, flowing southwards, with a tributary flowing from Scotsgrove Mill, north of the town of Thame (Oxon.). The main channel of the tributary forms the county boundary from Scotsgrove Mill westwards to the junction with a minor channel; it then follows this channel north-west to the junction of the channel with the Thame. (c) The River Thame between Shabbington and Ickford (both Bucks.), which here marks the county boundary with Oxfordshire.

represent a straightening of the less sinuous part, it is possible that flow instability introduced by an obstruction may have caused increased erosion and hence sinuosity, or that a straightened section will develop bends in order to achieve a reduced gradient. Two clear examples of sharp changes in sinuosity are included in Fig. 29. The first is from the Thames south-east of Bampton (Oxon.), close to Chimney. Here a wide meander loop of the Thames has been bypassed with the construction of the Shifford Lock Cut (Fig. 29a). The name of this watercourse identifies it as anthropogenic, and note also the inclusion of a lock at the downstream (eastern) end of this cut, presumably to reduce problems of increased flow rate that would have occurred had the lock been omitted. This cut is not present on the 1884 6-inch map, though the minor channel immediately to its north may have been more significant then.

Interestingly, a weir is included on the meandering loop section, presumably to help maintain suitable water depths for the lock. The meander loop itself comprises a series of tight bends as it heads south to the perhaps tautologically named Duxford Ford. On the return northward reach, the channel is much less sinuous, suggesting that this part has received the benefit of invasive engineering works which contrast with the westerly reach. These sections appear very similar in 1884. It is interesting to speculate as to the reasons for the construction of Shifford Lock Cut. The distance is reduced only by about 1,000 m by taking the new route, with the addition of the time costs of traversing a lock. Also, the lock would need to be constructed, maintained, and staffed, so this presumably represented a significant perceived advantage. Possibly the ford was difficult to navigate, or this section contained other weirs, and this may have been an added incentive to the construction of the cut. It would be interesting to investigate these possibilities using historical records. Within the meander loop is a series of drainage ditches, the southernmost of which appears possibly to represent an earlier bypass of just the tip of the large meander loop, similar in 1884. A minor channel to the south-west of the loop also hints at a wider loop at some time in the past.

The second example of pronounced sinuosity change is from the River Thame between Shabbington and Ickford (Fig. 29c). The upstream part of this reach is highly sinuous, while to the north a much straighter minor ditch parallels the main trend of the river. Suddenly, the river turns and appears to join this ditch, continuing towards Ickford Bridge in a series of arcuate stretches of around 200–500 m each. This section has a significantly lower sinuosity than the first. Before the bridge, the river turns sharply left through 90 degrees, after a straight stretch of around 400 m. After turning right through around 60 degrees, the river then follows an approximately straight course through Ickford Bridge. Another minor turn to the right brings the river into a remarkable section where seven short straight reaches are traversed, each at 90 degrees to the last. This gives the appearance of the river occupying what may have been constructed as field boundary ditches of a rectangular field system, or possibly water meadow system. This latter section certainly appears to be very far from any arrangement that such a river is likely to adopt in the absence of human interference. A casual inspection of this location on the ground, to see if remnants of earlier channels survive as slight depressions in the land surface, revealed no obvious channels, though these can become silted up extremely quickly under conditions prevailing in the UK. The relationship of these features to the historic Ickford Bridge, and to a small earthwork beside the causeway to the north of the bridge, might well be documented in local sources. Also marked on the OS 1 : 25,000 map is the name "Whirlpool Arch" at Ickford Bridge. This perhaps provides some independent, if weak, evidence of human channel engineering, creating (probably unintentionally) a whirlpool. The apparent high degree of manipulation of this reach of the Thame may represent an attempt to rectify this problem, perhaps originally caused by the construction of the bridge. The pattern of channels observed on the modern map is essentially the same as those observed in 1885.

Around Oxford, there are several features of interest in the arrangement of river channels and ditches (Fig. 30). Between higher ground to the east including Rose Hill and Headington Hill, and to the west including Hinksey Hill, Boar's Hill, Cumnor Hill, and Wytham Hill, is a large low-lying area, comprising the modern floodplain of the Thames, and the partially eroded remnants of Pleistocene terraces at slightly higher altitudes. Oxford was founded at the southern tip of a large "peninsula" of gravel formed by the Summertown–Radley terrace, with the River Cherwell to the east, and the larger River Thames to the west and south. In this setting, there can have been many different reasons for modifying the channel or channels, including defensive structures, navigation, positioning of mills and the creation of suitable heads of water to drive them, drainage for agriculture, water supply for watermeadows or industry, improving access across the floodplain with roads and tracks, establishing efficient and functioning river crossings, and controlling flood risk.

Perhaps the most striking aspect of this plan of the contemporary watercourses (see also Blair, below, p. 268 and Fig. 62) is the apparent alignment of Wytham Stream, also known as Seacourt Stream, to the north-west, with Hinksey Stream and the River Thames to the south-east, extending almost as far as Sandford Lock. This alignment has been highlighted with a broken line in Fig. 30. Whether this represents the remnants of a single channel, dug straight over several kilometres, is not clear, but it certainly seems highly unlikely that this alignment could occur coincidentally from the chance locations of different channel sections. Whether there is any evidence that the channel ever existed in what is now Osney Island would be an interesting hypothesis to test by excavation. This may represent part of an organized field system, bounded by ditches, perhaps constructed parallel or sub-parallel to a linear feature such as a former Roman road, perhaps now buried in alluvium.

Osney Island itself is striking, being bounded by a series of channel fragments which have the appearance of formerly forming an almost perfectly circular arrangement, known in places as the Bulstake Stream. The minor Roman road which passed between Boar's Hill and Cumnor Hill, and through North Hinksey,[10] presumably passed close to the centre of Osney Island. Possibly the Bulstake Stream represents a defensive structure built around a later settlement (Osney), although it is possible that an earlier structure was influential in the routing of the Roman road. These are fascinating speculations, generated by an inspection of the modern water courses. However, evidence to support these suggestions is likely only to come from documentary sources or archaeological excavation.

---

[10] see J. Blair, *Anglo-Saxon Oxfordshire* (Stroud, 1994), p. xvi.

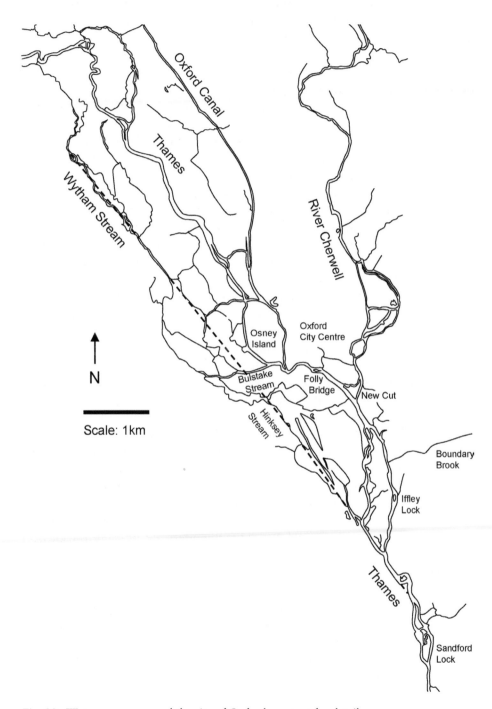

Fig. 30. Watercourses around the city of Oxford: see text for details.

After the late tenth century the main channel of the Thames was apparently redirected to its present course under Folly Bridge, south of Oxford city centre from a position further west, close to the present Hinksey Stream.[11] A series of rather wide linear sections take the present Thames to an early confluence with the Cherwell. Most of the latter river's flow is now taken by the New Cut constructed in the mid-nineteenth century. What may have been approximately the location of the former River Cherwell to the south of this position is represented today by a series of channels known as the Weirs Mill Stream. The Thames continues towards Iffley Lock. Examination of this reach of the river in the field provides excellent evidence of it representing a modified channel. The east side of the channel cuts into deposits of a higher terrace, forming a low grassy cliff up to around 1 to 1.5 m in height, while on the west side the channel is retained within a large bank. The level of the water surface is significantly above the level of the fields to the west of the channel, an area prone to flooding, just upstream of Iffley Lock. This is clearly not a situation that is stable, and is very unlikely to arise without human intervention. Sometimes, the original position of a river can be maintained, as the surrounding land surface is lowered, by the removal of an economic mineral such as sand and gravel or peat. This situation was presumably engineered at least in part to provide a good head of water to power Iffley Mill, formerly situated at Iffley Lock, though the location of this mill might equally represent the taking advantage of an existing situation.

Boundary Brook forms the administrative boundary between East Oxford and Cowley. From Fig. 30 it is clear that the lower parts of this brook are composed of two unnaturally straight sections, with a single rather large deviation to the south where the Oxford to Henley road (Iffley Road) crosses it. This water-course is directed towards the point known as Redbridge, where a minor Roman road, which crossed the top of Boar's Hill and ran down Hinksey Hill, crossed the Thames. The late Anglo-Saxon bounds of Kennington and Hinksey mention a 'stone ford' (*stanford*) in the vicinity of Redbridge,[12] and an ancient causeway survives which in some ways resembles the causeway at Grandpont, 2 km to the north. Possibly, Boundary Brook may represent a residual watercourse developed from one of the ditches of a former Roman road which joined that mentioned above. The alignment of the sections on either side of the deviation at the road crossing may suggest some antiquity to this ditch.

The final example presented here comes from the River Thames between Windsor (Berks.) and Runnymede Bridge (Surrey). Here a road follows the edge of the river. At two places, both the road and the river deviate towards the south. At both these locations the road appears to be in alignment on either

---

[11] Ibid. 104; cf. A. Dodd (ed.), *Oxford before the University* (Oxford, 2003), ch. 3.

[12] S. E. Kelly, *Charters of Abingdon Abbey*, ii, (*Anglo-Saxon Charters* VIII (Oxford, 2001), 238–41; Blair, *Anglo-Saxon Oxfordshire*, 87.

Table 6. Features which can act as indicators of human modification

These features can be found by
    Fieldwalking
or examining
    Maps
    Aerial photographs
    Satellite photographs

Straight channel sections
Silted former channels
Bank reinforcement
Building materials in channel
Channel not following *Thalweg* of valley
Stepped downstream profile
Sudden directional change
Pronounced change in sinuosity
Change in number of channels
Alignment of channel sections with other landscape features
    Roads
    Settlements
    Fords
    Bridges
    Ferries
Pairs of parallel channels

side of the diversion. This suggests that the two aligned sections of the road in each case may have been constructed as a single straight section, and that these deviations may have developed well after road construction. This relationship between the river and road may provide some indication of the rate and nature of channel modification, probably in this case taking place without significant human intervention, and gradually causing new routes for the road to have to be built.

Summary and Suggestions for Future Research

Table 5 summarizes some of the reasons why channel modification may have taken place in the past. Table 6 summarizes some of the features which can act as beacons to help identify such human modification. The examples provided above may be useful in illustrating the scale and nature of such modification. These examples do not necessarily reflect channels that have definitely been modified, and it would be very interesting to explore some of these using additional techniques. It seems rather likely that very few of the river channels in the UK have not been modified to a degree at some point in their history. The underlying causes of the common pattern for UK lowland rivers, comprising several channels, is still not completely clear, and a significant role for human input should not be discounted.

New scientific techniques exist that might be applied to assist in disentangling these complex landscape/human relationships. Palynology, or the study of pollen, allows landscapes to be reconstructed from the vegetation contributing to pollen at different levels within a core collected from lake sediments or peat. This can be complemented by the study of mollusca, coleoptera (beetles), chironimids (midges) and plant remains. Radiocarbon dating can be undertaken on plant remains to provide a firm date for events such as the initiation of sedimentation in an abandoned channel. Recent developments in luminescence dating,[13] in particular OSL dating of fluvial sediments, may allow significant clarification of the age relationships of different sedimentary components of fluvial systems, and possibly sediments deposited in anthropogenic structures such as ditches. This technique has the advantage of being applicable to the sediment itself, and does not require the presence of a special component (such as plant remains).

Through future interdisciplinary collaborations there is the possibility to complement more traditional studies of landscape evolution and development, and elucidate the magnitude and timing of human input. Interpreting the modification of channels in an archaeological or historical context may lead to significant new avenues for investigation, throwing light on the scale of regional and local political organization during time periods when specific documentary sources are relatively scarce.

[13] M. J. Aitken, *Introduction to Optical Dating* (Oxford, 1998).

7

Canal Construction in the Early Middle Ages: An Introductory Review

JAMES BOND

Popular literature on English canals usually presents two figures as the pioneers of inland navigation: Francis Egerton, third Duke of Bridgwater, who is supposed to have abandoned the fashionable London social circuit as a result of being crossed in love, instead diverting all his energies into the exploitation of his estates and coal mines at Worsley; and James Brindley, the millwright's apprentice from Tunstead in Derbyshire who, despite his humble background and limited literacy, by his own native genius provided the Duke with a means of water-borne transport. Brindley's engineering of the Duke of Bridgwater's Canal over the 16 km between Worsley and Manchester between 1759 and 1761, which included the construction of two embankments and a short aqueduct, was a genuinely courageous and innovative achievement, and it paved the way for the golden age of canals over the following eighty years.

Significant though the Duke of Bridgwater's Canal was for its influence upon the course of the Industrial Revolution, however, it was not the first artificial inland navigation canal to be constructed in England; nor was it even the first to eschew a course along a river valley. Several smaller canal projects had successfully been completed during the previous century and a half, including the ambitious Exeter Ship Canal, constructed between 1698 and 1701, 8 km in length, able to accommodate coasters and small seagoing vessels of up to 150 tonnes. This was itself the successor to an earlier artificial waterway constructed in 1564–6, containing three pound-locks, accommodating boats of up to 16 tonnes.[1] The sixteenth and seventeenth centuries had also seen significant improvements to the navigability of many rivers, involving the construction of pound-locks on the River Lea (1571), the Thames (1624–35), the Warwickshire Avon (1636–9), the Wey (1651–3), the Dick Brook (c.1653),

---

[1] P. C. De La Garde, 'On the Antiquity and Invention of the Lock Canal of Exeter', *Archaeologia*, 28 (1838); W. B. Stephens, 'The Exeter Lighter Canal, 1566–1698', *Jnl. of Transport History*, 3/1 (1957), 1–11; C. Hadfield, *The Canals of South-West England* (Newton Abbot, 1967), 19–25; K. R. Clew, *The Exeter Canal* (Chichester, 1984).

the Salwarpe (1660–2), the Worcestershire Stour (1665–7), the Hampshire Avon (1664), the Itchen (1665), the Trent (*c*.1699), and the Tone (1699 and 1717).[2]

Looking further back, into the middle ages, it has long been accepted that rivers were regularly used for the transport of bulky goods such as building stone and timber. At the end of the middle ages nowhere in England or Wales was more than 65 km from a coastal port or a navigable river. However, rivers had their limitations. Transfer of goods between barge and wagon was never especially convenient. Rivers were prone to flooding, silting, and drought, and there was also often a conflict between the needs of navigation and the construction of fisheries and mills. These problems could be resolved by making artificial channels to link the building site directly with the river, or to bypass obstacles. Yet the possibility that wholly artificial canals were constructed and used in the middle ages, let alone before the Norman Conquest, usually receives scant acknowledgement.

Perhaps this neglect is a result of the fact that canal histories have, by and large, been compiled by writers with an engineering or business history background who are more familiar with the economics and technology of the Industrial Revolution, rather than by medieval historians or archaeologists. Certainly, evidence of medieval undertakings is elusive. Documentary records tend to be scattered through a wide range of sources, and are often ambiguous. Even a very recent survey by one medieval historian concluded that river improvement and canal-building episodes were rare in the middle ages.[3] Incontrovertible archaeological evidence is even harder to find, though some archaeologists recently have taken a more optimistic view.[4] Nevertheless, isolated examples of medieval river diversion and canal construction had attracted comment from the earliest years of antiquarian writing. In the early sixteenth century John Leland had recognized diversions of the Thames near Abingdon and of the River Brue near Glastonbury, and had also noted the recent recutting of the Foss Dyke between Torksey and Lincoln.[5] Could Leland's information be the tip of a much larger iceberg? This is, undoubtedly, a topic ripe for re-examination.

    [2] A. Raven, 'Ware', in *VCH Herts.* iii (1912), 380, quoting *Statutes of Realm*, iv (1). 553; T. S. Willan, *River Navigation in England 1600–1750*, 2nd impression (London, 1964), 89; P. J. Huggins, 'A Medieval Bridge at Waltham Abbey, Essex', *MA* 14 (1970), 128; F. S. Thacker, *The Thames Highway: A History of the Inland Navigation* (London, 1914; repr. New York 1968), 67; C. Hadfield, *Canals of the West Midlands* (Newton Abbot, 1966), 16, 56–60; C. Hadfield, *Canals of Southern England* (London, 1955), 30–2, 36–7; L. T. C. Rolt, *The Inland Waterways of England* (London, 1950), 68–71; E. Course, 'The Itchen Navigation', *Proc. Hants Field Club & Archaeol. Soc.* 24 (1967), 113–26.

    [3] R. Holt, 'Medieval England's Water-Related Technologies', in P. Squatriti (ed.), *Working with Water in Medieval Europe: Technology and Resource-Use* (Leiden, 2000) (and see above, p. 1).

    [4] See, for example, S. Rippon, *The Transformation of Coastal Wetlands: Exploitation and Management of Marshland Landscapes in North-West Europe during the Roman and Medieval Periods* (Oxford, 2000), 209, 240.

    [5] *The Itinerary of John Leland in England and Wales, 1535–1543*, ed. L. T. Smith (London, 1910), i. 29, 148–9; v. 76.

Medieval River Navigations

The extent to which rivers were navigable in England during the middle ages and the extent to which inland water-borne trade contributed to economic growth and urban development have been the subject of lively debate in recent years; but there is general agreement that river traffic was more considerable and more extensive between the eleventh and thirteenth centuries than it became after the fourteenth century.[6]

The story of many rivers through the middle ages is of a constant struggle by bargemen and merchants to keep navigation open against the encroachments of riparian owners wishing to construct mills and fish-weirs. There appears to have been an assumption in law that the tidal reaches of rivers had the status of a common highway under the protection of the Crown, whereas above the tidal limit the river was the property of riparian landowners, and that a right of passage for boats existed only through ancient custom, or by agreement, or if confirmed by statute. An early study of the Thames by Thacker collected many references to disputes and inquiries over the state of the river during the middle ages, and the navigability of the river in the later middle ages has been investigated in greater detail by Robert Peberdy.[7] Similar conflicts have been documented along the Trent and the Great Ouse.[8]

It is difficult to divorce the study of canals in the middle ages from that of river navigations. Almost any navigational use of a river involves some sort of modification of its natural course, which, at its very least, may include occasional dredging, embankment, and the provision of waterfront structures such as quays or jetties.[9] A complaint about interference to the navigation of the Trent in 1346–7 described how the banks were reinforced by a firm construction of 'piles, poles and hurdles' fixed to preserve the passage for boats at any time of the year.[10] In 1438 there was a plan to dredge and widen the River Wissey and to build two stone jetties at a point within 30 m of its junction with the Great Ouse, though it is not clear whether this was ever done.[11]

---

[6] For opposing views see J. F. Edwards and B. P. Hindle, 'The Transportation System of Medieval England and Wales', *JHG* 17 (1991), 123–34; J. Langdon, 'Inland Water Transport in Medieval England', *JHG* 19 (1993), 1–11; J. F. Edwards and B. P. Hindle, 'Comment: Inland Water Transportation in Medieval England', *JHG* 19 (1993), 12–14. Some of the conflicting opinions are partly resolved by E. Jones, 'River Navigation in Medieval England,' *JHG* 26 (2000), 60–75.

[7] Thacker, *The Thames Highway*, 12–61; R. Peberdy, 'Navigation on the River Thames between London and Oxford in the Late Middle Ages: A Reconsideration', *Oxoniensia*, 61 (1996), 311–40.

[8] A. C. Wood, 'The History of Trade and Transport on the River Trent', *Trans. Thoroton Soc.* 54 (1950), 1–44; D. Summers, *The Great Ouse: The History of a River Navigation* (Newton Abbot, 1973).

[9] G. Milne and B. Hobley (eds.), *Waterfront Archaeology in Britain and Northern Europe*, CBA Research Rep. 41 (London, 1981); G. L. Good, R. H. Jones, and M. W. Ponsford (eds.), *Waterfront Archaeology: Proceedings of the Third International Conference on Waterfront Archaeology*, CBA Research Rep. 74 (London, 1991).

[10] *Cal. Pat. R. 1345–8*, 237, 398.

[11] W. Dugdale, *History of Imbanking and Drayning* (London, 1772), 295; Summers, *The Great Ouse*, 29.

From the deepening and widening of natural watercourses and the reinforcement of banks, it is but a small step to making short artificial cuts to bypass
meanders, shallows, or man-made obstructions such as mills or fish-weirs, or
to construct arms to give access from the river to some nearby settlement or
building site. David Pannett's work on the fish-weirs of the Severn in Shropshire
has shown that the weirs were bypassed by barge-gutters, some of which may
simply be modifications of naturally braided channels, while others seem to
have been deliberately dug. At Preston Weir in Shropshire (Fig. 63) the general
alignment and narrow width of the barge-gutter in relation to the broader
channel containing the fish-weir show that it was not the original river course,
yet it is of sufficient antiquity to be followed by the parish boundary. In a civil
court case in 1839 the owner of this weir was able to demonstrate that it had
been in continuous use since before the Domesday survey, that it had been
owned and maintained by Haughmond Abbey through much of the middle
ages, and that a previous legal challenge to its presence in 1422 had been
overthrown.[12] Documentary references show that many other fish-weirs on the
Severn, and on other rivers, date back to the middle ages.[13] Wherever these
rivers were also used by barges, considerable effort must have been expended
in digging bypass channels. In marshland areas also, there is usually a long and
complex history of human interference with the natural watercourses, in which
the requirements of drainage often played a greater part than navigation, yet
usage by boats remained a significant secondary purpose.

Along the coast, short artificial cuts might also be made to carry seagoing
vessels into sheltered estuaries. By the early middle ages the outlet of the River
Blyth into the North Sea had become deflected southwards for some 5 km by
the growth of the Kingsholme spit, which protected the harbour of Dunwich.
However, the spit was far from stable, and by 1249 the river had broken
through it to enter the sea further north at Walberswick. The rapid silting of
the Dunwich outlet prompted the men of that town to stop up the northern
breach and to reopen their own haven, but their work was undone by a storm
on New Year's night in 1287. Although the Dunwich haven was again, for
a time, reopened, it was becoming evident that this was a losing battle. A
severe north-easterly gale on 14 January 1328 finally blocked the Dunwich
mouth completely, and repeated attempts to reopen it met with no success. A
succession of natural breaches and artificial cuts was opened through the spit
at various points further north during the fourteenth and fifteenth centuries.
However, the approach to Dunwich up the silting channels behind the shingle
bank was becoming ever more difficult, and the more northerly trading ports
of Walberswick and Southwold, though not without access problems of their
own, were increasingly the beneficiaries of Dunwich's decline. In 1589 there

---

[12] D. Pannett, 'Fish Weirs on the River Severn', in M. Aston (ed.), *Medieval Fish, Fisheries and Fishponds in England*, BAR British Ser. 182 (2 vols., Oxford, 1988) 2, 371–89.
[13] C. J. Bond, 'Monastic Fisheries', in Aston (ed.), *Medieval Fish, Fisheries and Fishponds*.

was one last proposal to make a direct cut into Dunwich haven. However, in the following year the men of Southwold and Walberswick took matters into their own hands, making their own cut at the north end of the spit, which precipitated an ineffective legal campaign from Dunwich.[14]

Further north, in Norfolk, the River Bure originally flowed into the sea north of Winterton, along the present line of the River Thurne and Hundred Stream. Now it turns south at Thurne Mouth towards Acle, ultimately joining the Yare at Great Yarmouth, and the direction of drainage of the River Thurne has been reversed. Between Thurne Mouth and Wey Bridge at Acle, the Bure follows a relatively straight course which looks artificial, though of some antiquity, since it is followed by parish boundaries. Changes have also taken place to the course of the River Ant, which formerly flowed east along the Hundred Dyke into the River Thurne, and has been diverted southwards below Ludham Bridge along a 900-m-long cut into the River Bure. At the hub of this maze of waterways is the Benedictine abbey of St Benet, Hulme, and it seems possible that, with the natural silting of the Thurne outfall north of Winterton, the monks may have diverted both the Ant and the Bure, to reduce flooding and to keep access open for boats coming up to the abbey quay.[15]

Occasionally medieval river diversions on a much larger scale can also be documented. It is evident, for example, that the course of the River Don had been subjected to considerable alteration below Thorne, 14 km north-east of Doncaster, even before the well-documented construction of the Dutch River and other drainage works by Cornelius Vermuyden in 1626–7 (cf. Cole, above pp. 64–6). The original course of the river meandered eastwards across Hatfield Chase through Crowle to reach a confluence with the River Trent near Addingfleet. However, certainly by the early fifteenth century, and perhaps even before 1344, most of the river's flow had been canalized into a new artificial channel running northwards through New Bridge to a confluence with the River Aire near Eskhamhorn. The question of the precise date and purpose of this diversion remains unresolved; it almost certainly had the effect of diverting floodwater away from Hatfield Chase and Thorne Moors, but a more economically rewarding purpose for such a major undertaking was probably the opening of a direct navigable link between Doncaster and the medieval port of Airmyn.[16]

The primary aim of the present chapter is to explore the evidence for more ambitious artificial canal schemes in the early middle ages; but short river diversions, canalized rivers, and wholly artificial canals are, in reality, all part of the same process of improvement in water transport, and it is impossible to study any of them in isolation.

---

[14] N. Comfort, *The Lost City of Dunwich* (Lavenham, 1994), 59–66.

[15] T. Williamson, *The Norfolk Broads: A Landscape History* (Manchester, 1997), 76. I am grateful to Bob Malster for this reference and for other information on this area.

[16] G. D. Gaunt, 'The Artificial Nature of the River Don North of Thorne, Yorkshire', *YAJ* 47 (1975), 15–21.

The Classical Background: Europe and the Mediterranean

The new works undertaken during the middle ages have to be seen in the context of a much longer tradition of water management, going back to the fourth millennium BC, when the earliest irrigation canals were cut in Mesopotamia. As a more immediate background to the subject of canal construction in early medieval England we need to know something of what was happening in the Roman world, and whether it retained any lasting influence. It cannot be doubted that the Roman imperial, provincial, and civic governments and the Roman army possessed the technical ability and organizational skills, and commanded the labour resources, to undertake substantial civil engineering projects. The construction of Roman conduits, aqueducts, mill-leats, irrigation systems, and drainage works all required a practical ability to survey accurately along the contour and to conduct water within an artificial channel. Moreover, classical authors make a number of casual references to the existence of navigable canals, without expressing any particular surprise or admiration at these achievements, which suggests that the concept was familiar.[17] Between the time of the late Republic and the early second century AD at least fourteen canal schemes were successfully completed, in Italy, Gaul, Germany, Dacia, and Egypt, and at least another seven seriously proposed (Fig. 31, Table 7); only a couple of schemes devised by Nero, for a ship canal through the isthmus of Corinth (which would have necessitated a cutting 90 m deep) and another linking the mouth of the Tiber with the Gulf of Pozzuoli (a distance of 236 km), were derided by contemporary writers as overambitious.[18]

Canalization schemes proposed or completed within the classical world occur in a number of different contexts, but their overall distribution is predominantly coastal (Fig. 31). This group includes cuts linking seaways through a narrow isthmus; short canals associated with harbours; waterways providing access to the coast from military bases, cities, and production centres further inland; and drainage canals in coastal marshes which had a secondary use for the carriage of goods by boat. Links over inland watersheds to connect neighbouring navigable rivers presented a greater technical challenge, but examples can be recognized, most notably the ambitious scheme of Lucius Vetus around AD 55–6 to open a waterway link between the Mediterranean and the German frontier by having his troops construct a canal between the Saône and Moselle.[19]

---

[17] Brief reviews have been provided by N. A. F. Smith, 'Roman Canals', *Transactions of the Newcomen Society*, 49 (1978), 75–86; K. D. White, *Greek and Roman Technology* (London, 1984); and C. Wikander, 'Canals', in O. Wikander (ed.), *Handbook of Ancient Water Technology*, Technology and Change in History 2 (Leiden, 2000), 321–30.

[18] See, for example, A. Sneh, T. Weissbrod, and I. Perath, 'Evidence for an Ancient Egyptian Frontier Canal', *American Scientist*, 63 (1975), 542–8; and J. Sasel, 'Trajan's Canal at the Iron Gate', *Jnl. of Roman Studies*, 63 (1973), 80–5. Nero's projects are described by Suetonius, *The Twelve Caesars*, ed. R. Graves (London, 1957), *Nero*, 19, 31.3, and Tacitus, *The Annals*, ed. J. Jackson (3 vols., London, 1931–7), 14, 5 n, 15. 42.

[19] Tacitus, *Annals*, 13. 53; E. M. Wightman, *Gallia Belgica* (London, 1985), 60, 151; G. H. Allen, 'A Problem of Inland Navigation in Roman Gaul', *Classical Weekly*, 27 (11 Dec. 1933), 65–9.

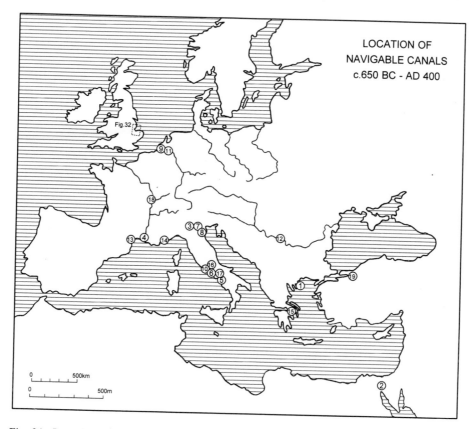

Fig. 31. Location of navigable canals, *c.650* BC–AD 400.

In a number of cases the initial motive for construction was purely military, prompted by the need to move troops, equipment, and supplies. In other cases, as in the Po valley or the Pontine marshes, navigation was an almost incidental benefit of drainage operations.

It is clear that the construction of canals for use by both warships and cargo vessels was not uncommon in the classical world. Labour was cheap, surveying skills were sufficient, and constructions over distances of 50 km or more were achieved on several occasions without undue difficulty (Table 7). Much longer distances involving changes of level were also seriously considered, which must imply either a capacity to construct locks or lifts, or a willingness to manhandle cargoes or vessels over a bar, perhaps by means of rollers. Failure to complete some of the more ambitious schemes, such as Vetus' Saône–Moselle canal, appears to be as much a matter of politics as a product of technical limitations.

Several of the constructions recorded as achieving their ends were, it has to be said, short-term expedients devised during military campaigns which were abandoned once the strategic need had passed. There is no evidence, for example, that the cut made by Xerxes through the isthmus of Actium to

Table 7. Navigable Canals: Continental Europe and the Mediterranean, *c.*650 BC–AD 400

No. on fig. 31	Date	Initiator	Location/terminal points	Approx. length
Completed schemes				
1	481 BC	Xerxes	Isthmus of Actium, Chalcidice	*c.*5 km
2	284–246 BC	Ptolemy II Philadelphus	?Bilbeis, R. Nile–Suez	*c.*115 km
3	109 BC	M. Aemilius Scaurus	Po Valley/Modena/ Parma	> *c.*55 km
4	101 BC	Gaius Marius	Arles–Fos-sur-Mer	*c.*42 km
5	37 BC	Marcus Agrippa	Lago Averno-Gulf of Pozzuoli	*c.*2 km
6	27 BC–AD 14	Augustus	Foro Appi–Terracina	26 km
7	27 BC–AD 14	Augustus	R. Adige	Unknown
8	27 BC–AD 14	Augustus	Ravenna–Po estuary	> *c.*55 km
9	12 BC	Drusus	Arnhem, R. Rhine –Doesburg, R. IJssel	*c.*14 km
10	AD 42–46	Claudius	Ostia–R. Tiber	2 km
11	AD 47	Corbulo	?Monster, Old R. Maas –Roomburg, Old R. Rhine	*c.*37 km
2	AD 89	Trajan	?Bilbeis, R. Nile –Suez	*c.*115 km
10	AD 98–117	Trajan	Ostia, harbour	2 km
12	AD 100–102	Trajan	Iron Gate, R. Danube	5 km
13	Unknown		Narbonne, R. Aude –Etang de Bages	21 km
14	Unknown		Fréjus, harbour	1 km
Uncompleted proposals				
15	*c.*628–588 BC	Periander of Corinth	Isthmus of Corinth	6.5 km
2	521–486 BC	Darius I of Persia	Bilbeis, R. Nile–Suez	*c.*115 km
16	60–44 BC	Julius Caesar	R. Tiber, Ostia–Rome	*c.*25 km
17	AD 54–68	Nero	Mouth of R. Tiber –Gulf of Pozzuoli	236 km
15	AD 67	Nero	Isthmus of Corinth	6.5 km
18	AD 55–56	Lucius Vetus	?Corre, R. Saône –Arches, R.Moselle	*c.*70 km
19	AD 111	Pliny the Younger	Lake Sophon–Sea of Marmara nr. Nicomedia, Asia Minor	16 km

facilitate naval support for his army during the Persian invasion of Greece in 481 BC was ever used again. Indeed, few of the military canals show any evidence of prolonged use. The *Fossa Mariana*, cut by Gaius Marius in 101 BC to provide his troops in Gaul with an alternative supply line avoiding the dangerous Rhône estuary, was subsequently handed over to the Massiliotes, who gained a considerable income from tolls from the canal. By the first century AD, however, it had silted up and fallen into disuse.[20] Whereas initial construction could call upon the resources of the army, long-term maintenance must have been much more of a problem for the civil authorities. The *Fossa Drusiana*, cut by Drusus from the Rhine to the IJssel in 12 BC, was used again for the carriage of troops and supplies in AD 15, AD 17, and probably also in AD 47, but the works were destroyed in the Batavian revolt of AD 70 and it was never reopened.[21]

On occasions the Romans successfully resurrected earlier canal schemes. One of the recurring ambitions of the ancient world was the construction of a ship canal to link the River Nile with the Red Sea, a minimum of 100 km by any route. According to Pliny the first man to propose such a scheme was the twelfth-dynasty pharaoh Senusret III (*c.*1878–1840 BC).[22] Herodotus describes another attempt to make a canal wide enough for two triremes to pass, undertaken by Necho, a pharaoh of the 26th dynasty (*c.*609–594 BC).[23] The Persian ruler Darius I (521–486 BC) also seems to have undertaken some construction, and although Strabo was sceptical of his achievement, during the construction of the modern Suez Canal a granite block was discovered bearing an inscription commemorating Darius' success in opening a way for ships to sail from Egypt to Persia.[24] Darius' canal was reopened and enlarged by Ptolemy II Philadelphus (284–246 BC), who cut a trench, reputedly about 30 m wide × 10 m deep and 56 km in length, from the Red Sea to the Bitter Lakes. According to Pliny, however, Ptolemy was deterred from going further by fears about the risk of flooding and pollution of the River Nile.[25]

It was left to the Romans to resurrect and complete the canal: it was reopened in AD 89 by Trajan, who hoped thereby to facilitate sea trade with the east. The first part of the route closely parallels the modern Suez Canal, linking the Gulf of Suez near the ancient port of Clysma (modern Qal'at el Qulzum near Suez) with the Great Bitter Lake; but the ancient canal then turned westwards through the Wadi Tumilat to join the easternmost branch of the Nile delta which enters the Mediterranean at Pelusium (modern Tell el Farama). A likely course between the Bitter Lakes and the Nile has been

---

[20] Plutarch, *Lives*, ed. B. Perrin (London, 1920), *Gaius Marius*, 15. 1–4.
[21] Suetonius, *Claudius*, 1. 2; Tacitus, *Annals*, 1. 9, 2. 8, 11. 18; Tacitus, *Histories*, 5. 19.
[22] Pliny the Elder, *Natural History*, ed. H. Rackham and W. H. S. Jones (10 vols., London, 1944–62), 5. 70.
[23] Herodotus, 2. 158–9.
[24] H. Last, 'A Roman Citizen Surveys the World', in J. H. Hammerton (ed.), *Universal History of the World*, iii (London, [*c.*1930]), 1989; the inscription is quoted by Wikander, 'Canals', 325, n. 15.
[25] Pliny, 5. 166.

traced by aerial photography, passing through Thaubasium (near Isma'iliya), Heroopolis (Tell el-Maskhuta), and Thou (Tell abu Suleiman).[26]

In fact the canal can only ever have been of limited value. Keeping its western end clear of the huge quantities of silt brought down by the Nile every year demanded constant dredging, while the Gulf of Suez was dangerous because of its shallows and reefs; moreover, access into the Gulf from the Red Sea was made difficult by the prevailing northerly winds. Although Pliny in the first century and Egeria in the fourth century both noted that ships plied to India from the port of Clysma, most traders preferred to land their cargoes at Berenice or one of the other ports along the west coast of the Red Sea for transport overland to the Nile higher upstream.[27]

Yet, despite the difficulties, efforts were made over many centuries, with at least intermittent success, to keep it open. The Spanish nun Egeria, who visited Egypt during a pilgrimage to the Holy Land in AD 381–4, describes the town of Heroopolis or Pithom (modern Tell el Maskhuta) in the Wadi Tumilat, 'with an arm of the Nile flowing through it', failing to realize that this was the artificial canal, not a natural branch of the river.[28] Very soon after Egypt came under Islamic rule the canal was restored by 'Amr Ibn el 'As in AD 640; and, during the early 760s a party of Irish monks and clergy travelling in Egypt included one Fidelis, who records that his party 'boarded ships in the River Nile and sailed to the entrance of the Red Sea'. Fidelis's account of his pilgrimage came into the hands of another Irish monk, Dicuil, who included it in his geographical text *De Mensura Orbis Terrae*, compiled in AD 825. The canal was finally blocked in AD 767 by the Abbasid Caliph Abu-Dja'afar al-Mansur, and thereafter we hear no more of it.[29]

The Roman achievement in building canals for transport was considerable, even if not all schemes commenced were ever brought to completion, and even if not all completed schemes served any long-term purpose. Nevertheless, despite formidable difficulties, the Nile–Red Sea canal remained at least intermittently in use into the second half of the eighth century. Did any other Roman canals remain in use, or survive in sufficiently recognizable form to be re-excavated in the early middle ages? Or, failing that, did knowledge of past achievements survive long enough to influence the ambitions of medieval rulers or landowners?

Roman Canals in Britain (Fig. 32)

The possibility that canals had been built in Britain during the Roman administration has been admitted since the eighteenth century. Suggestions that Roman

---

[26] Smith, 'Roman Canals', 80–2; A. Sneh, T. Weissbrod, and I. Perath, 'Evidence for an Ancient Egyptian Frontier Canal', *American Scientist*, 63 (1975), 542–8.

[27] Pliny, 6. 26–7; J. Wilkinson (ed.), *Egeria's Travels* (3rd edn. Warminster, 1999), 103.

[28] Wilkinson (ed.), *Egeria's Travels*, 117.

[29] A. Letronne, *Recherches géographiques et critiques sur le livre 'De Mensura Orbis Terrae'* (Paris, 1814), 20, 24; J. Wilkinson, *Jerusalem Pilgrims before the Crusades* (Warminster, 2002), 23, 31, 231–2.

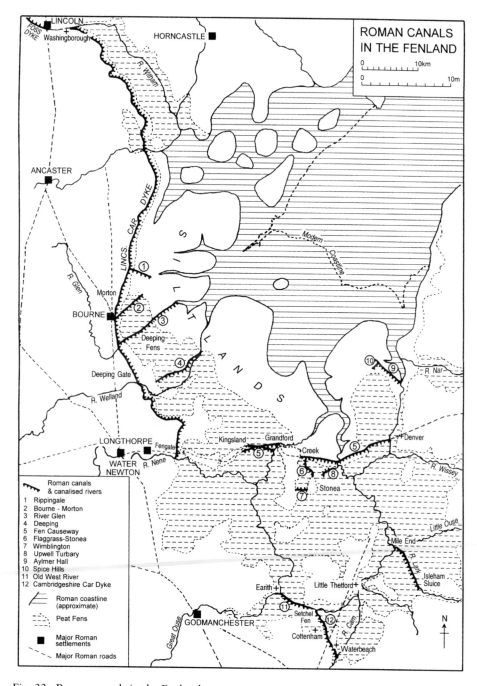

Fig. 32. Roman canals in the Fenland.

military authorities and civilian producers alike used canals for the transport of bulky goods in Britain are not intrinsically unreasonable in view of the continental evidence cited above. One writer has gone so far as to suggest that water, rather than road, was always the preferred mode of transport for heavy goods, not just in Britain but throughout the Roman empire; that canals must have been widespread; and that the Romans must, in consequence, have developed the use of the pound-lock (the so-called 'Piercebridge formula', propounded by Selkirk in 1983). Such extreme views have not met with general acceptance; but the baby should not be ejected with the bathwater, and enough evidence for canal construction in Roman Britain has been put forward in previous literature to require further serious investigation.[30]

In terms of sheer length, the most impressive construction in Britain is the Lincolnshire Car Dyke (Fig. 33), first identified as a Roman canal by Stukeley in the eighteenth century. This is an obviously artificial channel, approximately following the 3 m contour, cut from the River Witham at Washingborough, 5 km below Lincoln, and continuing southwards for over 90 km, cutting across the lines of natural drainage to the Wash, intersecting the Welland at Deeping Gate, and meeting the Nene at Fengate near Peterborough.[31] This is, in fact,

Fig. 33. The Roman Car Dyke, seen here at Metheringham (Lincs.).

[30] R. Selkirk, *The Piercebridge Formula* (Cambridge, 1983); for a refutation, see J. D. Anderson, *Roman Military Supply in North-East England: An Analysis of and an Alternative to the Piercebridge Formula*, BAR British Ser. 224 (Oxford, 1992).

[31] Ven. E. Trollope, *Sleaford and the Wapentakes of Flaxwell and Ashwardburn in the County of Lincoln* (Sleaford, 1872), 64–71; C. W. Phillips, 'The Present State of Archaeology in Lincolnshire,

significantly longer than any completed Roman navigation canal known from the Continent. Trollope's estimate of its original dimensions, a navigable width of 9–15 m and depth of 2.4 m, has been fairly closely matched by later observations quoting a width of 12 m and depth of 2 m.[32]

A further artificial watercourse bearing the same name links the Old West River at Setchel Fen with the Cam at Waterbeach, 8 km below Cambridge, providing a link between the Cam and the Ouse. Sections of this waterway have revealed dimensions of 20 m in width and 2 m in depth.[33] It has been suggested that the Cambridgeshire section of the Car Dyke would only make sense as a navigable canal if there was, at the time it was cut, no natural watercourse between Earith on the Ouse and Little Thetford on the Cam. If the present Old West River, which links those two places, had then been available, the effort of cutting a canal 8 km in length from Waterbeach to Setchel Fen would scarcely seem to be justified by the saving of 13 km, estimated as a mere four hours' work for a horse or six bowhaulyers towing four 20-tonne barges. The western part of the Old West River from Setchel Fen to Hermitage Sluice at Earith is fairly straight, and the suggestion that this is itself part of the canal system seems entirely plausible; if so, this increases the total length of the Cambridgeshire Car Dyke system to 17 km. The diversion of part of the flow of the Ouse down the Old West River to pass east of Ely may be a result of post-Roman subsidence and the blocking of earlier outfalls to the Wash.[34]

It has been suggested in the past that the Lincolnshire and Cambridgeshire Car Dykes were linked into a single waterway system between the Nene and Ouse by intervening rivers and further artificial cuts, but no proof of this has ever been forthcoming; indeed, it is not even certain whether both parts are of the same date.

The Roman date of the Cambridgeshire end of the Car Dyke seems secure. In 1926 a section near Waterbeach Lodge identified an early Saxon hut, the floor of which had been cut through the silt deposited on the bank to reach the gravel below. Rubbish from the hut overlay the infill of the dyke, indicating that the canal had been disused for some considerable time before the Anglo-Saxon settlement.[35] In 1947 a further section was cut several kilometres to the north-west of Waterbeach, at Bullock's Haste, Cottenham, at a point where

---

Part ii', *ArchJ* 91 (1934), 118–21; RCHME, *Peterborough New Town: A Survey of the Antiquities in the Areas of Development* (London, 1969), 40–3.

[32] D. Hall, *The Fenland Project 2: Fenland Landscapes and Settlement between Peterborough and March*, EAA 35 (Cambridge 1987), 28; D. Hall and J. M. Coles, *Fenland Survey: An Essay in Landscape and Persistence*, English Heritage Archaeological Rep. 1 (1994), 105; R. Thorpe and T. Zeffert, 'Excavation of the Lincolnshire Car Dyke, Baston', *Fenland Research*, 6 (1988–9), 10–15.

[33] J. G. D. Clark, 'Report on Excavations on the Cambridgeshire Car Dyke, 1947', *Antiquaries Journal*, 29 (1949), 145–63; S. Macaulay and T. Reynolds, 'Excavations and Site Management at the Cambridgeshire Car Dyke, Waterbeach', *Fenland Research*, 8 (1993), 63–9.

[34] G. Fowler, 'Fenland Waterways, Past and Present, South Levels District: I', *Proc. Cambridge Antiquarian Soc.* 33 (1933), 117.

[35] T. C. Lethbridge, 'An Anglo-Saxon Hut on the Car Dyke at Waterbeach', *Antiquaries Jnl.* 7 (1927), 141–6.

the dyke ran through a system of earlier droveways and 'Celtic' fields. Belgic sherds were found under its north bank, and pottery from the primary silts of the channel suggested to the excavator a construction date of AD 50–60.[36] However, re-examination of the critical pottery by Brian Hartley revealed nothing that need be earlier than the second century.[37] The construction of the Cambridgeshire Car Dyke was carried out with some ruthlessness, pursuing a contour course regardless of any pre-existing features in its way. This implies that it was directed by a powerful authority, and it has been viewed by some as part of a major programme of imperial planning associated with the settlement of the Fenland under Hadrian at the time of his visit to Britain around AD 120.[38] Evidence from the 1947 excavation at Cottenham initially suggested that the canal remained in use until the end of the second century, and was then blocked by a dump of gravel containing third- and fourth-century pottery carrying a droveway over it, but subsequent work has suggested that it remained in use until the middle or late fourth century.[39] The date of the Lincolnshire portion is less securely established, but several short branches from it into the Deeping Fens lie in conformity with the Romano-British field systems there.

William Stukeley believed that both sections of the Car Dyke were part of the same system of canals made by the Romans for the carriage of corn from the Fenland to Lincoln, York, and the northern military zone.[40] Even if Stukeley's view of a unified waterway can no longer be upheld, many later writers have accepted the primary function of both Car Dykes as transport.[41] Certainly the dimensions recorded would be compatible with the use of either section as a barge canal. Others have preferred to see them as catch-water drains, designed to intercept water from higher ground inland and to prevent it from inundating the summer pastures on the fens.[42] The two alternative interpretations are not necessarily in total conflict,[43] but arguments can be

[36] Clark, 'Report'; a 1st-century date is also accepted by S. J. Hallam, 'Settlement round the Wash', in C. W. Phillips (ed.), *The Fenland in Roman Times* (London, 1970), 74.

[37] Quoted in J. B. Whitwell, *Roman Lincolnshire*, History of Lincolnshire II (Lincoln, 1970), 94; and in J. Wacher, *The Towns of Roman Britain* (London, 1975), 136.

[38] Whitwell, *Roman Lincolnshire*, 57.

[39] Clark, 'Report'; Wacher, *Towns of Roman Britain*, 136; Macaulay and Reynolds, 'Excavations and Site Management', 63–9; Thorpe and Zeffert, 'Excavation of the Lincolnshire Car Dyke', 14.

[40] W. Stukeley, *Itinerarium Curiosum* (London, 1724), 1–34; S. Piggott, *William Stukeley: An Eighteenth-Century Antiquary* (New York, 1985), 39–40.

[41] Phillips, 'Present State', 118–22; Clark, 'Report'; Whitwell, *Roman Lincolnshire*, 57–8; Smith, 'Roman Canals', 77.

[42] S. B. J. Skertchly, *Geology of the Fenland*, Memoirs of the Geological Survey (London, 1877); B. B. Simmonds, *The Lincolnshire Car Dyke* (Swineshead, 1975); B. B. Simmonds, 'The Lincolnshire Car Dyke: Navigation or Drainage?', *Britannia*, 10 (1979), 183–96; B. B. Simmonds, 'Iron Age and Roman Coasts around the Wash', in F. H. Thompson (ed.), *Archaeology and Coastal Change*, Soc. of Antiquaries Occasional Paper NS 1 (London, 1980), 61–4; P. Salway, *The Oxford Illustrated History of Roman Britain* (Oxford, 1993), 384.

[43] See e.g. Trollope, *Sleaford*, 64–71; S.S. Frere, *Britannia: A History of Roman Britain* (London, 1967), 275–6; Wacher, *Towns of Roman Britain*, 136.

raised against both of them. The catch-water interpretation is reduced in credibility by the fact that the Lincolnshire section has 13-m-wide banks on both sides, still surviving to a height of 1 m, which would have prevented surface runoff from the higher ground entering directly into the canal,[44] while the Cambridgeshire section only intersects one stream, the Beach Ditch, and it too has a partial embankment on its western side.[45] The argument for navigational use has to overcome Pryor's demonstration that the Lincolnshire dyke does not consistently follow the contour, being subject to slight variations in altitude, and Simmonds's observation that it is crossed at several points by causeways made apparently of unexcavated natural gravel.[46] It is possible that the purpose of the causeways was actually to break the length of the canal into separate pounds at slightly different levels in order to overcome the difficulties of maintaining a navigable depth of water throughout the entire length all year round; the necessity of transferring goods from one boat to another or carrying the barges themselves over the obstructions would be a fairly cumbersome operation, but it might still be preferable to road transport. A third possibility put forward by Mackreth that the dykes may represent the bounds of an imperial estate made up of improvable lands in the Fens must also be considered, though the scale of construction seems excessive for this purpose.[47]

A third canal in eastern Britain which seems likely to be of Roman origin is the Foss Dyke, providing a water connection 17 km long between the Witham at Lincoln and the Trent at Torksey, and theoretically enabling military supplies to be taken by water to York. There must have been some form of tidal sluice where the Foss Dyke joined the Trent. The pottery kilns of Little London south of Torksey and those on Lincoln Racecourse were situated immediately adjacent to the canal, and their products may have been carried by water.[48] The boundary between Kesteven and Lindsey follows the Foss Dyke for 6 km west of Lincoln, and this would seem to underline its antiquity, though along this stretch the canal in part follows the original course of the River Till.[49] The navigation of the Foss Dyke was reopened in the middle ages (see further below) and again in the eighteenth century.

The Fen Causeway, the major Roman route crossing the fens in an east–west direction between Denver near Downham Market and Longthorpe near Peterborough, was constructed as a road in the later part of the first century. The western portion from Peterborough to Kingsland and the central portion between Grandford and Creek near March followed a natural watershed on slightly higher ground, but for some 5 km between Kingsland and Grandford

---

[44] Hall, *The Fenland Project 2*, 28; Hall and Coles, *Fenland Survey*, 105.

[45] Fowler, 'Fenland Waterways I', 117.

[46] A. Pryor, 'The Car Dyke', *Durobrivae*, 6 (1978), 24–5; Simmons, 'Lincolnshire Car Dyke', 10, 189.

[47] D. F. Mackreth, *Orton Hall Farm: A Roman and Early Anglo-Saxon Farmstead*, EAA 76 (Manchester, 1996), 233–4.

[48] Whitwell, *Roman Lincolnshire*, 58.   [49] Phillips, 'Present State', 117–18.

and for 15 km of the eastern half from Creek to Denver the causeway crossed the peat fen. Subsequently, probably during the second century, a canal 10–15 m wide and 1.7 m deep was dug immediately to the south of the road in the eastern section, with salterns along its banks, and a similar construction appears to have been made west of Grandford, where the road follows the north bank. Road and canal were both buried by silt deposits caused by one or more episodes of marine inundation in the later third century, which may also have put an end to the salt-working. The road was then re-established over the south bank of the earlier canal. As the peat contracted through later drainage operations, the silt fill of the canal was left as a causeway up to 40 m wide and 2 m high with the road along its centre.[50] The Fen Causeway canals contributed nothing to the local drainage, and it is difficult to find any explanation for them other than as an alternative transport link between the upland margins. Other straight lengths of waterway near March connected the north end of Stonea Island with the Fen Causeway and Wimblington with the Old River Nene, and these too appear to be of Roman origin, but were silted up before the end of the Roman period.[51]

Other putative Roman canals or river canalizations in the Fens remain unproven, though plausible reasons for their construction can be put forward. A section of the River Lark has clearly been canalized between Isleham Sluice and Mile End, and the antiquity of this modification is indicated both by the fact that it is followed by the county boundary between Cambridgeshire and Suffolk for 7 km, as far downstream as Spooner's Farm, and by Romano-British pottery reported from dredgings. The need to transport clunch from Romano-British quarries at Isleham may have provided a motive for this improvement.[52] Aerial photography has produced evidence of a canal apparently associated with Romano-British turbaries south of the Fen Causeway and east of Stonea Island.[53]

Other Fenland waterways seem to serve urban or rural communities rather than being related to the exploitation of particular resources. Between Lincoln and the coast, the River Witham now hugs the north-eastern edge of its alluvial valley. It includes many straight stretches which are clearly artificial, while aerial photographs show traces of an older estuarine creek system through the middle of the valley.[54] At least some parts of its course, including in Lincoln

---

[50] Hall, *The Fenland Project 2*, 41–2; R. J. Silvester, *The Fenland Project 4: The Wissey Embayment and the Fen Causeway, Norfolk*, EAA 52 (Cambridge, 1991), 97–115; M. Leah, 'The Fenland Management Project, Norfolk', *Fenland Research*, 7 (1992), 53–4; A. Crowson, 'Excavations on the Fen Causeway at Straw Hill Farm, Downham West', *Fenland Research*, 9 (1994), 25.

[51] D. Hall, *The Fenland Project 6: The South-Western Cambridgeshire Fenlands*, EAA 56 (1992), 71.

[52] G. Fowler, 'Fenland Waterways, Past and Present, South Levels District: II', *Proc. Cambridge Antiquarian Soc.* 34 (1934), 26–7.

[53] D. Hall, *The Fenland Project 10: Cambridgeshire Survey, the Isle of Ely and Wisbech*, EAA 79 (Cambridge 1996), fig. 96; Rippon, *Transformation*, fig. 25b, plate 39.

[54] T. J. Wilkinson, 'Palaeoenvironments of the Upper Witham Fen: A Preliminary View', *Fenland Research*, 4 (1987), 52–6; Simmonds, 'Iron Age and Roman Coasts around the Wash', 70–1;

itself the 0.8 km between Brayford Pool and Stamp End Lock, which may have replaced an original course along Sincil Dyke, and the 14 km between Lincoln and Bardney, have been suggested as Roman cuts.[55]

Elsewhere in Lincolnshire, several further canals of probable Roman date extend from the Car Dyke out over the peat towards the settled siltlands. The longest ran for some 6 km from the Car Dyke at Bourne north-westwards to the head of a creek in Morton Fen, and was 13 m wide × 3 m deep and scoured by tidal flow.[56] The canalized outfall of the River Glen between Thetford on the Car Dyke and Pinchbeck Fen, and shorter canals in similar alignments at Rippingale and Deeping, may also be of Roman date.[57] Although these all connect with tidal creeks, the lack of any sea wall indicates that flood control was not their primary function, and Rippon suggests that improved communications across the Fenlands between the coastal communities and the fen-edge settlements may have been a more important consideration.[58] In the Norfolk siltlands a canal at least 5.6 km long, 12 m wide, and *c*.1.1 m deep, and scoured by the tide, linked an area of Romano-British coastal settlement around Aylmer Hall in Tilney St Lawrence with the Great Ouse estuary with a branch towards Spice Hills. Neither survived into the early Anglo-Saxon period.[59]

Further inland, other straight canal arms meeting the River Cam below Cambridge, once interpreted as Roman, now seem more likely to belong to the early middle ages, and will be discussed further below.

Evidence for Roman canals in Britain is very heavily concentrated around the shores of the Wash, but possibilities need to be considered in other parts of the country. There are some problems with the traditional interpretation of the Raw Dykes at Leicester as part of the town aqueduct, not least because it was cut at a level 6 m below that of the baths. It is by no means impossible that there was some means of raising the water up into the town; but Sheppard Frere suggested that the shape of the earthwork bore more resemblance to a navigable canal, perhaps leading to docks on the river.[60] Only a short length survives today.

P. Chitwood, 'Lincoln's Ancient Docklands: The Search Continues', in G. L. Good, R. H. Jones, and M. W. Ponsford (eds.), *Waterfront Archaeology: Proceedings of the Third International Conference on Waterfront Archaeology*, CBA Research Rep. 74 (London, 1991), 170–1.

[55] F. Hill, *Medieval Lincoln* (Cambridge, 1948), 6, 12.

[56] Hallam, in Phillips, *The Fenland in Roman Times*, 255–6; P. P. Hayes and T. W. Lane, *The Fenland Project 5: Lincolnshire Survey, the South-West Fens*, EAA 55 (Cambridge, 1992); T. W. Lane, 'The Fenland Project in Lincolnshire: Recent Evaluations', *Fenland Research*, 8 (1993), 42; P. Murphy, 'Environmental Archaeology: Second Progress Report', *Fenland Research*, 8 (1993), 38; P. Murphy, 'Environmental Archaeology: Third Progress Report', *Fenland Research*, 9 (1994), 28; Rippon, *Transformation*, 70.

[57] Hayes and Lane, *The Fenland Project 5*, 84, 190; Rippon, *Transformation*, 70.

[58] Rippon, *Transformation*, 71.

[59] M. Leah and A. Crowson, 'Norfolk Archaeological Unit, the Fenland Management Project', *Fenland Research*, 8 (1993), 44–5; P. Murphy, 'Environmental Archaeology: A Review of Progress', *Fenland Research*, 7 (1992), 38; R. J. Silvester, *The Fenland Project 3: Marshland and the Nar Valley, Norfolk*, EAA 45 (1988), 54, 100, 104, 111, 156; Rippon, *Transformation*, 71.

[60] Frere, *Britannia*, 245.

Canals and River Diversions on the Continent, 400–1250 (Fig. 34)

On present evidence the Roman impetus to construct canals had largely spent itself by the middle of the second century. Certainly after the middle of the third century, increasing political turmoil and waning imperial power did little to encourage the initiation or maintenance of large-scale engineering schemes, except perhaps during a few brief periods. Apart from one abortive episode in Asia Minor, where a scheme first mooted by the younger Pliny soon after AD 111 to construct a canal from Lake Sophon to the Sea of Marmara was briefly resurrected by Justinian in the mid-sixth century, nothing more was done for nearly six centuries.[61]

The revival of classical technology appears to begin in Francia in the late eighth century, with Charlemagne's ambitious scheme to make a navigable connection between the Rhine and Danube near Weissenburg in Bavaria (Fig. 1). In 792 Charlemagne was faced with two simultaneous threats, a Saxon rebellion in the north and a threat from the Avars in the east, and the prospect of a

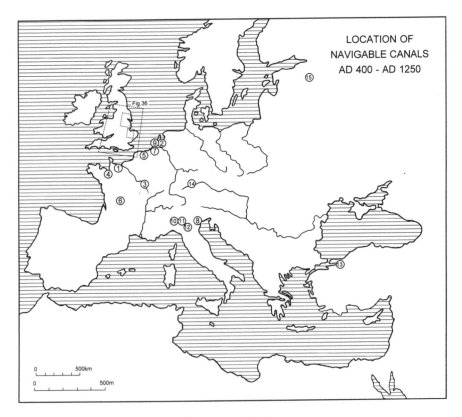

Fig. 34. Location of navigable canals, AD 400–1250.

[61] F. G. Moore, 'Three Canal Projects, Roman and Byzantine', *American Journal of Archeology*, 54 (1950), 108–10.

Table 8. Navigable Canals: Continental Europe and the Mediterranean, AD 400–AD 1250

No. on fig. 34	Date	Initiator	Location/terminal points	Approx. length
Completed schemes				
1	1066–83	Lanfranc	Nouvel Odon, Caen	1 km
1	1104	Robert Curthose	Robert Canal, Caen	1 km
2	c.1150	Unknown	Oude Gracht, Utrecht	c.12 km
3	1157–74	Count Henry	R. Seine, Troyes	2 km
4	?1158	King Henry II	R. Couesnon, Pontorson	c.7 km
5	1175	Benedictine Abbey of Anchin	R. Scarpe, Anchin, nr. Douai	Unknown
6	1175–1275	Unknown	Poitou	Unknown
7	c.1180	City of Bruges	Bruges–Damme–Sluys	c.15 km
8	1190–1201	Unknown	Padua–Monselice	18 km
9	1203	Count Willem II	Gouwe Canal, Alphen–Gouda	c.20 km
10	Late 12th cent. and 1269	City of Milan	R. Ticino–Milan	?c.50 km
11	13th cent.	Unknown	Muzza Canal, Cassano d'Adda	Unknown
12	13th cent.	City of Cremona	Taleata Canal	>90 km
Uncompleted proposals				
13	AD 527–65	Justinian	Lake Sophon–Sea of Marmara nr. Nicomedia, Asia Minor	16 km
14	AD 792–3	Charlemagne	R. Rezat–R. Altmühl nr. Weissenburg, Bavaria	1.6 km
15	12th cent.	Unknown	Novgorod–R. Volga	Unknown

prolonged conflict on two fronts. An efficient means of transferring troops and supplies across his realm would have been of great strategic value, both immediately and in the longer term, and there was also a considerable potential benefit to trade. How far this was inspired by Charlemagne's general admiration for the achievements of imperial Rome, and whether he was consciously imitating any particular model, can never be known; but as an example of an attempted linkage of two river systems over an inland watershed, it rivalled the plan of Lucius Vetus to connect the Saône and Moselle over seven centuries before. Construction of the canal 1.6 km long, over 27 m wide, and up to 6 m deep was commenced in the late spring of 793, and intended to link two headstreams, the rivers Rezat and Altmühl. Since the Altmühl was at a lower level, the two rivers could not have been connected without an intervening dam, graded for the winching or manhandling of vessels over it; otherwise the Rezat would simply have drained away towards the Danube, emptying the upper part of the connection towards the Rhine. In the event the project was defeated by heavy rainfall during the autumn of that year, and by geological problems. The sides of the cutting were undermined, constantly slumping into the excavated bed, and the project was abandoned just before Christmas. Considerable traces of the works, comprising a great cutting *c.*1100 m long and 45–75 m wide, can still be seen.[62]

A number of river diversions were undertaken in France between the ninth and twelfth centuries, for a variety of motives. The threat of Norman invasion seems to have prompted Charles the Bald (840–77) to cut new courses for the rivers at Senlis, Noyon, and Chartres. At Senlis part of the waters of the River Nonette south of the city was diverted into a contour channel, named as the *Nouvelle Rivière* by 1141, which seems to have served a dual function of draining marshy land to the east of the city while at the same time making it possible to flood the main valley in time of military emergency. At Chartres the River Eure was partly diverted into a second channel known as the *Vieux Fossés* to distinguish it from a further series of defensive moats cut around the walls after 1350; this had certainly taken place before 1050, and a dry moat is also documented by 1030. Similar works were later carried out at Châlons-sur-Marne, Beauvais, and Étampes. At Beauvais branches of the River Thérain were diverted through the city in the early eleventh century, apparently primarily to serve mills. At Troyes the River Seine was similarly diverted into a series of branches between the eleventh and thirteenth centuries, which ultimately worked some twenty-five mills, and similar works were undertaken at Rouen, Chartres, and Lille. At Provins the River Durtein and other watercourses were diverted during the first half of the twelfth century to drain a marshy area and to work mills.[63] None of these works was intended primarily, if at all, for navigation, but they all helped to keep alive and to develop the techniques and methods of water management on a large scale, skills essential for canal construction.

Following the selection of Caen by William of Normandy and his wife Matilda as a principal ducal residence, and their foundation of the two great abbeys of Saint-Étienne and La Trinité there in 1062-3, Lanfranc, then Abbot of Saint-Étienne, began dredging the River Orne and improving its tributary, the Odon, in order to develop a port suitable for seagoing ships. The Odon was diverted into a canalized course, the *Nouvel Odon*, cutting off a large meander loop along the southern side of the town between the abbey and the bridge of Saint-Pierre, an operation probably completed between 1066 and

---

[62] H.H Hofmann, 'Fossa Carolina: Versuch einer Zusammenschau', in W. Braunfels (ed.), *Karl der Grosse: Lebenswerk und Hachleben*, i: *Persönlichkeit und Geschichte*, ed. H. Beumann (3rd edn. Düsseldorf, 1967), 437–53; R. Koch, 'Neue Beobachtungen und Forschungen zum Karlsgraben', *Jahrbuch die Historischen Vereins für Mittelfranken*, 97 (1994–5), 1–16; J. Boyes, 'Inland Waterways', in I. McNeil (ed.), *An Encyclopaedia of the History of Technology* (London, 1990), 494; F. Gies and J. Gies, *Cathedral, Forge and Waterwheel: Technology and Invention in the Middle Ages* (New York, 1994), 69; J. Haywood, *Dark Age Naval Power: A Reassessment of Frankish and Anglo-Saxon Seafaring Activity* (rev. edn. Hockwold-cum-Wilton, 1999), 147–52 and map 5. See also above, pp. 2–3.

[63] A. E. Guillerme, *The Age of Water: The Urban Environment in the North of France, AD 300–1800*, trans. of *Les Temps de l'eau* (College Station, Tex., 1988), 26–32, 57–60, 65–7; M. Daumas, *Histoire générale des techniques* (Paris, 1962); trans. E. B. Hennesy, as *A History of Technology and Invention* (London, 1969), 550.

1083. Further works were undertaken under Robert Curthose, including the diversion of the Orne into two separate channels and the construction of the Robert Canal along the eastern side of the marshy Île Saint-Jean on the southern side of the town, completed in 1104. The Robert Canal (later enlarged into what is now the Bassin Saint-Pierre) provided further dock space in the twelfth century, and also helped to drain the marshes of the island.[64]

By the second half of the twelfth century river diversions and canalizations were beginning to appear in France in some numbers, though few of them were yet of any great length. Between 1157 and 1174 Count Henry of Troyes diverted part of the waters of the River Seine along a canal 2 km long, passing along the valley of the River Tréffoir, apparently crossing the Vienne by a wooden aqueduct into the city, passing through the count's palace, and exiting by the Cordé channel. In 1175 the Countess of Flanders granted the Benedictine monks of Anchin near Douai the right to dig a new bed for the River Scarpe, on condition that it was made navigable. Henry II of England seems to have undertaken a diversion of the River Couesnon at Pontorson in connection with the building of a new frontier castle on the borders of Normandy and Brittany. Marshland drainage again sometimes produced navigational benefits: renewed attempts to drain the marshes of Poitou between 1175 and 1275 involved the use of movable gates or sluices, permitting boats to travel along some streams.[65]

In the Low Countries, as in the English Fenland, it is especially difficult to disentangle drainage from navigation works, and ancient cuts from more recent works and the maze of natural waterways. The Rhine and Meuse break up into a braided series of distributaries, the most important of which are the Oude Rijn (Old Rhine), IJssel, Lek, and Waal. Natural tidal processes in the North Sea have always tended to clog up the more northerly branches while scouring out more southerly channels, and the outlet of the Oude Rijn became blocked by silt deposits during the post-Roman period. Until the late twelfth century merchants from the north aiming to reach the Rhine travelled over the Zuider Zee to the River Vecht to the dyke at Otterspoor. From there an artificial canal, made around 1150, allowed vessels to pass through the city of Utrecht and on southwards to the River IJssel, a distributary of the Lek and thence of the Rhine itself, at Het Gein, where another dyke had to be crossed. The main canal through Utrecht, the Oude Gracht, is still essentially the construction of the mid-twelfth century.[66] This route was to be challenged by the opening of new canals during the thirteenth century.

In Flanders and Holland canals with some form of chamber-lock or lift-lock are known from the late twelfth century onwards. In 1180 a lock gate was

---

[64] D. Bates (ed.), *Regesta Regum Anglo-Normannorum: The Acta of William I* (Oxford, 1998), nos. 45, 50, 53–4; Guillerme, *The Age of Water*, 38–9, 54–6.
[65] Guillerme, *The Age of Water*, 63–5; Daumas, *History of Technology and Invention*, 549–50.
[66] S. J. Fockema Andreae, 'The Canal Communications of Central Holland', *Jnl. of Transport History*, 4/3 (1960), 174–5.

constructed at Damme, on the canal from Bruges to the sea.[67] Wooden lift-locks and sluices were installed in the small River Reie allowing vessels access to Bruges by 1236; they were rebuilt in stone in 1394–6, with a sea-gate, a river-gate, and a masonry floor. The gates operated vertically, guillotine-fashion, and were counterbalanced with lead weights. By the fifteenth century, however, Bruges was fighting a losing battle against the silting of its canals, and increasingly goods were carried overland to and from the outport of Sluis.[68]

Soon after 1203 Count Willem I of Holland opened up the Gouwe Canal from the Oude Rijn near Alphen to the IJssel at Gouda, where a single-door sluice was constructed to be let down at high water. This provided an alternative passage into Flanders and the Rhine system for traders coming from the north, bypassing the older route through Utrecht. The Gouwe Canal was subsequently extended northwards by further waterways to link with the River IJ north of Haarlem. In 1220–6 the sea embankment was completed in the vicinity of the Spaarne Lake, and this must already have included some sort of lock. By 1253 the Spaarndam lock is said to have been 7.3 m wide, sufficient for the passage of seagoing vessels. Destroyed in the great flood of 1277, the lock was reconstructed on several occasions through the later middle ages.[69] Further development of the waterway system in the Low Countries continued throughout the later middle ages, with immense implications for trade and the prosperity of towns, but this story cannot be followed here.

In Italy works resumed in the Po basin in the twelfth century, beginning with the diversion of the tributary River Ticino above Pavia. Another canal, 18 km long, built between 1190 and 1201, linked Padua with Monselice. In the late twelfth century a water supply and irrigation canal was constructed from the Ticino below its outlet from Lake Maggiore. In 1269 this was enlarged and made navigable as the Naviglio Grande, and was used for carrying marble for the building of Milan Cathedral after 1385. A lock was built at Milan near the Porta Ticinese on the southern side of the city to allow boats into the city moat, and the late fourteenth-century lock chamber there is preserved. East of Milan the River Adda was diverted at Cassano d'Adda to form the Muzza Canal in the 1220s. Another thirteenth-century venture was the Taleata Canal, over 90 km long, built by the merchants of Cremona to escape the tolls taken on the River Po by their rivals at Mantua. Irrigation and navigation works continued in Lombardy and Piedmont through the fourteenth and fifteenth centuries.[70]

In medieval Russia attempts to link Novgorod by artificial waterway with the River Volga to the east are reported in the twelfth century, but nothing

---

[67] C. Singer, E. J. Holmyard, A. R. Hall, and T. I. Williams (eds.), *A History of Technology*, ii (Oxford, 1956), 657.

[68] D. Nicholas, *The Later Medieval City* (London, 1997), 40.

[69] Singer et al. (eds.), *History of Technology*, ii. 688–9; Fockema Andreae, 'Canal Communications', 175.

[70] Boyes, in McNeil (ed.), *Encyclopaedia of the History of Technology*, 500; R. Magnusson and P. Squatriti, 'The Technologies of Water in Medieval Italy', in Squatriti (ed.), *Working with Water*, 228–31.

seems to have been achieved.[71] However, rivers such as the Volkhov, Lovat, and Dnieper continued to carry much trade between the Baltic and Black Sea.

From the examples assembled above (and, almost certainly, more remain to be discovered) it is clear that the benefits and technical principles of modifying natural waterways and constructing artificial ones were well understood in continental Europe in the early middle ages. It should, therefore, come as less of a surprise to find evidence of the same processes taking place in Britain.

The Use and Restoration of Roman Canals in Medieval England

There seems to be little evidence from the Continent that any Roman canals there remained open. By contrast, and perhaps unexpectedly, at least a couple of the Roman canals in Britain show signs of use in the early middle ages. Either they had remained passable throughout the intervening six or seven centuries, or, perhaps more likely, had survived in a sufficiently visible form to encourage re-excavation when circumstances prompted it.

The clearest archaeological and documentary evidence pertains to the Foss Dyke in Lincolnshire (Fig. 40). From the quantity of coins and ornamental metalwork found at Torksey, this place had clearly become an important trading centre by the eighth and ninth centuries. Much of the water-borne trade was doubtless coming up the Trent, but there may also be an implication that the Roman canal had been reopened. By contrast, fewer tenth-century coins have been found at Torksey and very little tenth-century Torksey pottery was reaching Lincoln, perhaps because the dyke had again become blocked through lack of maintenance. Substantial quantities of Torksey wares were once more finding their way to Lincoln by the early eleventh century, suggesting that the waterway had again been reopened; but the use of Torksey pottery in Lincoln dwindled sharply after the Norman Conquest, and another period of neglect of the dyke may in part explain the considerable decline of Torksey recorded in the Domesday survey.[72]

According to Symeon of Durham the Foss Dyke was once more recut by order of Henry I and reopened for traffic in 1121.[73] Its renewal helped to revive the prosperity of both Torksey and Lincoln in the twelfth century. Wool produced in the English midlands and destined for export to Flanders regularly travelled down the Trent and then along the Foss Dyke to Lincoln, which became a port and staple town on the strength of its transport facilities.[74] A complaint against Robert of Donham, bailiff of William of Valence, recorded in the Hundred Rolls, states that he was levying excessive tolls of a halfpenny on every ship passing along the Foss Dyke to the Trent and then upstream to

---

[71] P. Hanson, 'Soviet Inland Waterways', *Jnl. of Transport History*, 3/1 (1963), 3.
[72] P. Sawyer, *Anglo-Saxon Lincolnshire*, A History of Lincolnshire III (Lincoln, 1998), 197.
[73] Symeon of Durham, *Historia Regum*, ed. T. Arnold, Rolls Ser. 75, ii (London, 1885), 260.
[74] Hill, *Medieval Lincoln*, 14.

Dunham, and that his receipts for the year amounted to half a mark.[75] This would imply some 160 ships a year travelling along the Foss Dyke to go up the Trent, and there must have been as many or more going downstream also.

By the end of the thirteenth century, however, the passage seems to have deteriorated once again. In the years after 1299 goods bought by the bursar of Durham Priory at Boston fair were being taken from Lincoln to Torksey by cart rather than water.[76] It still carried occasional traffic into the early fourteenth century, as in 1319, when a party of Cambridge clerics travelled along it on their way to York; but it may be significant that this journey took place in winter, when the canal would have contained a greater depth of water.[77]

The deteriorating condition of the Foss Dyke caused serious concern in Lincoln, and in 1335 a commission was appointed to investigate and resolve the problem; but funds collected for the clearance of the waterway were said to have been embezzled.[78] A second commission was appointed in 1365, since it was claimed that the canal was so obstructed by silting and grass growth and by the actions of men who had lands, meadows, and pastures on either side driving their cattle over it, that boats could no longer pass through it to and from Lincoln.[79] Clearly nothing was achieved, since a third commission was appointed in 1376, following further presentments about the impassability of the dyke.[80] Though there is one final reference in April 1395 to the 'scouring of a canal (*fossatum*) whereby boats come to the city with divers victuals in greater numbers than they used to',[81] the task eventually proved too great a burden for the burgesses of Lincoln; but the loss of this important transport link contributed to the commercial stagnation of their city throughout the later middle ages, while Torksey went into terminal decline, and even Boston was affected.[82] Despite one abortive attempt in 1518 the Foss Dyke was not reopened until 1672.[83]

Portions of the Lincolnshire Car Dyke were also probably still occasionally used for transport in the middle ages. A forged writ of Edward the Confessor containing possibly genuine material records an agreement whereby the monks of Ramsey gave Peterborough Abbey 4,000 eels during Lent in exchange for stone from Peterborough's Barnack quarries.[84] The most likely route for the

---

[75] J. Caley and W. Illingworth (eds.), *Rotuli Hundredorum Temp. Hen. III & Ed.I.*, Record Commissioners (London, 1818), i. 320a.

[76] J. T. Fowler (ed.), *Extracts from the Account Rolls of the Abbey of Durham*, ii, Surtees Soc. 100 (Durham, 1898), 495–6, 512, 532; Hill, *Medieval Lincoln*, 311.

[77] F. M. Stenton, 'The Road System of Medieval England', *EcHR* 7 (1936), 1–21 (reprinted in D. M. Stenton (ed.), *Preparatory to Anglo-Saxon England* (Oxford, 1970), 234–52), 20; Langdon, 'Inland Water Transport'.

[78] *Cal. Pat. R. 1334–8*, 148, 203.   [79] *Cal. Pat. R. 1364–7*, 138.

[80] *Cal. Pat. R. 1374–7*, 322.   [81] *Cal. Pat. R. 1392–6*, 414.

[82] G. Platts, *Land and People in Medieval Lincolnshire*, History of Lincolnshire IV (London, 1985), 219–20, 228.

[83] M. W. Barley, 'Lincolnshire Rivers in the Middle Ages', *Lincolnshire Architectural & Archaeological Soc. Reports & Papers*, NS 1 (1938), 10–11; Jones, 'River Navigation', 64–9.

[84] S 1110.

carriage of the Barnack stone would have been down the River Welland to Deeping and then southwards along the Car Dyke to Peterborough, and thence by the River Nene to Ramsey. A lost load of dressed Barnack stone was found in the bed of the cut at Morton, 17 km north of the junction with the Welland at Deeping, and there is an unverified tradition that Great Tom, Lincoln Cathedral's largest bell, was carried there by canal from Peterborough; even if untrue, this may preserve a folk-memory of local transport along the Car Dyke. However, other sections had become impassable by the twelfth century. Just to the east of Sempringham Hall Farm the construction of a moated enclosure, possibly a grange of Sempringham Priory, encroached right across the line of the dyke.[85]

New Canals in England before 1066

During the early middle ages monastic houses developed a particularly formidable range of skills and experience in many aspects of water management, including water supply, sewerage, land drainage, and the operation of mills and fishponds.[86] As permanent corporations their expectation of long-term security of landholding and their ability to look to a future beyond the lifespan of the individual encouraged investment in large-scale projects in a way that not even the Crown could match. Many of the greater English monasteries reformed under Benedictine rule in the tenth century by Dunstan, Æthelwold, and Oswald lay close to rivers or marshlands. It is in this period that we find the monasteries emerging as new pioneers in the diversion of rivers and the construction of canals.

Dunstan's own monastery at Glastonbury stood in a typical marshland-edge situation. Surviving earthworks and antiquarian accounts suggest that an artificial watercourse once connected the abbey and town with the River Brue, 1.5 km away, following the contour along the north flank of Wearyall Hill. The turnpike road to Street was built over part of its course in 1821. Excavation on this alignment by Charles and Nancy Hollinrake (below, pp. 235–8) produced twelfth- and thirteenth-century pottery from the upper fill, and a timber revetment stake gave a radiocarbon date centred in the mid-tenth century, the very period when Dunstan was reforming and rebuilding the abbey. This watercourse may well have been deliberately filled when it was incorporated into the abbot's deer park of Wirrall in the early thirteenth century.[87]

[85] C. W. Phillips, 'The Car Dyke, Lincolnshire', *Antiquity*, 5 (1931), 106–9; Platts, *Land and People in Medieval Lincolnshire*, 154.

[86] Summarized most recently in C. J. Bond, 'Monastic Water Management in Great Britain: A Review', in G. Keevill, M. Aston, and T. Hall (eds.), *Monastic Archaeology: Papers on the Study of Medieval Monasteries* (Oxford, 2001), 88–136.

[87] C. Hollinrake and N. Hollinrake, 'A Late Saxon Monastic Enclosure Ditch and Canal, Glastonbury, Somerset', *Antiquity*, 65 (1991), 117–18; C. Hollinrake and N. Hollinrake, 'The Abbey Enclosure Ditch and a Late-Saxon Canal: Rescue Excavations at Glastonbury, 1984–1988', *Proc. Somerset Archaeol. and Nat. Hist. Soc.* 136 (1992), 73–94.

Table 9. Navigation Canals, AD 400–AD 1250: England

Date	Initiator	Location/terminal points	Length
Completed schemes			
Reopened		Foss Dyke, 　Lincoln–Torksey	17 km
Reopened		? part of Lincs. Car Dyke	Unknown
Mid-10th cent.	?Dunstan	Glastonbury, R. Brue 　–Abbey	1.5 km
?10th cent.	?Ramsey Abbey	Cnut's Dyke	>10 km
c.1052–65	Abbot Orderic	?Swift Ditch, Abingdon	2 km
1097–1123	Abbot Hugh	Selby–Monk Fryston	12 km
1100–17	Abbot Faricius	R. Thames, Abingdon	?c.2.5 km
?c.1135–40	Rievaulx Abbey	R. Rye–Rievaulx Abbey	?c.0.6 km
?post-1150	?Bishop of Ely	Ely Cut, Great Ouse	c.9 km
c.1154–89	Tupholme Abbey	Tupholme–R. Witham	2 km
c.1160–82	Meaux Abbey	Eschedike	2 km
c.1175	Sawtry Abbey	Monks' Lode, 　Sawtry–R. Nene	c.8 km
c. 1190	Abbot Walter	Cornmill Stream, 　Waltham Abbey	c.2.5 km
c.1200	Bullington Priory	Bullington–R. Witham	
1210–20	Meaux Abbey	Monk Dyke	6 km
1221–35	Meaux Abbey	Forth Dyke	5 km
?c.1230–50	Glastonbury/Wells	R. Brue, 　Street–Highbridge	c.25 km
c.1230	Glastonbury Abbey	Pilrow Cut	c.12 km
1240	Bristol corporation	R. Frome	c.0.3 km
1249–69	Meaux Abbey	Skerndike	3 km
1277	King Edward I	R. Clwyd, Rhuddlan	c.4.5 km
?13th cent.	Unknown	R. Ouse, Littleport– 　Brandon Creek	c.6 km
Unknown	Castle Acre Priory	Castle Acre Priory 　–R. Nar	c.0.2 km
Unknown	Butley Priory	Butley River–Butley 　Priory	c.2 km
Unknown	Barlings Abbey	Barlings Eau	1.1 km
Unknown	Bardney Abbey	Bardney–R. Witham	0.6 km
Unknown	Norwich Cathedral	R. Wensum–Cathedral	0.25 km
Unknown	York Minster	Bishop Dyke, Sherburn- 　in-Elmet–Cawood	10 km
Unknown	Unknown	R. Cam, Bottisham Lode	3.6 km
Unknown	Unknown	R. Cam, Swaffham 　Bulbeck Lode	5.1 km
Unknown	Unknown	R. Cam, Reach Lode	4.6 km
Unknown	Unknown	Cottenham Lode	c.6.5 km
Unknown	Unknown	Little Ouse, Botany 　Bay–Brandon Crk.	c.10 km
Unknown	Unknown	Portions of R. Axe	Unknown

Three pioneer water management projects have been attributed to Dunstan's friend and colleague Æthelwold, successively Abbot of Abingdon (*c.*954–63) and Bishop of Winchester (963–84). At Abingdon he is said to have diverted part of the Thames into a mill-leat and dug a sewerage channel from the abbey reredorter, while at Winchester he carried out a more ambitious scheme, diverting the River Itchen into a series of artificial channels running across the city.[88] While none of these schemes was intended for navigation, they may have initiated an interest and expertise in the manipulation of watercourses which bore fruit in later years. Certainly Abingdon Abbey was involved in at least one, and possibly more than one local diversion and alteration of the River Thames between the mid-eleventh and early twelfth centuries. According to the abbey's chronicle, in the time of Abbot Orderic (1052–65) the citizens of Oxford petitioned for the abbey's consent to divert the navigation channel through the abbey's meadow to the south, in order to allow their vessels more convenient passage upstream, bypassing the dangerous shallows near Barton Court. The watercourse made for this purpose is almost certainly the backwater now known as the Swift Ditch (Figs 35, 61). In exchange for this facility 100 herrings a year were to be paid to the abbey cellarer, and in 1110 the abbot successfully sued some Oxford boatmen who had attempted to evade payment of the agreed toll. In an earlier investigation it was suggested that the account in the chronicle implied a second river diversion just upstream near Thrupp. This view was partly influenced by the fact that the Thrupp backwater today is a minor ditch, whereas a sixteenth-century map shows it as a much more considerable stream.[89] However, John Blair argues below (pp. 266–8) that the reference to Thrupp served only to locate the difficult stretch of the 'natural' river, and that only a single canalization, that of the Swift Ditch, need be postulated.[90]

Martin Biddle and Derek Keene have drawn attention to a charter of 1045 by which Edward the Confessor granted land at South Stoneham to the Old Minster at Winchester. The charter-bounds run 'Of þære ealden Icenan on ufwyrd þonæ orcerd on þa niwan ea,' and they suggest that the 'new river' may have been constructed to improve navigation.[91] It may have been a relatively recent feature of the landscape, since it is not named on an earlier charter of 990 × 992 which covers the same ground. Chris Currie (below, pp. 246–50) has identified this watercourse with a broad channel up to 15 m wide to the south of the present river, bypassing both Wood Mill and Gater's Mill, still

[88] J. Stevenson (ed.), *Chronicon Monasterii de Abingdon*, Rolls Ser. 2, (London, 1858), i. 480–1, ii. 270, 278–80, 282, 285; M. Biddle and D. J. Keene, 'Winchester in the Eleventh and Twelfth Centuries', in M. Biddle (ed.), *Winchester in the Early Middle Ages*, Winchester Studies I (Oxford, 1976), 282–4.

[89] *Chron. Abingdon*, i. 480–1, ii. 282; C. J. Bond, 'The Reconstruction of the Medieval Landscape: The Estates of Abingdon Abbey', *Landscape History*, 1 (1979), 59–75.

[90] Personal communication, 4 August 2005.

[91] S 1012; Biddle and Keene, 'Winchester in the Eleventh and Twelfth Centuries', 270–1; C. K. Currie, 'Saxon Charters and Landscape Evolution in the South-Central Hampshire Basin', *Proc. Hants. Field Club and Archaeol. Soc.* 50 (1995), 111–15.

Fig. 35. Abingdon (Berks.): Swift Ditch, probably to be identified with the canal built by Abbot Orderic just before the Norman Conquest (see also p. 268). Visible in the background is the pound-lock put in by the Oxford-Burcot Commission in or shortly before 1628.

visible in 1940, but since filled and levelled.[92] One purpose of the channel may have been to permit salmon access to the higher reaches of the river for spawning, but it seems an extravagant width for that purpose alone. If it was used by boats, it had been forgotten by the 1270s, when the navigation of the Itchen and its obstruction by mills was the subject of a royal inquisition.

In the north-east midlands the River Idle formerly flowed north into the old course of the Don on Hatfield Chase, but before the end of the eleventh century it had been diverted eastwards below Misson to join the Trent at West Stockwith. The new cut, the Bykers Dyke, was some 8.5 km in length.[93]

The most extensive network of waterways during the late Anglo-Saxon period was, once again, in the Fenland. Of the three main estuaries on the southern shore of the Wash, the central estuary below Wisbech at that time provided the principal drainage outlet from the Fens for the waters of the Nene, Great Ouse, Cam, Lark, Little Ouse, and Wissey. By contrast, the Lynn estuary to the east provided an outlet only for the Nar and its tributary the Gay.[94] Within this maze of natural watercourses, improved watercourses, canals, and drains of all dates, isolating those cut primarily for navigation and determining

[92] C. K. Currie, 'A Possible Ancient Water Channel around Woodmill and Gater's Mill in the Historic Manor of South Stoneham', *Proc. Hants. Field Club and Archaeol. Soc.* 52 (1997), 89–106.
[93] See Cole, above, p. 65.
[94] H. C. Darby, *The Medieval Fenland* (Cambridge, 1940; repr. Newton Abbot, 1974), 94–6.

their precise date and who initiated their construction is a particularly difficult problem. However, some cuts do seem to originate in the period before the Norman Conquest.

The most substantial works appear to have been along the course of the River Nene. Cnut's Dyke extends for over 16 km across the Fens to the south-east of Peterborough. A Roman origin has been claimed for this in the past, primarily because of the straightness of much of its course.[95] Certainly it was not designed as a drainage work, because it cuts right across the natural lines of drainage of the Bedford Middle Level. Nor, however, is there any real evidence to support a Roman date; no significant Roman sites are known along any part of its length or near its terminus. The most likely explanation is that it was made to facilitate the bringing in of stone from the direction of Peterborough for construction work at Ramsey Abbey, perhaps at the time of the abbey's first foundation in the tenth century.[96] Cnut's Dyke bypassed the meandering course of the Nene through Whittlesey Mere. The King's Dyke, probably of similar date, provided a more northerly cut across the great loop of the Nene, but may have been more important as a drain. Lower downstream the Nene originally flowed around the northern margins of the gravel island at March, but David Hall's survey has suggested that the short cut through the gravels was made in late Saxon times, possibly by Ely Abbey for drainage purposes.[97]

The River Cam shows signs of canalization at Cambridge which Jeremy Haslam has associated with the foundation of Edward the Elder's borough in 917.[98] Below Cambridge three straight artificial canals, Bottisham Lode (3.6 km long), Swaffham Bulbeck Lode (5.1 km long), and Reach Lode (4.6 km long), link the uplands to the east with the River Cam, cutting across the line of the natural drainage. Claims have also been made for a Roman origin for these canals in the past, along with Wicken Lode, Monks' Lode, and Burwell Old Lode, partly because of their straightness and partly because the Reach Lode in particular has produced quantities of Romano-British pottery. The fact that the early post-Roman Devil's Dyke subtly changes direction in order to terminate at the head of Reach Lode has also been put forward as evidence that the latter was already there when the dyke was constructed.[99] However, David Hall's survey has found no satisfactory evidence for a Roman date for any of these waterways; there are no Roman sites at their landward ends, and the Reach Lode actually seems to have cut through a Roman villa. Again a late Anglo-Saxon or early medieval construction date now seems more likely. All had quays at their landward end in the middle ages, and it seems likely that their primary function was for transport, though they would also assist local

---

[95] Phillips (ed.), *The Fenland in Roman Times*, 186.

[96] Hall, *The Fenland Project 6*, 42.   [97] Hall, *The Fenland Project 2*, 46.

[98] J. Haslam, 'The Development and Topography of Saxon Cambridge', *Proc. Cambridge Antiquarian Soc.* 72 (1983).

[99] Fowler, 'Fenland Waterways I', 113–14; RCHME, *An Inventory of Historical Monuments in the County of Cambridge*, ii. *North-East Cambridgeshire* (London, 1972), pp. xxvii, liv–lv, lxv.

drainage by carrying water from the higher ground, which would originally have drained through the buried channels in the fen, more directly into the Cam.[100]

West of the Cam the Cottenham Lode, which intersects the Roman Car Dyke, also looks largely artificial. An ancient legend that the builders of Cottenham church repeatedly found their stone being removed by the Devil from the old village centre to the place where the church now stands, 1 km to the north-east, may enshrine a genuine memory of the replacement of the original wooden church in stone and the decision to relocate it to a place to which stone could be brought by water.[101] Cottenham Lode certainly existed by the time of the Norman Conquest, when it was used by William the Conqueror in his assault on Ely.[102]

Numerous small hythes are recorded elsewhere in the Fens, sometimes at a distance from major waterways, which implies that many smaller streams were navigable in the early middle ages (see also Cole and Gardiner above). Little Downham hythe was connected to the Old Croft River by the Darcy Lode, a canalized watercourse largely followed by parish boundaries. This can only have functioned while the Old Croft River formed the main outlet of the Great Ouse.[103] A hythe at Willey near Chatteris, recorded in 1251, implies that Fenton's Lode was navigable, providing an artificial waterway linking the March island with the Great Ouse to the south.[104]

New Canals in England after 1066 (Fig. 36)

Did the Norman Conquest represent a significant new stage in the development of inland water transport? Lanfranc's works at Caen, mentioned earlier, demonstrate the Norman ability to undertake major works of this type, and the new programme of building stone castles and rebuilding monasteries and churches on a much more massive scale, which continued through the twelfth century and beyond, provided a recognized need. Even so, the quantity and quality of direct documentation remains limited until the very end of the period under discussion here.

Canals in the Fenland after the Conquest (Fig. 37)

The quarries of Barnack continued to supply stone for many buildings in East Anglia, particularly during the eleventh and twelfth centuries, and continuing on a smaller scale into the fifteenth century. The transport of building stone from this and other east midland quarries over the Fenland rivers has been

---

[100] Hall, *The Fenland Project 10*, 112.
[101] J. R. Ravensdale, *Liable to Floods: Village Landscape on the Edge of the Fens, AD 450–1850* (Cambridge, 1974), 123.
[102] E. O. Blake (ed.), *Liber Eliensis*, Camden Soc., 3rd Ser. 92 (London, 1962), 182.
[103] Hall, *The Fenland Project 6*, 81.   [104] Ibid. 94.

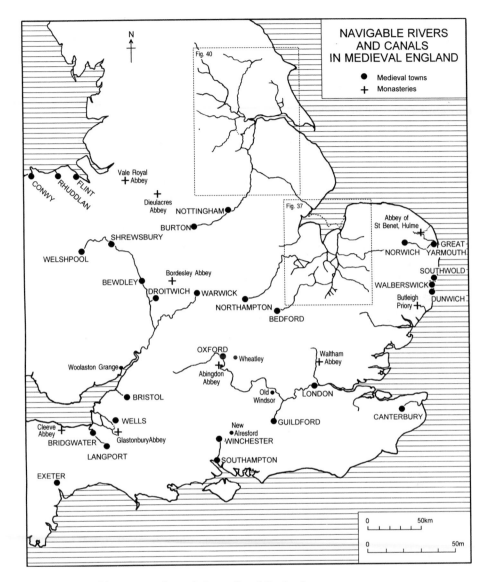

Fig. 36. Navigable rivers and canals in medieval England.

examined by Jennifer Alexander.[105] Discussion here will be confined to the evidence for the use of artificial waterways.

The Cam lodes previously described were used by barge traffic throughout the middle ages and after. They are of particular interest for the small ports which developed on the edge of the fen at their heads. Reach Lode, recorded in a document of the late eleventh century, seems to have been the most

---

[105] J. S. Alexander, 'Building Stone from the East Midlands Quarries: Sources, Transportation and Uses', *MA* 39 (1995), 107–35, esp. 124–7.

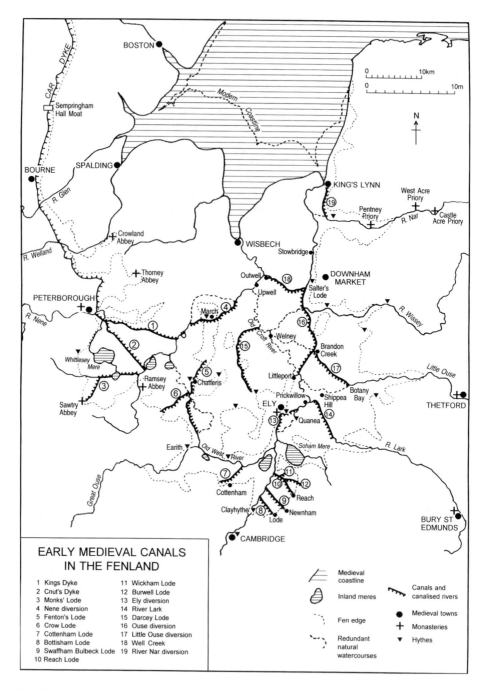

Fig. 37. Early medieval canals in the Fenland. Compare Figs. 15 and 20.

important. The small interchange port and village of Reach developed astride the Devil's Dyke which served as the parish boundary between Burwell and Swaffham Prior at the head of the lode. By 1131 Henry I had granted to the burgesses of Cambridge a monopoly of water-borne trade on the Cam and freedom from tolls, and a fair at Reach was confirmed to the burgesses in 1201.[106] By the early fourteenth century the small port at Reach had become of more than local significance, shipping local agricultural produce, timber, iron, and clunch. Clunch from Reach was used at Cambridge Castle in 1295, at Ely Cathedral, and at a number of Cambridge colleges.[107] Improvements to the waterway may have been carried out after the end of the middle ages, since boats continued to come up to the quays at Reach into the nineteenth century.

Of the other Cambridgeshire canals, Bottisham Lode and Swaffham Bulbeck Lode are both recorded by name in the Hundred Rolls of 1279, but there is little evidence that either of them carried more than local traffic during the middle ages.[108] Bottisham Lode developed a small interchange port called simply Lode at its head, but this never rivalled Reach. Similarly, a new medieval hamlet called Newnham (now Commercial End) developed near the head of Swaffham Bulbeck Lode, possibly as a small port, beyond the northern end of Swaffham Bulbeck village. Significant lengths of both canals were recut on new alignments immediately to the north-east, probably around the middle of the seventeenth century, but in both cases considerable traces of the earlier cuts, followed by the parish boundaries, can still be seen alongside the current watercourses.[109]

East of Ely the Great Ouse had originally followed a meandering course through Middle Fen, past the low islands of Stuntney and Quanea, now represented by Rolls Lode. The Ely Cut was dug some time in the early middle ages between Cawdle Fen to the south and Prickwillow to the north, in order to bring the waters of the river closer to the town. A map of 1826 shows its course prior to the further alterations carried out in 1830 for improved drainage,[110] and its abandoned lower course can still be traced on the ground. The diversion could have been undertaken to bring building stone into the abbey as early as the time of the first abbot, Brihtnoth (*c*.970–999), for it is recorded that his steward was digging new ditches elsewhere to fix the bounds of the abbey's property; but a passage in the *Liber Eliensis* implies that access to the island by boat had become very difficult by the middle of the twelfth century. The elevation of Ely to a see in 1109 increased the importance of the place and provided a succession of bishops as able as their monastic predecessors to command the resources for such an undertaking. On balance the most likely period for the diversion of the river would seem to be between 1150 and 1300.

[106] H. M. Cam, 'The City of Cambridge', in *VCH Cambs.* iii (1959), 6, 32, 91.
[107] F. R. Chapman (ed.), *The Sacrist Rolls of Ely* (Cambridge, 1907), ii; F. M. Page, 'Industries', in *VCH. Cambs.* ii (1948), 366; *RCHME Cambs.*, ii: *North-East Cambs.*, pp. lv, lxv–lxvi, 85–6, 89–90.
[108] *Rotuli Hundredorum*, ii. 484.
[109] *RCHME Cambs.*, ii: *North-East Cambs.*, pp. lxv–lxvi, 74, 81–3, 96–7, 113.
[110] Cambridge RO, Q/Rum2; Hall, *The Fenland Project 10*, 40.

By the later middle ages the economic focus of the town had moved eastwards, away from the older market places west and north of the abbey, and down to the Waterside quays below the further end of Fore Hill.[111] Excavations undertaken in 2000 by the Cambridge Archaeological Unit under the direction of Alison Dickens have revealed three broad channels cut at right-angles to the river some time after 1400 to provide additional wharf and dock space, abandoned and infilled in the seventeenth century. Despite the diversion of the river, the old course also still carried sufficient water for the hythe at Quanea to continue functioning long after the twelfth century.[112] Early medieval pottery found on the roddon of Rolls Lode near Quanea seems more likely to represent material dropped from boats than evidence of settlement.[113]

By the thirteenth century the silting of the Nene estuary below Wisbech had caused profound changes to the pattern of Fenland drainage, and also prompted a series of cuts to provide alternative routes to the sea. First the Little Ouse, which had originally flowed west past Shippea Hill Farm to join the Great Ouse near Littleport, was diverted north-westwards from Botany Bay to follow a canalized course through Brandon Creek, passing west of the Southery and Hillgay island to Salters' Lode, intercepting the River Wissey on its way, and ultimately reaching the sea at King's Lynn. The abandoned original course of the Little Ouse can still be followed for much of its length by a meandering line of ditches followed by parish boundaries.[114] The Great Ouse, which had originally flowed down the Old Croft River course through Welney to join the Old River Nene at Upwell above Wisbech, was then diverted north-eastwards from Littleport to Brandon Creek, there joining the Little Ouse. This provided a much more direct connection between Ely and King's Lynn.[115] A description of the episcopal manor of Ely in the Old Coucher Book of Ely in 1250 already refers to the 'Old Welney' River, which implies that it was no longer the main channel. Dugdale quotes a late thirteenth-century source which seems to imply that some part of the water formerly draining northwards through Wisbech was now diverted down the Well Creek into the new Ouse system at Salters' Lode.[116]

By the late thirteenth century the Well Creek had become an important route between the River Nene and King's Lynn. However, it was reported in 1301 that it had become obstructed near Outwell, so that boats were forced to take

[111] Fowler, 'Fenland Waterways II', 23; C. Taylor, *The Cambridgeshire Landscape* (London, 1973), 248–9.

[112] Fowler, 'Fenland Waterways II', 18; Summers, *The Great Ouse*, 32; Hall, *The Fenland Project 10*, 40.

[113] T. C. Lethbridge and G. Fowler, 'Excavations in the Bed of the Old Cam at Quanea (Rollers Lode)', *Proc. Cambridge Antiquarian Soc.* 33 (1933), 129–32.

[114] Fowler, 'Fenland Waterways II', 20.

[115] R. J. Silvester, '"The Addition of More-or-Less Undifferentiated Dots to a Distribution Map"? The Fenland Project in Retrospect', in J. Gardiner, *Flatlands and Wetlands: Current Themes in East Anglian Archaeology*, EAA 50 (1993), 34–5.

[116] Dugdale, *History of Imbanking and Drayning*, 393–5; Fowler, 'Fenland Waterways I', 112; Summers, *The Great Ouse*, 9–18; Darby, *Medieval Fenland*, 94–6.

Fig. 38. The Monks' Lode, seen here north of Hingney Grange (Hunts.).

the more circuitous route up the Old Croft River by Welney to Littleport, there joining the diverted Great Ouse. Jurors at the resulting inquiry gave witness that the Nene had formerly flowed from Peterborough through March and the Well Creek to Lynn, and, moreover, that there had even been a direct navigation from Crowland to Lynn.[117] The carriage of weighty goods such as building material continued to provide a motive for construction of artificial waterways. In 1176 the pope confirmed the right of the Cistercian monks of Sawtry to the watercourse which they had made at their own cost to carry building stone to their church. This was the canal now known as the Monks' Lode (Fig. 38), extending for nearly 8 km from the River Nene to the abbey. The stone came from the Barnack quarries, and after the first couple of kilometres the whole journey could then have been made by water down the Welland, along the Roman Car Dyke to the Nene, and finally up the canal to the abbey, where a complex of earthworks interpreted as docks and quays has been reported.[118] The construction occasioned a long dispute with the Benedictines of Ramsey, 8 km to the east, which was finally resolved in 1192 when the Ramsey monks conceded to those of Sawtry the right to use the canal and to build a rest-house for the men working the stone barges.[119] Four large blocks of Barnack stone recovered from a barge sunk in Whittlesey Mere, one of them bearing a chiselled destination symbol, may have been on their

[117] Dugdale, *History of Imbanking and Drayning*, 301; Darby, *Medieval Fenland*, 96.
[118] S. Inskipp Ladds, 'Ancient Earthworks', in *VCH. Hunts.* i (1926), 302–3.
[119] *Cart. Mon. de Rameseia*, i. 166, ii. 347.

way to either Ramsey or Sawtry Abbey.[120] In later years the canal may have been used for the carriage of peat and corn.[121] By 1342, however, an inquiry into the obstruction of various watercourses in the fens of Sawtry, Conington, and Wood Walton found that the Monks' Lode had become blocked through neglect of repair and a season of drought.[122]

One factor in the diversion of the Little Ouse might have been the demand for better access for boats carrying stone up to the Cluniac priory of Thetford. A similar case might be made for the lower portion of the River Lark, previously put forward as a possible Roman canalization, and for the northward diversion of the River Nar below Setchey Hythe: the former would have provided improved access towards the Benedictine abbey of Bury St Edmunds, while the latter served the Augustinian priories of Pentney and West Acre and the Cluniac priory of Castle Acre. It is of some interest that all those rivers of the Fens whose courses reveal major medieval alterations were linked with important Benedictine, Cluniac, or Augustinian monasteries founded before the middle of the twelfth century. Only the River Wissey, which had no early monastic houses along its course, escaped significant modification.[123]

Canals in the Somerset Moors after the Conquest

The Somerset moors were dominated by two great landowners, Glastonbury Abbey and the bishops of Wells. Here a series of river diversions, canalizations, and artificial cuts were undertaken during the thirteenth century, and three of these in particular had significant transport implications: the canalization of the River Axe, the diversion of the River Brue, and the construction of the Pilrow Cut. These schemes are discussed in detail by Stephen Rippon below, and will not be considered further here.

Monastic Canal Construction in England, 1066–1250

There are numerous references to the use of rivers for transport by monastic houses in the post-Conquest period, and the remains of riverside quays survive on several sites, for example at the remote Benedictine abbey of St Benet Hulme (Norfolk). Three massive soleplates of a timber quay, dated by dendrochronology to the mid-twelfth century, have been recorded in the intertidal zone of the Severn estuary at the mouth of Grange Pill below Woolaston. Early in the thirteenth century this quay was extended seawards using stone blocks and upright timber posts, so that it would then have accommodated seagoing ships as well as river craft, at all stages of the tide. Tintern Abbey had acquired the manor of Woolaston in 1131 and established five granges within its bounds,

---

[120] D. Purcell, *Cambridge Stone* (London, 1967), 98, pl. 47.
[121] Hall and Coles, *Fenland Survey*, 137.   [122] *Cart. Mon. de Rameseia*, i. 174–80.
[123] A. K. Astbury, *The Black Fens* (1958; repr. Wakefield, 1970), 157; Silvester, in Gardiner, *Flatlands*, 37.

which would have produced marketable surpluses of grain, livestock, dairy products, wool, hides, fish, timber, and probably iron ore. It seems likely that the abbey built the quay in order to export this produce to the markets of Bristol, Gloucester, and south Wales. The quay was modified after serious damage, perhaps in the early fifteenth century, when the lower part was abandoned, but it remained in use for some decades after the Dissolution.[124] Rolt quotes a tradition that Caen stone for Tewkesbury Abbey was carried up the Severn and then up the diminutive Swilgate Brook on spring tides.[125] In 1233 the monks of Canterbury Cathedral Priory were permitted to carry six shiploads of stone down the Thames for the completion of their refectory.[126] Abingdon Abbey had a quarry at Wheatley from which stone was carted to Sandford-on-Thames and then taken downriver to Abingdon by boat.[127] In 1276 the Abbot of St Mary's, York, complained that Edmund, Earl of Cornwall, had prevented the free passage of his ships and boats carrying victuals and other goods to the abbey along the Yorkshire Ouse and Ure (Fig. 40), a right which he claimed the abbey had enjoyed from time immemorial.[128] Rievaulx Abbey also claimed a right to have ships in the River Ouse. The sacrist of Ely regularly travelled by boat along the Ouse and Cam to Barnwell and Lynn during the fourteenth century, and when the central tower of the abbey collapsed in 1322, the stone and timber needed to build its successor, the great octagonal lantern, was brought in by water.[129] Rivers were also widely used for the export of the produce of monastic estates, such as wool: in the early thirteenth century Rievaulx, Byland, and other Yorkshire abbeys regularly brought their wool to Clifton, on the outskirts of York, a place not far from the extensive sheep pastures of the North Yorkshire Moors and Yorkshire Wolds, but also convenient for continental merchants, with direct water access down the Ouse and Humber to the port of Hull.[130]

Several of those ancient Benedictine houses which had been involved in river diversions and canal constructions before the Conquest continued to alter watercourses for the improvement of transport into the twelfth century and after. The role of Glastonbury Abbey in the Somerset moors has already been noted. John Leland, who visited Abingdon in the early sixteenth century, stated that the present main stream of the Thames on the north-western side of Andersey Island was the product of another diversion carried out after

---

[124] M. G. Fulford, S. Rippon, J. R. L. Allen, and J. Hillam, 'The Medieval Quay at Woolaston Grange, Gloucestershire', *Trans. Bristol & Gloucs. Archaeol. Soc.* 110 (1992), 101–22.

[125] Rolt, *Inland Waterways*, 16.   [126] *Cal. Close R., 1231–34*, 205.

[127] R. E. G. Kirk (ed.), *Accounts of the Obedientiars of Abingdon Abbey*, Camden Soc., NS 51 (London, 1892), 29, 48–9.

[128] W. P. Baildon (ed.), *Notes on the Religious and Secular Houses of Yorkshire*, i, Yorks. Archaeological Soc., Record Ser. 17 (1894), 232–3.

[129] Chapman (ed.), *The Sacrist Rolls of Ely*, ii. 32, 94, 169, 179.

[130] B. Waites, 'Monasteries and the Wool Trade in North and East Yorkshire during the Thirteenth and Fourteenth Centuries', *YAJ* 52 (1980), 111–21; B. Waites, *Monasteries and Landscape in North-East England* (Oakham, 1997), 191.

the time of Abbot Faricius (1100–17) to bring the navigation channel closer to the abbey. Does Leland's comment, written long after the event described, genuinely reflect a new venture, or was it, as John Blair has suggested,[131] simply a muddled and misattributed memory of the works of Abbot Orderic? After the twelfth century the Swift Ditch does seem to have been the lesser stream (though it would be brought back into use again when the first pound-locks were built on the Thames in the seventeenth century). If Leland was correct, the implication seems to be that both of the streams which surround Andersey Island are at least partly artificial. Leland goes on to state that in wet periods the old course of the river still flooded so that there were three streams. Indeed, an abandoned meandering watercourse can still be followed in part across the middle of the island in the lowest part of the valley floor.[132]

Benedictine houses founded after the Conquest continued the tradition of making canals to serve their needs. Works connected with the rebuilding of Selby Abbey on a new site under the second abbot, Hugh de Lacy (1097–1123), appear to have included the construction of a canal to bring stone from the quarries at Monk Fryston, 12 km to the west.[133] The line of this waterway may be perpetuated by a ditch alongside the modern road between the two places, which joins the Selby Dam at Thorpe Willoughby.

The new monastic orders coming into England in the twelfth century made further contributions. Amongst the charters of the Augustinian abbey of Waltham is a writ of William Longchamp, Bishop of Ely and royal chancellor, to the sheriff of Essex, dated to 1190 or 1191, announcing that he had authorized Abbot Walter of Waltham Holy Cross to divert the course of the River Lea in Waltham as he wished, without harming anyone and respecting the convenience of navigation.[134] The Cornmill Stream at Waltham is an artificial leat following the edge of the gravel terrace. Its primary purpose was to provide a head of 2 m of water for the second abbey mill, just north-west of the church, but it is likely also to have fed water into the fishponds on Veresmead and through the claustral buildings to flush the reredorter. However, the headrace of the mill was also used by boats bringing grain and other produce both to the mill and to the grange yard to the north-east of the abbey, where remains of a dock and wharf were discovered in 1970–1. The dock may have been built as early as 1200. It had an opposed slipway with a sill so that a floating depth of water could be retained when the use of the mill caused the level in the leat to fall, and it would have been capable of accommodating flat-bottomed boats up to 3.7 m long × 1.2 m beam. The timber-framed wharf which superseded the dock appears to have remained in use till about 1700. An inquiry into a complaint that the abbot was taking too much water from the river for his

[131] Pers. comm., 4 August 2005.   [132] Leland, *Itin.* v. 76; Bond, 'Reconstruction', 69–70.

[133] H. Farrar, *Selby Abbey* (Selby, 1989), 5; J. T. Fowler (ed.), *The Coucher Book of Selby*, i, Yorks. Archaeological & Topographical Association, Record Ser. 10 (1890), [3], [22].

[134] R. Ransford (ed.), *The Early Charters of the Augustinian Canons of Waltham Abbey, Essex, 1062–1230* (Woodbridge, 1989), 200–1, no. 303.

Fig. 39. Canal terminus (now dry) serving the Cluniac priory of Castle Acre, Norfolk.

mill in 1482 described the entry to the Cornmill Stream from the river as 16 ft (4.9 m) wide where it should only be 4 ft (1.2 m) wide. Following the Act of 1571 this mill-leat became the principal navigation channel of the Lea, with a pound-lock near the mill taking boats down to the level of the old river.[135]

Where rivers were used for transport, it was a relatively simple matter to construct a short artificial arm in order to bring boats closer to the point where their contents were needed. The Cluniac priory of Castle Acre (Norfolk) was constructed on made-up ground over marshland on the north side of the River Nar. The main stone-lined drain, after passing through the reredorter, kitchen, and mill, broadens to a width suitable for boats along the north side of the grain storage and processing buildings in the outer court (Fig. 39), continuing on down to the River Nar as a canal for a further 200 m. The sewer outfall from the reredorter did not flow back into this drain, but was led off southwards to an earlier junction with the river. The Nar was navigable downstream from the abbey, and there may be further wharfage along the riverside, while upstream small boats could probably have reached the castle.[136] There may have been many more relatively short-distance diversions or improvements of this sort.

[135] P. J. Huggins, 'Excavation of a Medieval Bridge at Waltham Abbey, Essex, in 1968', *MA* 14 (1970), 126–9; P. J. Huggins, 'Monastic Grange and Outer Close Excavations, Waltham Abbey, Essex, 1970–1972', *Trans. Essex Archaeological Soc.* 4 (1972), 81–9.

[136] R. Wilcox, 'Castle Acre Excavations, 1972–76', *Norfolk Archaeology*, 37 (1980), 231–76; R. Wilcox, 'Excavation of a Grain Handling Complex at Castle Acre Priory, Norfolk' (forthcoming).

A rather longer artificial channel, about 2 km long and 3 m wide, was cut across Stonebridge Marshes from the tidal Butley River up to a wharf within 180 m of the Augustinian priory of Butley in Suffolk. The construction of the church, claustral buildings, and gatehouse during the thirteenth and fourteenth centuries employed stone from the Yonne valley similar to that used in Auxerre Cathedral, as well as Purbeck marble. The French stone must have been carried down the Yonne to the Seine and across the English Channel to Orford, then up the Butley River. Excavations in 1933 located remains of the wharf and landing stage at the end of the canal. A shallow creek had been filled in with successive layers of faggots, earth, clay, and mortar, creating a platform of reclaimed land 7–15 m wide providing access to the edge of a deep-water channel. A timber revetment was traced for a length of some 27 m along the northern edge of the deep water. This first construction was interpreted as dating back at least to the early part of the thirteenth century, possibly even to the foundation of the priory in 1171. Subsequently the timber revetment was replaced with a buttressed stone wall some 18 m long, with a jetty 5.5 m to the south constructed of a line of stout oak baulks infilled behind with building debris, including ashlar fragments, mortared crag, and tile. Above the jetty a road roughly surfaced with flint and tile gave access to a platform behind the wall which may have contained storage sheds. The construction of the wall intersected the line of an earthenware drain of thirteenth- or fourteenth-century date at its western end, and there was no pottery from the mud in front of the jetty dating from before the second half of the fourteenth century; the reconstruction of the wharf may, therefore, be connected with the rebuilding of the east end of the church and the gatehouse in the fourteenth century. Pottery from the silts in front of the jetty suggests that the deep-water channel remained in use at least up to the seventeenth century, but by the eighteenth century it had been abandoned and was used as a dump for rubbish.[137]

During the twelfth century Cistercian abbeys became especially prominent in the field of water engineering. As a strongly centralized organization, they may have had access to technical expertise on a scale unavailable to older, more autonomous houses. Arriving as latecomers and settling in generally remote, underdeveloped regions, they had both opportunities and incentives to modify the local landscape on a substantial scale. The river diversions undertaken by the Cistercians were not invariably designed for transport. At Bordesley Abbey (Worcs.) the River Arrow was diverted around the edge of the valley so that the old valley floor could accommodate fishponds and a watermill, an operation now dated to within a couple of decades of the first monastic settlement.[138] At Byland Abbey (Yorks.) several streams rising on the Hambleton Hills which originally flowed eastwards through a waterlogged morass towards the River

[137] J. B. Ward-Perkins, 'The Priory Wharf or Landing-Stage', in J. N. L. Myres, W. O. Caroe, and J. B. Ward-Perkins, 'Butley Priory, Suffolk', *ArchJ* 90 (1933), 260–4.

[138] G. G. Astill, *A Medieval Industrial Complex and its Landscape: The Metalworking Watermills and Workshops of Bordesley Abbey*, CBA Research Rep. 92 (York, 1993).

Rye were intercepted by an artificial cut, the Long Beck, constructed during the later twelfth century, which drained the water south-westwards towards the River Swale. This had the effect both of removing surplus water from the new site of Byland Abbey and of providing a supply to the fishpond and mill of Newburgh Priory, and the work may have been undertaken co-operatively by both monastic houses, since it was to the benefit of both.[139] In some cases natural watercourses were realigned simply in order to provide more space for building. At Kirkstall Abbey (Yorks.) the old course of the River Aire was deliberately filled in during the thirteenth century and the flow of the river diverted into a new channel 60–70 m further south, in order to extend the precinct and its buildings.[140] Diversion of rivers and streams from their original beds in order to accommodate buildings or to enlarge precincts is also evident at Dieulacres Abbey (Staffs.) and Cleeve Abbey (Somerset).[141] While none of these schemes was undertaken with navigation in mind, they all reflect the reservoir of technical ability available to the Cistercians, and it comes as no surprise to find them carrying out works more specifically for transport elsewhere.

The Rye valley at Rievaulx contains evidence of a succession of natural and artificial watercourses both above and below the abbey (Fig. 40). These were first described at the beginning of the last century by Henry Rye, whose assessment of them stands out as a bold and creditable piece of interpretation well in advance of its time. Rye drew attention to a series of land grants made to the abbey by local landowners around 1142–6, 1160–70, and 1193–1203 which, he suggested, had been made in order to allow the abbey to cut artificial canals in order to bring in building stone from its quarries upstream at Penny Piece and downstream at Hollins Wood.[142] Fifty years later the evidence was reassessed by Weatherill, who put forward the view that the land grants were designed to give the monks a compact block of meadow or grazing land on their own (i.e. the eastern) side of the river. He accepted that the earthworks represented canals for the transport of stone, but suggested that the canal system antedated the land grants, and was constructed *c*.1135–40, when major building works were in progress.[143] In more recent descriptions of the site the canal interpretation has been ignored in favour of the view that the real purpose of the successive river diversions was simply to give the monks of Rievaulx complete control of the valley floor, partly in order to permit them to

[139] J. McDonnell and M. R. Everest, 'The "Waterworks" of Byland Abbey', *Ryedale Historian*, 1 (1965), 32–9.

[140] S. Moorhouse and S. Wrathmell, *Kirkstall Abbey*, i: *The 1950–64 Excavations: A Reassessment* (Wakefield, 1987), 33–6.

[141] M. J. C. Fisher, *Dieulacres Abbey, Staffordshire* (Stafford, 1969), 13; R. Gilyard-Beer, *Cleeve Abbey, Somerset* (London, 1960), 16.

[142] H. Rye, 'Rievaulx Abbey, its Canals and Building Stones', *ArchJ* 57 (1900), 69–88.

[143] J. Weatherill, 'Rievaulx Abbey: The Stone used in its Building, with Notes on the Means of Transport, and a New Study of the Diversion of the River Rye in the Twelfth Century', *YAJ* 38 (1954), 333–54.

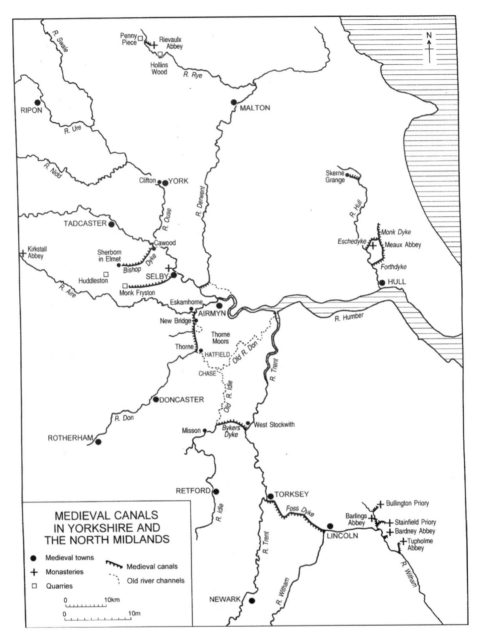

Fig. 40. Medieval canals in Yorkshire and the north midlands.

utilize the old course as a leat for a chain of three successive mills and partly
in order to have more meadowland on their side of the river.[144] Yet in some

---

[144] G. Coppack, *Abbeys and Priories* (London, 1990), 95–6; P. Fergusson and S. Harrison, *Rievaulx
Abbey: Community, Architecture, Memory* (New Haven, 1999).

respects this interpretation also remains unconvincing. In particular, access to meadows could surely have been obtained more easily and cheaply simply by building a bridge or two rather than by diverting the entire river. Perhaps the lowest portion of the 'canal' system does make more sense as a set of supply leats and tail-races serving the fulling mill and smithy; Weatherill recognized that there was a fall of nearly 3 m from the abbey down to the level of the Hollins Wood quarry, and his suggestion that boats could still have brought stone up to the abbey by means of some sort of lock and balancing reservoir may be a little extravagant. Nevertheless, there is no such obstacle to Rye's original interpretation of the channel *above* the abbey as a canal from the Penny Piece quarry. From the point where it diverges from the present river just below the quarry, the now dry course runs along the eastern flank of the valley in an almost straight line at a slightly higher elevation than the valley floor for a distance of nearly 500 m, terminating below the site of the abbey's corn-mill. It is embanked on its lower side, and its scale appears excessive for a mill supply leat. The line of the present river above the abbey may also be artificial, since it too follows a straight course. Between the two a wide abandoned meander curves round in the old valley bottom. More work is needed on this highly complex site before it can fully be understood.

The Cistercian abbey of Meaux was heavily involved in operations to drain and reclaim the low-lying boulder clay and peatlands of the Hull valley in the twelfth and thirteenth centuries, but it is evident that some of the new watercourses cut were more than mere drainage channels.[145] Some time between 1160 and 1182 a canal called the Eschedike was cut for just over 2 km in a straight alignment across the natural drainage line and through the boulder clay from the abbey to the River Hull.[146] Clearly drainage was not its primary function, and a more likely purpose was to provide a means of transporting building stone and produce up to the abbey. In 1210–20 the abbey negotiated with its neighbours the right to divert part of the waters of the Lambwath Stream to make a further canal, 22 feet (6 m) wide, known as the Monk Dyke, passing in a north–south direction for 6 km from west of Arnold Grange to west of Benningholme Grange, and linking with the Eschedike.[147] Then, between 1221 and 1235, arrangements were made for the cutting of the Forthdike, 16 ft (5 m) wide, linking with the Monk Dyke at Fairholme and directing the main flow of the Lambwath Stream in two straight lengths of canal totalling 5 km around to the east and south of Wawne, providing a lower access to the River Hull.[148] The lower 3.5 km of the Forthdike also intersects a boulder clay watershed, and cannot have been designed primarily as a drain. Finally, further up the valley, the Skerndike, 3 km long and 20 feet (6 m) wide was cut between 1249 and 1269 to provide access by boat between Skerne

---

[145] J. A. Sheppard, *The Draining of the Hull Valley* (York, 1958), 3.
[146] E. A. Bond (ed.), *Chronicon Monasterii de Melsa*, Rolls Ser. 43.i (London, 1866), 60.
[147] Ibid. 354–6.   [148] Ibid. 409–12.

Grange and the River Hull.[149] During the later fourteenth century several complaints were made against the abbey over its inadequate maintenance of the Eschedike, Monk Dyke, and Withdike, though by this time the concerns seem to be drainage rather than navigability.[150] The pattern of water channels in this area has been much altered by later drainage works, but the basic alignments of the monastic canals still remain.

The dense cluster of monastic houses along the east flank of the Witham valley below Lincoln (Fig. 40) clearly made considerable use of the river, and the courses of many of the tributary streams show signs of diversion or canalization. At the Gilbertine priory of Bullington the construction of a canal is documented around the end of the twelfth century 'for the easy transit of ships' from the river. Three kilometres to the south, at the Premonstratensian abbey of Barlings, a tributary of the Witham, the Barlings Eau, shows clear signs of diversion. The old course, followed by the parish boundaries sweeping round to the east of the abbey, has been shortened by a straight cut about 1.1 km long, passing close to the abbey buildings. Less than 2 km to the east of Barlings at the Benedictine nunnery of Stainfield the Stainfield Beck was diverted into a straightened cut, the original meandering course being reused in part for fishponds. Three kilometres lower down the Witham the Benedictine abbey of Bardney was served by a navigable cut extending eastwards from the river for some 600 m.[151] A further 4 km downstream is the Premonstratensian abbey of Tupholme, where Henry II granted the canons a watercourse (*fossatum*) large enough to allow vessels to come and go from the River Witham to allow carriage of stone to the abbey buildings, a distance of nearly 2 km.[152] Presumably this was the stream now forming the parish boundary below the abbey site which was straightened and improved.

Water transport was obviously preferable to road transport for the carriage of stone, timber, and other heavy or bulky materials, and monastic houses clearly had a particular incentive to undertake works of this sort at times when major building projects were in hand. Nevertheless, where no water route was available, considerable quantities could still be carried by road. The construction of the Cistercian abbey of Vale Royal in Cheshire demanded some 35,000 cartloads of stone between 1278 and 1281, brought from quarries on Eddisbury Hill, 8 km away.[153] The excavation of the Pilrow Cut by Glastonbury Abbey during the first half of the thirteenth century seems to mark the end of any major capital investment in canalization by monastic houses. Once the major building projects were completed the need to maintain

---

[149] *Chron. Melsa*, Rolls Ser. 43.ii (London, 1867), 110.

[150] *Chron. Melsa*, Rolls Ser. 43.iii (London, 1868), 165, 172.

[151] P. L. Everson, C. C. Taylor, and C. J. Dunn, *Change and Continuity: Rural Settlement in North-West Lincolnshire*, RCHME (London, 1991), 47, 66–8.

[152] H. M. Colvin, *The White Canons in England* (Oxford, 1951), 101.

[153] J. Brownbill (ed.), *The Ledger-Book of Vale Royal Abbey*, Record Soc. of Lancs. and Cheshire 68 (Edinburgh, 1914), 198–203.

the canals may often have receded, though some were still used for other purposes. Certainly the Pilrow Cut was still passable by boats as late as 1500, when the churchwardens' accounts of St John's church in Glastonbury record that this was the route taken by boats bringing new stalls from Bristol.[154]

Episcopal Canal Construction in England, 1066–1250

The bishops, deans, and chapters of cathedrals were also great builders in the post-Conquest period, and their needs were similar to those of the monasteries. The likelihood that the bishops of Ely had a hand in the diversion of the Great Ouse in the early twelfth century has already been mentioned.

During the construction of the Norman cathedral at Norwich much freestone from Caen and Barnack was brought into the city along the River Wensum. It was probably in order to facilitate the carriage of building stone that a canal was cut from the Wensum at Pull's or Sandling's Ferry, east of the cathedral, passing through the close wall by a watergate and penetrating the Lower Close immediately inside the southern wall of the precinct for a distance of some 250 m. Stone was still being brought from Caen in 1288, when the communar rolls of the cathedral priory record its shipment to the port of Yarmouth, each shipload then being transferred to between four and six river barges for carriage up to Norwich. The line of the canal remained open until around 1780, when it was filled in to make the 'Prebends' Walk'. Its remains have from time to time been observed during the laying of drains and sewers.[155]

In Yorkshire an artificial watercourse known as the Bishop Dyke runs for over 10 km from Sherburn-in-Elmet to Cawood on the River Ouse (Fig. 40), and it has been suggested that this was made in order to bring magnesian limestone from the quarries at Huddleston, 4 km west of Sherburn, for building works at the cathedral.[156] If this is correct, its construction and subsequent abandonment must date from the early middle ages. The relatively high costs of transporting stone between Huddleston and Cawood and the explicitly stated use of carts and sleds recorded in the York Minster fabric rolls during the fifteenth and early sixteenth centuries argue against the use of water transport for this leg of

[154] W. E. Daniel (ed.), 'Churchwardens' Accounts of St John's, Glastonbury', *Somerset & Dorset Notes and Queries*, 4 (1894–5), 322; Hadfield, *Canals of South-West England*, 76; V. E. J. Russett, 'Hythes and Bows: Aspects of River Transport in Somerset', in Good, Jones, and Ponsford (eds.), *Waterfront Archaeology*, 62.

[155] E. Fernie and A. B. Whittingham (ed.), *The Early Communar and Pittancer Rolls of Norwich Cathedral Priory*, Norfolk Record Soc., 41 (Norwich, 1972), 54–5; I. Atherton, 'The Close', in I. Atherton, E. Fernie, C. Harper-Bill, and H. Smith (eds.), *Norwich Cathedral: Church, City and Diocese, 1096–1996* (London, 1996), 644; B. S. Ayers, 'Building a Fine City: the Provision of Flint, Mortar and Freestone in Medieval Norwich', in D. Parsons, (ed.), *Stone Quarrying and Building in England, AD 43–1525*, Royal Archaeological Institute (Chichester, 1990), 223; B. S. Ayers, 'From Cloth to Creel: Riverside Industries in Norwich', in Good, Jones, and Ponsford (eds.), *Waterfront Archaeology*, 3.

[156] J. S. Miller and E. A. Gee, 'Bishop Dyke and Huddleston Quarry', *YAJ* 60 (1983), 167–8.

the journey at this period, though there is no doubt that boats were still being used to take the stone from Cawood up the Ouse to York.[157]

Some of the antiquarian traditions of episcopal canal construction do not stand up to critical re-examination. According to Dugdale, Godfrey de Lucy, Bishop of Winchester (1189–1204), 'restored the navigation of the River Itchen, not only from the port of Southampton as far as Winchester, but also to the very head of that river, where he constructed a dyke'.[158] Two documents appeared to support this statement: a charter of King John licensing the bishop to levy tolls on all hides, leather, and other goods entering the river by the canal ('per trancheam') he had made; and an entry in the Pipe Rolls of the bishopric in 1208–9 recording carriage of wool 'per aquam' between Bishops Sutton, near the source of the River Alre (a headstream of the Itchen), and Beaulieu.[159] It has been suggested that the purpose of continuing navigation works up the Itchen above Winchester and on up the River Alre was to facilitate the carriage of supplies up to the episcopal palace at Bishops Sutton. In order to ensure sufficient depth of water for boats in the diminutive River Alre, it has further been suggested that Bishop Godfrey constructed the great dam, 6 m high, 370 m long, just south of the village of Old Alresford, in order to create an artificial reservoir originally covering some 80 hectares (now reduced to 12 hectares through silting) at the head of navigation. From near the southern end of the dam, the bishop then laid out a broad rectangular market place extending up to the London road, the nucleus of the new town of Alresford Forum or New Alresford, for which he acquired a charter for a Thursday market in 1200, and an annual fair a couple of years later.[160]

However, archaeological surveys have found no trace of any long-distance navigation works antedating the scheme of the late seventeenth century.[161] Moreover, neither Derek Keene nor Edward Roberts finds any support for the existence of Bishop Lucy's canal in their reassessment of the documentary evidence.[162] No mention is made of any canal, either in the detailed mid-thirteenth-century rental and custumal of Alresford, or in the long series of bishopric pipe roll entries. Indeed, where the pipe rolls identify any means of

[157] D. Parsons, 'Stone', in J. Blair and N. Ramsay (eds.), *English Medieval Industries: Craftsmen, Techniques, Products* (London, 1991), 22 n.

[158] W. Dugdale, *Monasticon Anglicanum*, ed. J. Caley, H. Ellis, and B. Bandinell, Record Commissioners (London, 1817–30), i. 196.

[159] C. Deedes (ed.), *Registrum Johannis de Pontissara* (Oxford, 1924), ii, 741–3; H. Hall (ed.), *The Pipe Roll of the Bishopric of Winchester, 1208–9* (London, 1903), p. xix.

[160] F. Brough, 'Liberty of Alresford', in *VCH Hants. and I.of Wight,* iii (1908), 350; E. M. Hewitt, 'Industries: Introduction', in *VCH Hants. and I.of Wight,* v (1912), 451; Hadfield, *Canals of Southern England,* 30; M. W. Beresford, *New Towns of the Middle Ages* (London, 1967), 109–10, 177, 442–3.

[161] R. T. Schadla-Hall, *The Winchester District: The Archaeological Potential* (Winchester, 1977), 44; M. F. Hughes, 'Settlement and Landscape in Medieval Hampshire', in S. J. Shennan and R. T. Schadla-Hall (eds), *The Archaeology of Hampshire*, Hants Field Club, Monograph 1 (1981), 66–77.

[162] D. Keene, *Survey of Medieval Winchester*, Winchester Studies 2.i (Oxford, 1985), 57–9; E. V. Roberts, 'Alresford Pond, a Medieval Canal Reservoir: A Tradition Assessed', *Proc. Hants Field Club & Archaeol. Soc.* 41 (1985), 127–38.

transport between Alresford, Winchester, and Southampton, it is invariably by road, using horse-drawn carts. Roberts has argued that the carriage 'by water' between Bishops Sutton and Beaulieu mentioned in 1208–9 is more likely to refer to goods being ferried across Southampton Water and the Solent between the Hamble and Beaulieu estuaries at the southern end of the journey. The first record of King John's reputed charter giving Bishop Lucy the right to extract tolls on goods carried by canal between Winchester and Southampton actually appears in the register of Bishop John de Pontoise (1282–1304), and the editor of the register believed this charter to be spurious. At that time proposals to create a navigation between Winchester and the sea certainly were under active discussion (see further below), and the forgery may have been designed either to encourage the bishop to sacrifice his income from the mills and fisheries to support the scheme, or to deter the burgesses of Winchester from promoting their own canal scheme independently. The very fact that the bishop had continued to enjoy an uninterrupted income from his mills on the Itchen between the late eleventh and late thirteenth centuries suggests that any navigational use of the river during this period was only local. Moreover, there is no mention of Bishop Lucy's canalization in the inquiry into the scheme of the later thirteenth century. The pipe roll records of Alresford Pond show that it was primarily a fishpond, though it may also have served the local mills. The pond was drained and left dry throughout the summer and autumn of 1253 in order to repair the weir and sluices and to remove accumulated silt from its bed; it was refilled and restocked with fish in the following year. Had its function been to maintain a head of water in a navigable canal or canalized river, this interruption would have serious implications, but the pipe rolls mention no adverse consequences.

The spurious charter of King John was quoted in the 1620s by John Trussell, mayor of Winchester, who was a keen supporter of the contemporary Itchen Navigation proposals, bitterly opposed in Southampton. Trussell's propagandist stance was followed uncritically by local antiquarian writers, and the notion of an early thirteenth-century episcopal canalization of the Alre and Itchen subsequently passed into the realm of historical orthodoxy.

Civic Undertakings

Only one successful major undertaking is known to have been carried out by any civic authority. This was in Bristol, where the expansion of trade during the early thirteenth century was placing severe pressure upon the quayside facilities.[163] In March 1240 the citizens purchased the eastern part of St Augustine's Marsh from St Augustine's Abbey, and began the task of digging a new channel for the River Frome to accommodate ships able to carry up

---

[163] M. D. Lobel and E. M. Carus-Wilson, 'Bristol', in M. D. Lobel (ed.), *The Atlas of Historic Towns*, ii (Baltimore, 1975).

to 200 tuns of wine.[164] In April 1240 the king ordered the men of Redcliffe, the suburb south of the Avon, to assist the burgesses, since its completion was 'for the common good of all', but it could not be completed 'without great costs'.[165] A new stone-revetted quay for seagoing ships was made on the north bank of the new channel, opposite St Stephen's church. Smaller coastal vessels continued to use the Avon, tying up on Welsh Back below Bristol Bridge. The bridge itself was rebuilt in stone at the same time, a task which involved the temporary damming and diversion of the Avon while work was in progress. The cutting of a moat around the outer side of the new Port Wall around the Redcliffe and Temple suburbs in the 1240s may have accommodated the diverted flow of the river.

The original course of the River Frome had been described by William Worcestre as flowing close under the inner western walls of the city, and William Smith's map of 1568 shows what appears to be a vestigial portion of this watercourse surviving between Marsh Street and Back Street on the southern side of Baldwin Street. On this evidence it has been proposed that the new cut diverged from the old course of the Frome above St Stephen's church and that the Frome originally flowed east of St Stephen's, round the western and southern sides of Baldwin Street, to join the Avon near St John's chapel on Welsh Back.[166] However, a reassessment of the evidence by Roger Leech has shown that the Marsh suburb was already being settled in the twelfth century, and that neither Henry II's charter to the men of this suburb, nor the deeds of medieval properties along either side of Baldwin Street, make any mention of the Frome in this alignment. Leech regards the area of water shown outside the line of Baldwin Street on Smith's map as highly suspect, since no hint of a former river course here is given by any subsequent map of the city. In Leech's view the old course of the Frome lay further to the south-west, beyond the line of the Marsh Wall of *c*.1240, following the line of King Street. The point of diversion of the new cut from the old course of the Frome would then have been some 366 m downstream from the location originally proposed, and the confluence with the Avon some 300 m downstream.[167]

The burgesses of Winchester proposed a scheme to make the River Itchen navigable from the city down to the sea in the late thirteenth century, but there may have been an opposing faction within the city. Edward I ordered an inquisition, and in 1276–7 the jurors reported that the river could be deepened and made navigable up to Bishopstoke, but only by the removal of the bishop's mills of Wood Mill, Stoke, Brambridge, and North and South Twyford, along

---

[164] E. W. W. Veale (ed.), *The Great Red Book of Bristol*, i, Bristol Record Soc. 4 (Bristol, 1933), 89–90; *Cal. Pat. R. 1216–25*, 540.

[165] N. Dermott Harding (ed.), *Bristol Charters, 1155–1373*, Bristol Record Soc. 1 (Bristol, 1930), 18–19.

[166] Lobel and Carus-Wilson, 'Bristol'.

[167] R. Leech, 'The Medieval Defences of Bristol Revisited', in L. Keen (ed.), '*Almost the Richest City': Bristol in the Middle Ages*, British Archaeological Association, Conference Trans. 19 (1997), 26–8.

with a salmon fishery and the two fulling mills at Priors Barton.[168] Given the possibility that Wood Mill at least may have been bypassed by a navigable channel in the early eleventh century, as mentioned above, the obstacles may have been political rather than practical; but no more is heard of proposals for navigation works on the Itchen before the seventeenth century.

Crown Undertakings

The Crown appears conspicuously absent from the roll of initiators of water transport projects in the middle ages. Only one example of a royal involvement in canal construction is identified in the comprehensive survey of the medieval kings' works provided by Brown, Colvin, and Taylor. This was Edward I's canalization of the River Clwyd at Rhuddlan, undertaken during his first campaign against Llywelyn ap Gruffydd in 1277. In strict terms this took place after the end of the period which is the main concern of this chapter. However, it is difficult to ignore it, simply because the quality of information available on the organization of the work is so superior to anything that has gone before.

Edward's overall strategy for the conquest of Gwynedd had demanded the movement of troops and provisions along the north coast of Wales by ship, and all but one of the new Edwardian castles built within the borders of Gwynedd during the campaigns of 1277–95 were sited in coastal or estuarine locations. The exception was Rhuddlan, the second major Edwardian castle to be built in north Wales, which was located some 4.5 km inland, just below the highest tidal point on the River Clwyd. In 1255–6 Llywelyn had recovered the border cantref of Tegeingl and in 1263 he was able to capture and destroy its principal castle at Dyserth. Llywelyn's gains had been recognized by the Treaty of Montgomery in 1267, but Edward's new assault in 1277 brought Tegeingl back under his control, and necessitated the building of a replacement for Dyserth. The site chosen was just downstream from the long-abandoned Anglo-Saxon *burh* of *Cledemutha* and Norman castle and borough of Rhuddlan. To make possible the import of building materials and the subsequent provisioning of the castle by sea, it was decided to straighten out the lower reaches of the River Clwyd between the castle and the sea. In July 1277 a man named William of Boston was deputed to recruit ditch-diggers in Lincolnshire, and 300 men were marched up to north Wales from the Fens under armed guard. By mid-September, 968 ditch-diggers appear on the royal payroll at Rhuddlan, and while some of these were probably digging the castle moat, the canalization of the river seems to have been the major expenditure over the next couple of years (see further below).[169] When in October 1282 Edward I granted the cantrefs of Rhos and Rhufoniog and the commote of Dinmael to Henry de Lacy, Earl of Lincoln, he reserved certain lands to himself, including 'all the

---

[168] Hewitt, in *VCH Hants and I. of Wight*, v. 451–2.
[169] R. A. Brown, H. M. Colvin, and A. J. Taylor, *The History of the King's Works*, i. *The Middle Ages* (London, 1963), 319–31.

marsh that is within the new course of the river Cloyt and the old course of that river, which marsh used anciently to pertain to the town of Rothelan'.[170] The present direct course of the river below the castle contrasts strongly with the meandering course upstream, and traces of the old river channel are clearly visible on either side. A short arm was led off the river immediately below the south wall of the castle to form a dock, accessible from the outer ward by means of a postern defended by a tower. A similar dock, capable of accommodating vessels of 40 tonnes at high tide, was provided at Beaumaris in 1296, while Flint, Conwy, and Caernarfon were served by open quays and Harlech's access to the sea was protected by an outer enclosure and a watergate with a dock just outside it.

Conclusion

Earlier writers concerned with the canals of the eighteenth and nineteenth centuries have, with a few honourable exceptions, shown little awareness of any medieval background to their field of study. Yet there is, without question, a long tradition of stream and river diversions and canalizations during the centuries between the collapse of the Roman empire and the late middle ages. However, there remain many difficulties in the investigation of early waterways. Local antiquarian anecdotes and traditions, while often containing grains of truth, cannot be accepted uncritically. Documentary records are numerous, but scattered, and references within them are often oblique and incidental rather than specific. Field evidence is often equally ambiguous and difficult to interpret. In marshland areas it is often very hard to distinguish works undertaken for drainage from those undertaken for navigation; indeed, there was often no inherent conflict between the two aims. Waterways which have remained in use, or have been restored in later centuries, pose particular problems: eighteenth- and nineteenth-century improvements to waterways such as the Foss Dyke may well have destroyed evidence of medieval and Roman use and construction. Equally, canals which are not maintained rapidly fall into decay, their embankments erode away, their beds silt up, and they can merge back into the landscape to leave little or no surface trace. If there has been relatively little detailed archaeological fieldwork, there has been even less scientific excavation. Much more investigation is needed, both on the documentation and on the ground by means of close-contour surveying and selective excavation, before the overall chronology, technology, and economic rationale of medieval canal construction in Britain can fully be understood. In conclusion, a dozen fundamental questions need to be posed:

1. Much remains to be done on the basic questions of chronology, location and dimensions of transport canals. On present evidence some work seems to have been undertaken in England around the middle of the tenth and

---

[170] *Cal. Chancery Rolls, 1277–1326*, 241.

middle of the eleventh centuries, with a prolonged period of more intensive activity throughout the second half of the twelfth and the whole of the thirteenth centuries, tailing off in the later middle ages; but the date of many of the works detectable in the landscape either remains entirely unknown or rests upon very uncertain evidence. The geographical distribution seems to be predominantly eastern and southern, with particular concentrations both before and after the Norman Conquest in the Fenland. Apart from a few undertakings in Yorkshire after the Conquest, little has yet been discovered in the midlands or the north. River diversions and arms led off rivers appear to be the commonest forms of construction, and no attempts were made to cross major watersheds. In terms of length, few medieval schemes rivalled the greater achievements of classical antiquity, most of them running for no more than a few kilometres. The few that exceeded this limit were generally modified natural watercourses. Whereas the length of a canal or river diversion is usually ascertainable, at least as an approximation, very little information is available on width and depth, yet these dimensions are just as important, in terms both of the amount of labour involved in construction and of the capacity of vessels the waterway was designed to accommodate.

2. Further work is needed on the nature and purpose of Roman canals, and in particular the extent to which they survived physically or remained in use into the early middle ages. Did the recognition of their remains prompt renewed experiments? Did any knowledge of the classical literature on water management, for example, the works of Frontinus and Vitruvius, play any part in the development of new canals?

3. To what extent were developments in England influenced by what was happening on the Continent during the early middle ages? We clearly need to know a great deal more about contemporary undertakings in France, in particular in Normandy and Anjou, and in the Low Countries.

4. Who were the prime movers in medieval canal construction in England, and what were their motives? Edward I's canalization of the Clwyd at Rhuddlan seems to stand alone as an undertaking made primarily for military building and supply purposes. Sometimes the Crown supported civic ventures, such as that at Bristol, where the needs of commerce were the driving force. On present knowledge, however, monastic and ecclesiastical corporations seem to dominate the earliest records of work in England, through their need to facilitate the import of building materials. Are we being taken in by the sometimes dubious evidence of monastic chroniclers concerned to bolster the reputations of their heroes? Is the prominence of the monasteries merely a result of their reputation for an ability to undertake large-scale capital projects, or the superior survival of monastic documentation, or their genuine need for bulk transport, or simply the bias of investigation so far?

5. Under what circumstances was water transport preferable to road transport? Clearly speed was not one of its advantages. In 1319 two parties
of scholars from King's Hall, Cambridge, travelled to York to attend
Edward II's Christmas feast. One group travelled overland by hackneys,
completing the 240-km journey in five days. The other group travelled
by boat to Spalding, then by horse to Boston, then again by boat up the
Witham, along the Foss Dyke to Torksey, then by another boat down the
Trent and up the Ouse to York. This took nine days, and they arrived
three days too late.[171] The chief attraction of water transport was its low
cost: one estimate suggests that the expense of carting goods overland
could be ten times that of carriage by water over the same distance.[172]
Water also has obvious advantages for the carriage of bulky, cumbersome,
or heavy material and for fragile goods. Economies could be made by
delaying departure until a sufficiently large cargo had accumulated to fill
a boat; but this might be offset by storage costs in the meantime, while
the size of boat was itself limited by the capacity of the waterway. In
addition to the many documentary records of building stone and timber
being carried by water, there is also indirect archaeological evidence. Chris
Gerrard has noted that the use in Somerset churches of the distinctive
stone from Ham Hill, near the headwaters of the River Parrett, shows
a strong relationship to the course of the river itself, suggesting that the
stone was often transported by water.[173] Occasionally in later periods we
do get some idea of comparative costs. The churchwardens' accounts of
St John's church in Glastonbury record the costs of the carriage of the
seats purchased in 1500 in Bristol, which included a journey by water over
some 70 km and a final overland journey of a little over 1 km. The hire of
two 'great boats' to carry them from Temple Back on the Bristol Avon to
Rooks Bridge on the Pilrow Cut cost 34s. 6d. They were there transferred
to thirteen smaller vessels which must have taken them by means of the
Pilrow Cut and the River Brue to Maydelode Bridge near Glastonbury,
at a cost of 15s 1d. The cost of a horse and cart moving them from the
landing place to the church came to 9s.[174] Thus, in this particular case, the
99 per cent of the journey which was by boat represented only 85 per cent
of the total transport costs. Salzman quotes a case from the fabric rolls of
York Minster where the carting of stone from the Thevesdale quarry to
Tadcaster cost 4s. per load and the carriage of the stone by water over a
distance of some 37 km down the River Wharfe from Tadcaster and then
up the River Ouse to York came to the same and concludes that 'so far as
possible water carriage was employed for the transport of stone and heavy

---

[171] F. M. Stenton, 'The Road System of Medieval England', *EcHR* 7 (1936), 1–21 (reprinted in
D. M. Stenton (ed.), *Preparatory to Anglo-Saxon England* (Oxford, 1970), 234–52), 20.
[172] Jones, 'River Navigation', 61.   [173] Quoted by Russett, 'Hythes and Bows', 62.
[174] *Glastonbury Churchwardens' Accounts*, 322.

timber'.[175] Yet there are also numerous records of building timber being carried overland over prodigious distances, so road transport was by no means impossible. Indeed, while inland water transport in general appears to be declining by the later middle ages, the greater flexibility of overland transport gave more incentive for its improvement by the construction of bridges and the replacement of oxen by horses for haulage.[176]

6. How was the construction of canals financed in the middle ages? The royal building accounts for Rhuddlan Castle between November 1277 and November 1280 record a total expenditure of £755 5s. 3d. specifically upon the 'great ditch from the sea up to the castle'. This implies the employment of an average of sixty-six diggers working six days a week at a daily wage of 3d. throughout the three years, though the actual numbers would certainly have been subject to considerable variation through the period.[177] Sources of funds for the maintenance of the Foss Dyke included private donations encouraged by offers of indulgences, and local taxes.

7. What technical skills were required to survey, construct, and maintain canals? In the absence of the bubble spirit level, not invented until 1666, the most obvious practical problem is how to follow the contour. Two means of determining a horizontal plane had come through from classical antiquity. The most basic method, known from ancient Egypt, was to use an A-frame with a plumb-bob, where the cross-bar of the frame gave a horizontal sighting when the cord of the plumb-bob coincided with a vertical mark down its centre. Various types of level employing a surface of still water had also been developed by Hero of Alexandria and by Vitruvius. Direct evidence for the use of such devices in the middle ages seems extremely elusive, yet clearly something of this sort must have been used for the construction of mill-leats and fishpond leats. How accurate could such methods be over longer distances, and did their limitations impose any constraint over the length of canals? Once the line was surveyed, the canal had to be dug out and the bed puddled with clay where the subsoil was permeable. Even a modest canal may require significant engineering works such as embankments and cuttings; how were they constructed and stabilized? How was the water level maintained? Is there evidence for balancing reservoirs, such as occur at the summit level of many eighteenth-century canals? How was surplus water removed to avoid bank erosion and flooding? Given the evidence for some forms of lock in the Low Countries by the late twelfth century, when did they first appear in England?

---

[175] L. F. Salzman, *Building in England down to 1540: A Documentary History* (Oxford, 1952), 348–54.

[176] Holt, 'Medieval England's Water-Related Technologies', 56–7.

[177] Brown, Colvin, and Taylor, *History of the King's Works*, i. *The Middle Ages*, 319–21.

8. Where were the necessary technical skills of surveying, levelling, and lining nurtured? To what extent did other economic activities, such as marshland drainage, the excavation of leats for watermills, or fishpond construction provide a reservoir of skilled labour? In the eighteenth century James Brindley himself began as a millwright's apprentice, and the making of a mill-leat demands much the same surveying skills and organization of muscle power as a canal. Watermills were a familiar feature of the landscape long before the Norman Conquest, and even in the ninth century the construction of a royal mill at Old Windsor had involved a considerable diversion of part of the flow of the Thames.[178]

9. Can the chronology of canal construction, itself still inadequately understood, be related to periods of economic prosperity and decline? On present evidence it appears that little work was undertaken in Britain between the withdrawal of the Roman imperial government and the tenth century. The Norman Conquest gave further encouragement to large-scale construction projects, and probably prompted a new period of waterway improvement for carrying building materials; but the impetus for improving water transport by canalization seems to have been on the wane by the end of the thirteenth century, and does not begin to revive again until the fifteenth century.

10. Constructing a canal is one thing, maintaining it over a prolonged period quite another. What proportion of medieval canals were viewed simply as short-term measures, useful only for the duration of a particular building project, and then abandoned? Where a longer period of use was envisaged or achieved (as appears to have been the case with the Pilrow Cut), what provision was made for dredging and repairs?

11. What impact did canal construction have upon patterns of trade? In particular, can any influence be detected upon the distribution of bulky goods, such as building stone, or fragile goods such as pottery and glass from identifiable sources?

12. What impact did canal construction have upon the settlement pattern? The development of canal ports in the eighteenth and nineteenth centuries at Runcorn, Stourport, Ellesmere Port, and Goole, along with the growth of smaller canalside settlements such as Brimscombe Port, is well documented.[179] Several small new ports and urban promotions dating from the middle ages—Rhuddlan, Rackley, Weare, Rooksbridge, Reach, Lode, Newnham in Swaffham Bulbeck—also owed their origins at least in part to the canalization of rivers and the construction of artificial waterways.

---

[178] Interim report in *MA* 2 (1958), 184–5; R. Holt, *The Mills of Medieval England* (Oxford, 1988), 3, 5.

[179] J. D. Porteous, *Canal Ports: The Urban Achievements of the Canal Age* (London, 1977).

8

Waterways and Water Transport
on Reclaimed Coastal Marshlands:
The Somerset Levels and Beyond

STEPHEN RIPPON

Introduction

The reclamation of coastal wetlands—the marshes and fens that fringe most
of England's estuaries—was an enormous undertaking, and created a series
of uniquely handcrafted landscapes. Central to the rationale and success of
these reclamations was the need to control water: an unwanted cause of flooding, but also an advantageous resource that, in addition to its importance in
agriculture, could be used for communication (navigation), power (milling),
and as a source of food (fish). Studies of reclaimed landscapes have tended to
focus upon the history of reclamation—drainage, enclosure, settlement, and
agriculture—while the potential multiple function of the artificial watercourses
so created has been neglected. It is easy to assume that the needs of communication, power, and drainage were united in single structures and systems, but
was this really the case?

These landscapes have their origin in reclamation, which was a costly
undertaking in terms of both the initial capital investment, and the subsequent
maintenance that the drainage and flood defence systems required. The constant
risk of flooding, both tidal and freshwater runoff from the adjacent 'uplands',
meant that living and farming in these areas was also a high-risk approach to
landscape utilization. In order to justify the costs and the risks, the returns from
marshland farming must have been considerable. Medieval estate records show
that land, particularly meadow, on reclaimed marshland was often as highly
valued as, or in some cases more highly valued than, that on adjacent 'drylands',
while recent work has shown that even by the thirteenth century specialized
regional economies had developed in certain marshes based around raising

livestock.[1] Around the Thames estuary, for example, the marshland economy had a strong bias towards sheep farming and dairy production,[2] whilst on the Somerset Levels cattle, pigs, and horses were raised as part of a pastoral economy with extensive meadows and over a third of demesnes sown with beans.[3] A critical factor in the success of these local economies was the ability to dispose of surplus produce for which there were two basic requirements: the presence of a market(s) and an efficient means of transporting goods there. Located beside tidal rivers and estuaries, these marshland landscapes inherently lay within the hinterland of major port towns, and were served by potentially navigable rivers, and one theme of this chapter examines how the needs of navigation often outstripped the capacity of these natural rivers, leading to the construction of artificial canals.

Reclaimed coastal wetlands are dominated by the need to manage water. In the historic period this consisted of a complex hierarchy of channels, ranging from plough-ridges ('furrows') and spade-dug gullies ('gripes') on the surface of fields, through minor field boundary ditches, to major drainage channels known variously as rhynes/rhines/reens (e.g. the Severn estuary), sewers (e.g. Romney Marsh and Fenland), drains/dykes (e.g. Fenland), and occasionally lakes (e.g. Fenland, Severn estuary).[4] This complex drainage system was designed to deal with an undesirable surplus of water: in this situation water was a problem. In other ways, however, water was an asset: if there were systems of dams and sluice-gates within these watercourses, as known on the Continent from the end of the Iron Age, then during summer months the water level in these ditches could have been kept artificially high in order for them to serve as 'wet fences' to control the movement of livestock, whilst also giving rise to lush pasture and meadow (the ditches could then be emptied in the winter to accommodate flood waters).[5] Major artificial watercourses could also be used for navigation and powering mills, while any wetland environment will be rich in fish and wildfowl. Overall, wetlands were remarkably productive landscapes if the water was managed effectively.

---

[1] S. Rippon, *The Transformation of Coastal Wetlands: Exploitation and Management of Marshland Landscapes in North-West Europe during the Roman and Medieval Periods* (Oxford, 2000), 220–40; S. Rippon, 'Adaptation to a Changing Environment: The Response of Marshland Communities to the Late Medieval "Crisis"', *Journal of Wetland Archaeology*, 1 (2001), 15–39.

[2] B. E. Cracknell, *Canvey Island: The History of a Marshland Community* (Leicester, 1959); J. Ward, 'Richer in Land than Inhabitants: South Essex in the Middle Ages, *c*.1066–*c*.1340', in K. Neale (ed.), *An Essex Tribute: Essays Presented to Frederick Emmison* (London, 1987), 97–108.

[3] I. J. E. Keil, 'The Estates of Glastonbury Abbey in the Later Middle Ages' (thesis, University of Bristol, 1965); S. Rippon, 'Making the Most of a Bad Situation? Glastonbury Abbey, Meare, and the Medieval Exploitation of Wetland Resources in the Somerset Levels', *MA* 48 (2004), 91–130.

[4] A. E. B. Owen, *The Records of a Commission of Sewers for Wiggenhall 1319–1324*, Norfolk Record Society 48 (Norwich, 1981); A. E. B. Owen, *The Medieval Lindsey Marsh: Select Documents*, Lincoln Record Society 85 (Lincoln, 1996); A. M. Kirkus, *The Records of the Commission of Sewers in the Parts of Holland 1547–1603*, Lincoln Record Society 54 (Lincoln, 1959); S. Rippon, *The Gwent Levels: The Evolution of a Wetland Landscape*, CBA Research Rep. 105 (York, 1996), 50–8; S. Rippon, *The Severn Estuary: Landscape Evolution and Wetland Reclamation* (London, 1997), 19–23.

[5] Rippon, *Transformation*, 84–90.

The aim of this chapter is to consider the origins and function of the largest artificial watercourses in these reclaimed landscapes: canals. After a brief consideration of the Roman period, to see whether medieval water management technology was based upon earlier endeavours, a number of reclaimed wetlands in Somerset are examined with particular regard to the multiple functions of the major artificial watercourses and the conflicts of interest that resulted.

The Roman Inheritance

Though the focus of this chapter is the medieval period, that was not the first occasion when human communities transformed coastal wetlands in Britain. The planned management of water on a landscape scale was a Roman innovation, occurring in three distinct contexts: 'unsystematic reclamation', where the construction of a sea wall was followed by piecemeal enclosure and drainage; 'systematic reclamation', where a sea wall was constructed and the area now protected was enclosed and drained in a planned fashion; and the 'modification' of marshlands, for example through the digging of localized ditched enclosure systems in an environment that was still open to occasional tidal inundation.[6] The best examples of 'unsystematic reclamation' are known on the North Somerset Levels, where survey and excavation at several locations has revealed landscapes comprising localized ditched enclosure complexes associated with individual farmsteads, drained through a hierarchy of channels ranging from small spade-cut gullies (very similar to modern 'gripes'), through to field boundary ditches. The wholly freshwater palaeo-environmental assemblages show that these field systems must have been protected from tidal inundation by sea walls,[7] in contrast to the ditched enclosure systems on the Caldicot Level (Gwent) just across the Bristol Channel,[8] and in Fenland,[9] that lay in an intertidal environment (an example of landscape 'modification'). The construction of the sea walls to protect the North Somerset Levels from tidal inundation must have been a collaborative/communal effort, though the subsequent enclosure of scattered areas of marsh was a piecemeal affair, and whilst individual enclosure complexes show some degree of planning in their layout, they were separated by areas of open, unenclosed landscape: there was no systematic attempt to drain the whole of the North Somerset Levels.

There is no evidence for any canals in the North Somerset Levels, though the area is crossed by a major tidal river, the Congresbury Yeo, beside which was

---

[6] S. Rippon, 'Romano-British Reclamation of Coastal Wetlands', in H. F. Cook and T. Williamson, *Water Management in the English Landscape: Field, Marsh and Meadow* (Edinburgh, 1999), 101–21; Rippon, *Transformation*, 52–3, 54–95.

[7] S. Rippon, 'The Romano-British Exploitation of Coastal Wetlands: Survey and Excavation in the North Somerset Levels, 1993–7', *Britannia*, 31 (2000), 69–200.

[8] F. M. Meddens and M. Beasley, 'Roman Seasonal Wetland Pasture Exploitation near Nash, on the Gwent Levels, Wales', *Britannia*, 32 (2001), 143–84.

[9] D. Hall, *The Fenland Project 10: Cambridgeshire Survey, the Isle of Ely and Wisbech*, EAA 79 (Cambridge, 1996); Rippon, *Transformation*, 65–79.

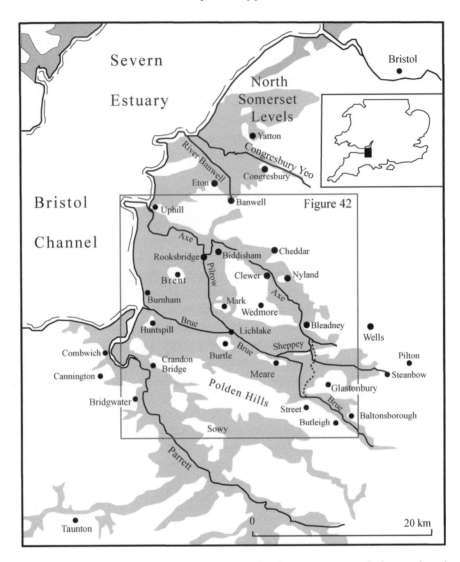

Fig. 41. The Somerset Levels, with the major canalized watercourses and places referred to in this chapter.

located a local coarse-ware pottery industry at Congresbury itself (Fig. 41).[10] 'Congresbury Ware' is known to have reached to the south of Mendip,[11] and its transportation via the Yeo and Bristol Channel is most likely. The major rivers that cross the main Somerset Levels must similarly have been used for navigation, with major settlements and possible ports at Cheddar (Ptolemy's

---

[10] Rippon, *The Severn Estuary*, 88–90.

[11] e.g. Brean Down: A. ApSimon, 'The Roman Temple on Brean Down, Somerset', *Proc. Univ. Bristol Spelaeological Soc*, 10/3 (1965), 195–258; Cheddar: P. Rahtz, *The Saxon and Medieval Palaces at Cheddar: Excavations 1960–62*, BAR British Ser. 65 (Oxford, 1979); the Axe valley: Somerset SMR 10418, 10338.

Ischalis?) beside the Axe, and Combwich and Crandon Bridge beside the Parrett.[12] These rivers certainly maintained their importance into the early medieval period, with high-status early medieval sites at Cannington (near Combwich), Cheddar, and Cadbury Congresbury beside the Parrett, Axe, and Congresbury Yeo respectively (Fig. 41). Though no coastal trading sites have been identified, equivalent to the south Devon site at Bantham,[13] one or more presumably exists in the region.

The use of the Axe for navigation during the early medieval period is suggested by the reference to a 'hythe' at Bleadney in 712 (Fig. 41).[14] South of Cheddar, in the Axe valley, there are also the extensive earthworks of another Romano-British relict landscape, similar to those on the North Somerset Levels, and this complex does include a possible canal *c*.11–12 m wide bank-top to bank-top, *c*.4–6 m wide at the base, and *c*.1.4 m deep,[15] remarkably similar in scale to the Aylmer Hall, Fen Causeway, and Bourne–Morton canals in the Norfolk and Lincolnshire Fenland.[16] There is, however, no direct dating of this feature, and whilst in places it appears integrated with the Romano-British landscape, elsewhere it is clearly stratigraphically later and so may relate to the medieval improvements of the Axe (see below).

All of these Romano-British landscapes on the Somerset Levels were abandoned during the late/post-Roman period, and so made no contribution to the medieval landscape and its system of water management. Across the Bristol Channel, however, on the Wentlooge Level, in Gwent, a very different Roman drainage system still survives in use today.[17] The carefully planned morphology of this landscape, with large blocks of long, narrow fields, is unique in Roman Britain in representing a 'systematic reclamation', and lying close to the legionary fortress at Caerleon, is likely to be of military origin. Palaeo-environmental evidence once again indicates a freshwater (i.e. reclaimed) environment, and the military connection makes it possible that water levels within the ditches were controlled through a system of dams and sluice-gates (hollowed tree trunks with hinged flap valves), as was the case close to the Roman frontier in what is now the Netherlands.[18] The survival of this landscape implies some degree of continued maintenance ever since, although the rigidly planned pattern of fields is totally different from the landscapes created in a piecemeal fashion when the

---

[12] Rippon, *The Severn Estuary*, 53–4.

[13] F. M. Griffith and S. J. Reed, 'Rescue Recording at Bantham Bay, South Devon, in 1997', *Devon Archaeological Society Proceedings*, 56 (1998), 109–32.

[14] S 1253; and see Cole, above p. 71.

[15] J. Grove, 'Reclamation and Utilization of the Upper Axe Valley during the Roman Period', *Archaeology in the Severn Estuary*, 13 (2002), 65–87. R. McDonnell, 'The Upper Axe Valley, an Interim Statement', *Proc. Somerset Archaeol. Nat. Hist. Soc.* 123 (1979), 75–82.

[16] A. Crowson, 'Excavations on the Fen Causeway at Straw Hill Farm, Downham West', *Fenland Research*, 9 (1994), 20–4. A. Crowson, T. Lane, and J. Reeve, *Fenland Management Project Excavations 1991–1995* (Sleaford, 2000), figs. 44 and 59.

[17] M. G. Fulford, J. R. L. Allen, and S. J. Rippon, 'The Settlement and Drainage of the Wentlooge Level, Gwent: Excavation and Survey at Rumney Great Wharf, 1992', *Britannia*, 25 (1994), 175–211.

[18] Rippon, *Transformation*, frontispiece, 84–90.

other coastal wetlands around the Severn were recolonized during the medieval period; the Roman approach to drainage in both Wentlooge and the other marshland areas appears to have had little if any influence over subsequent, medieval, attempts to reclaim the Severn wetlands.

The Severn estuary appears to be the only part of Roman Britain to see reclamation, though other coastal wetlands were certainly extensively settled and the environment 'modified' through the construction of localized ditched enclosure systems in what remained an intertidal environment (e.g. the Caldicot Level, Gwent: see above). The best-known example, and the most extensive, is Fenland, where there were also a series of major canals that linked fen-edge settlements with the major coastal estuaries.[19] In running between two areas of high ground, or through areas of backfen that were undrained and not settled, these canals were clearly concerned with navigation rather than drainage. The vast majority silted up long before the area was recolonized during the medieval period, and the handful of medieval canals that may follow a Roman line (e.g. the River Glen) are likely to have been re-excavated.

A number of key conclusions can be drawn from this rapid overview of water management in coastal wetlands during the Roman period. First, there was both reclamation and canal construction in Roman Britain, but very little evidence that this ability to modify and transform wetland environments survived into the medieval period. Secondly, it was possible in the Roman period to have reclamation without canals, and canals without reclamation: it was only in the medieval period that the two means of increasing landscape productivity commonly occur in the same place, though the question is whether artificial watercourses performed both functions.

The Medieval Period

Following a rise in relative sea level during the late/post-Roman period, there was renewed interest in coastal wetlands during the late first millennium AD, with extensive areas being embanked and reclaimed, notably around the Severn estuary, Romney Marsh, Thames estuary, and in Fenland.[20] Major artificial watercourses became important features of these reclaimed landscapes and the rest of this chapter will consider their different functions as navigational canals, to help prevent flooding by carrying upland rivers across low-lying areas, and using water as a resource in countering the effects of estuaries silting up, and powering mills. It is argued that these forms of water management inherently played little part in the drainage of those areas through which they passed.

As with the case of river canalization (see Bond, above pp. 188–97 and Blair, below pp. 267–8, 278–83), monastic houses were often at the forefront of

[19] Crowson et al., *Fenland Management Project*, 129–34, 202–5; Rippon, 'Romano-British Reclamation of Coastal Wetlands', in Cook and Williamson, *Water Management*, 101–21; Rippon *Transformation*, 69–71; and see Bond, above pp. 162–9.
[20] Rippon *Transformation*, 152–82.

attempts to transform wetland landscapes and certainly have the best surviving records. The Somerset Levels were dominated by Glastonbury Abbey, a particularly progressive estate manager whose archives are both extensive and accessible, and that area will be used as a case study. Hollinrake and Hollinrake (below pp. 235–43) consider the abbey's management of water in the immediate vicinity of its precinct, whereas this chapter will focus on the Axe and Brue valleys further to the west (Fig. 41). First, though, it is worth briefly examining the range of water management issues that faced medieval landowners and their tenants generally.

One major issue was the need to link inland/fen-edge settlements with major tidal estuaries often many kilometres away. As in the Roman period, in Fenland this could be achieved through the construction of canals that traversed the extensive unenclosed freshwater backfens. Either wholly artificial, or straightening stretches of natural rivers, such canals were mainly to improve communication rather than drainage: the backfens through which they passed were lower-lying than the areas towards the coast with the result that canals often had to be embanked and raised. However, although they did not drain waters from the areas through which they passed, these canals did help prevent flooding by carrying freshwater discharge from rivers flowing off the adjacent uplands to the sea. Such canals were most common in Fenland (see Bond, above pp. 180–8), while other examples include the works of Meaux Abbey in the Hull valley.[21] In these cases the canals were consciously created as landscape features, in contrast to the Pevensey Levels where a series of embanked/constrained rivers were created as the land on either side was separately embanked as part of the process of individual reclamations: the result was canal-like features which were never deliberately created as such.[22]

Another reason why major artificial watercourses had to be constructed was due to changes in coastal geomorphology, notably the silting of estuaries. In Fenland, for example, the gradual closing of the inlet at Wisbech led to a series of rivers being diverted to the Nar estuary at King's Lynn (see Bond, above pp. 186–7), while in Romney Marsh a major artificial watercourse, the 'Rhee Wall', was constructed across an already reclaimed landscape, simply to divert waters from the river Rother across the reclaimed marshes in order to flush out the estuary at New Romney.[23] Its sole purpose was to clear the estuary of silts: being completely embanked, it could not drain the land through which it passed (this function was carried out by a separate set of sewers). The Rhee Wall was constructed and maintained by the Cinque Port Liberty of New

---

[21] J. Sheppard, *The Draining of the Hull Valley* (York, 1958).

[22] A. J. F. Dully, 'The Level and Port of Pevensey in the Middle Ages', *SxAC* 104 (1966), 26–45; Rippon, *Transformation*, 187–90.

[23] J. Eddison, 'The Purpose, Construction and Operation of a Thirteenth-Century Watercourse: The Rhee, Romney Marsh, Kent', in A. Long, S. Hipkin, and H. Clarke (eds.), *Romney Marsh: Coastal and Landscape Change through the Ages* (Oxford, 2002), 127–39; Rippon *Transformation*, 197–8; E. Vollans, 'New Romney and the "River Newenden" in the Later Middle Ages', in J. Eddison and C. Green (eds.), *Romney Marsh: Evolution, Occupation, Reclamation*' (Oxford, 1988), 128–41.

Romney, at a time when several Royal Commissions reflect growing concern over the maintenance of navigational watercourses, but there is no evidence that it ever served any navigational purpose itself: it simply channelled water to where it was needed.

Another form of medieval water management is best illustrated by examples on the Caldicot Level in south-east Wales. A number of streams flow off the adjacent uplands into tidal creeks that for several millennia appear to have been used as landing places, such as Magor Pill, where a thirteenth-century boat has recently been recovered from the silted-up creek upon which a small port called *Abergwaitha* lay.[24] The character of the Romano-British pottery assemblage from a site just upstream suggests that it functioned as a small landing place from the late first/second century AD, whilst the earliest occupation beside the creek dates back to the Iron Age, an example of the long-term significance that these tidal inlets had in coastal landscapes.[25] Extending inland from the creek, a major embanked, artificial watercourse, 'Mill Reen', runs to the fen-edge at the village of Magor. A very similar embanked channel, the 'Monksditch', runs from a tidal creek at Goldcliff to the fen-edge village of Llanwern, while a third artificial watercourse, Elver Pill Reen, runs from Redwick to the fen-edge at Bishton. In each case the primary function of these channels was to carry a freshwater stream across the Levels, so avoiding flooding, with, in the case of Mill Reen and possibly Monksditch, the additional role of feeding watermills located both on the fen-edge and the coast. None of these channels served as canals, in the sense of being used for navigation, and indeed, a thirteenth-century description of Monksditch describes various sluice-gates along its line making it impossible for boats to have passed.[26] These embanked reens nevertheless demonstrate how controlling water, in order to prevent flooding, could also be used as a positive resource, in this case powering mills, and can be seen alongside reclamation as part of the policy of Anglo-Norman Marcher lords and their newly founded monastic houses of improving the productivity of their newly acquired estates in the late eleventh and twelfth centuries.[27]

[24] J. R. L. Allen, 'Magor Pill (Gwent) Multiperiod Site: Post-Medieval Pottery and Shipping Trade', *Archaeology in the Severn Estuary*, 10 (1999), 75–97; J. R. L. Allen and S. J. Rippon, 'Iron Age to Early Modern Activity and Palaeochannels at Magor Pill, Gwent: An Exercise in Lowland Coastal Zone Geoarchaeology', *Antiquaries Journal*, 77 (1997), 327–70; N. Nayling, *The Magor Pill Medieval Wreck*, CBA Research Rep. 115 (York, 1998).

[25] J. R. L. Allen, 'Magor Pill Multiperiod Site: The Romano-British Pottery and Status as a Port', *Archaeology in the Severn Estuary*, 9 (1998), 45–60; J. R. L. Allen, 'Magor Pill Multiperiod Site: The Romano-British Pottery and Status as a Port. A Postscript', *Archaeology in the Severn Estuary*, 10 (1999), 75–97; J. R. L. Allen and M. G. Fulford, 'The Distribution of South-East Dorset Black Burnished Category 1 Pottery in South West Britain', *Britannia*, 27 (1997), 223–81; A. Whittle, 'Two Later Bronze Age Occupations and an Iron Age Channel on the Gwent Foreshore', *Bulletin of the Board of Celtic Studies*, 36 (1989), 200–23.

[26] Rippon, *The Gwent Levels*, 68–71, 82, 94.

[27] Rippon, *The Gwent Levels*; S. Rippon, 'Reclamation and Regional Economies of Medieval Marshland in Britain', in B. Raftery and J. Hickey (ed.), *Recent Developments in Wetland Research* (Dublin, 2001), 139–58.

From this rapid overview it should be clear that a wide variety of substantial artificial watercourses were constructed within coastal wetland landscapes, both as part of and after reclamation. At first sight, most of these channels appear similar and could be regarded as canals: in practice they served a multiplicity of purposes and not all functioned as canals.

The Somerset Levels: The 'Old Brue', Axe, Sheppey, and Hartlake

The canals of Glastonbury Abbey, and to a lesser extent the bishops of Wells, are the best-documented examples in Somerset, and represent a major reorganization of the river systems of the Somerset Levels. Traditionally they are ascribed to the thirteenth century;[28] a re-evaluation of the documentary evidence, and its integration with physical evidence contained within the fabric of the historic landscape (the modern pattern of fields, roads, settlements and watercourses), suggests that Glastonbury was manipulating its water resources from at least the eleventh century. Glastonbury was initially linked directly to the Bristol Channel via the old course of the River Brue, which enters the Levels south of Baltonsborough, flowing through South Moor to Street (Figs. 41–2: some of its waters were later diverted to form a mill-leet, powering mills at Baltonsborough and Northover).[29] From Street, the Brue originally turned north and flowed along a meandering channel past Beckery and along what is now Great Withy Rhyne (the Meare–Glastonbury parish boundary), through the Godney–Garslade Gap and northwards through the Panborough–Bleadney Gap, where it was joined by the River Axe which rises in Wookey Hole (Fig. 42): the Brue could not flow directly west from Glastonbury as this area was blocked by the raised peat-bog around Meare. From where the old course of the Brue was joined by the Axe, they flowed along an increasingly meandering channel past Cheddar reaching the sea at Uphill.

In what is now known as the Axe valley, the course of the former Brue between Bleadney and Cheddar was subject to a series of improvements. By 1242/3 the abbot had built banks on either side of the river to increase its depth (Fig. 44), leading to a dispute with local commoners. The jury found that 'the bank is of old time, and that by the water channel alongside the bank, the abbot and his predecessors had a thoroughfare by their boats to go from their stone and their lime to the abbey'.[30] In 1273, it is recorded that the Axe was 'adequate for the Abbot to take stone and lime and corn from his manors and from other places in those parts to his abbey at Glastonbury'.[31] In 1316 the Axe was diverted into a wholly artificial canal between Clewer Bridge and Lower Weare

[28] e.g. M. Williams, *The Draining of the Somerest Levels* (Cambridge, 1970), 65–73.
[29] G. Aalbersberg, 'The Alluvial Fringes of the Somerset Levels' (thesis, University of Exeter, 1999), 97–173; Hollinrake and Hollinrake below.
[30] C. E. Chadwyke-Healey (ed.), *Somersetshire Pleas*, Somerset Records Society XI (Taunton, 1897–1929). no. 818.
[31] P. J. Helm, 'The Somerset Levels in the Middle Ages', *JBAA* 12 (1949), 37; Williams, *Draining*, 65.

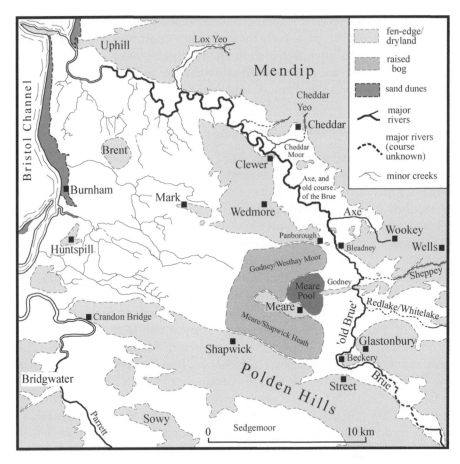

Fig. 42. The natural drainage system in the Somerset Levels during the early medieval period. Note how the Brue, Redlake/Whitelake and Sheppey rivers all drained northwards into the Axe valley as their more direct route westwards to the Bristol Channel was blocked by a raised peat-bog. These old courses of the Brue and Axe are reconstructed from earthworks, and the pattern of field and parish boundaries. The old course of the Parrett is shown before the meander south of Crandon Bridge was broken through in the seventeenth century.

in order to improve the flow of water (slicing through a series of meanders) and to avoid a mill.[32] A short stretch of canal cutting through the neck of a large meander at Barrrow's Hams may also date to this period (Fig. 43).

In addition to the main course of the Axe, there were several subsidiary canals (Fig. 43). The Cheddar Yeo takes the water flowing out of Cheddar Gorge that used to flow through a river that meandered across Cheddar Moor. The Yeo must have been canalized by 1212, when Hythe is first recorded.[33]

---

[32] W. H. B. Bird (ed.), *Calendar of the Manuscripts of the Dean and Chapter of Wells*, vol. i, Historic Manuscripts Commission (London, 1907), 188–9; C. D. Ross (ed.), *Cartulary of St Mark's Hospital, Bristol*, Bristol Record Society 31 (Bristol, 1959), nos. 209, 243; B. Hobhouse (ed.), *Register of Bishop Drokensford*, Somerset Record Society 1 (Taunton,1877), 7; Helm, 'The Somerset Levels', 45.

[33] See Cole, above p.71.

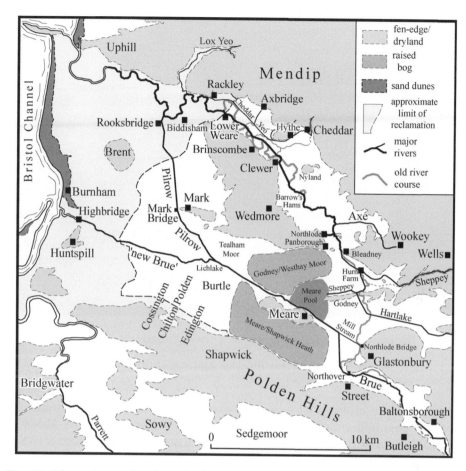

Fig. 43. The major medieval artificial watercourses in the Somerset Levels. Note how the Brue, Hartlake (Redlake/Whitelake), and Sheppey rivers were all diverted westwards through an artificial cut (the 'new Brue') through the raised bog west of Meare. Pilrow linked Glastonbury with its port and watermill at Rooksbridge. Compare Figs. 17 and 49.

Another tributary canal of the Axe is suggested by the place name 'Northlode', in Theale, recorded in 1308 (and see below).[34] By the thirteenth century there were a series of small ports and landing places in the Axe valley, suggesting that the amount of traffic was not inconsiderable (Bleadney, Northlode near Wedmore, Clewer, Brinscombe, Hythe near Cheddar, Axbridge, Lower Weare, Rackley, and Rooksbridge).[35]

South of Bleadney, the old course of the Brue was straightened and widened (to 12 feet, 3.7 m) in 1326 between Bleadney and Monkenmead (modern Hurn

[34] *Wells I*, 219; see Cole, above p.78.
[35] A. Major, 'Report on Excavation Work at Brinscombe, Weare, Somerset', *Proceedings of the Somerset Archaeological and Natural History Society*, 57 (1911), 110–13; V. E. J. Russett, 'Hythes and Bows: Aspects of River Transport in Somerset', in G. L. Good, R. H. Jones, and M. W. Ponsford (eds.), *Waterfront Archaeology: Proceedings of the Third International Conference on Waterfront Archaeology*, CBA Research Rep. 74 (London,1991), 60–6; and see Cole, above pp. 73–4.

Fig. 44. The heavily embanked River Axe as it passes through the Panborough–Bleaney Gap, looking north from ST 4804 4550. The watercourse is higher than the surrounding land, and so cannot have served any drainage function apart from carrying freshwater discharge across this low-lying area.

Farm), and two bridges that were built over the channel were high enough for boats to pass underneath. The work was carried out jointly by the Abbot of Glastonbury and the Bishop of Wells,[36] and the now-abandoned course of this old river still survives as a palaeochannel (a slight, silt-filled ridge between 50 and 100 m wide).[37]

Two tributaries of the 'old Brue' were also canalized. The Sheppey drains the area south of Wells, and enters the north-east corner of the Somerset Levels at Hay Moor in the episcopal manor of Wookey. It used to meander across Hay and North Moors before joining the 'old Brue' at Hurn Farm, but in the early fourteenth century it was diverted to a raised and embanked channel which follows the contours of Bower's Hill at Fenny Castle.[38] By the fourteenth century, the Sheppey was diverted down the 'old Brue' past Hurn Farm, Lineacre (Nineacre), and Godney into Meare Pool, and thence to the new course of the Brue (see below).[39] In 1326 another canal, the Hartlake, was created in order to carry the waters of two upland streams (the Redlake and Whitelake) through the moors north of Glastonbury, joining the new course

[36] D. A. Watkins (ed.), *Glastonbury Chartulary*, vol. i, Somerset Records Society 59 (Taunton, 1944), no. 161.

[37] Aalbersberg, 'The Alluvial Fringes', fig. 4.1.

[38] T. S. Holmes, *A History of the Parish and Manor of Wookey* (Bristol, 1886), 12.

[39] *Glastonbury Chartulary*, i, no. 161; D. A. Watkins (ed.), *Glastonbury Chartulary*, vol. ii, Somerset Record Society 63 (Taunton, 1948), no. 648.

of the Sheppey at Higher Bridge Farm in Godney.[40] These watercourses were certainly used for navigation, as in the early thirteenth century it is recorded that wine was carried by boat from Glastonbury's vineyard at Pilton (Fig. 41: between the Redlake and Whitelake rivers in the drylands east of Glastonbury) to the abbey.[41] As both the Sheppey and the Hartlake were embanked they cannot have actively drained the land through which they passed, though they did carry upland waters across the lowest-lying parts of the Levels so reducing their chances of flooding.

The Somerset Levels: The 'New Brue', Pilrow Cut, and Rooksbridge (Fig. 43)

As late as the sixteenth century, some of the waters of the Brue still flowed along the old course past Godney and Bleadney,[42] but most of the waters were carried in an artificial canal which left the old course of the Brue at Northover, taking a direct route north to Waterlease before turning west, past Meare Pool and the Abbot's Fish House (Fig. 48) as far as Lichlake in Tealham Moor (ST 4012 4472).[43] From here, the waters took two courses: the Brue itself, continuing west to Highbridge and thence the Bristol Channel,[44] while a second canal, the Pilrow Cut, went north through Mark to join the Axe at Rooksbridge.[45] At some stage a more direct route between Glastonbury and Meare was established through the digging of a further canal from Northload meeting the Brue at Waterlease; now called 'Mill Stream', this is presumably the *lād* that gave 'Northlode', first documented in *c*.1180, its name.[46]

It is unclear when the Brue's diversion occurred, but it was certainly before the thirteenth-century date previously attributed to it.[47] That stretch as far

---

[40] *Glastonbury Chartulary*, i, no. 157; *Wells*, i. 324; W. P. Baildon (ed.), *Calendar of the Manuscripts of the Dean and Chapter of Wells*, vol. ii, Historic Manuscripts Commission (London, 1914), 617.

[41] C. J. Elton (ed.), *'Rentalia et Custumaria' of Glastonbury Abbey*, Somerset Records Society 5 (Taunton, 1891), 176–8.

[42] L. Toulmin Smith (ed.), *The Itinerary of John Leland in or about the Years 1535–43, Part 1* (London, 1906), 148–9.

[43] This is the point where the raised peat-bogs of the central Brue valley give way to the alluvial claylands of Mark and Huntspill Moors.

[44] Between Lichlake and Liberty Moor the meandering line taken by the Brue suggests that it followed a natural stream that flowed out of the raised bogs of the central Brue valley. At ST 371 451 this natural stream turns south (RAF air photograph CPE/UK 1724 3022). The Brue, however, continues west in a wholly artificial line across Huntspill Moor, skirting around the northern edge of the medieval reclamations at Hackney (East Huntspill). This stretch of the Brue appears to pre-date the historic landscape (i.e. the enclosure and drainage of this area, which led to the present pattern of fields and roads). At Churchland Farm (ST 340 459) it takes a rather more sinuous line that at first sight suggests the artificial canal joined an existing creek that flowed into the tidal inlet south of Burnham and Highbridge. This stretch of the Brue, however, must also have been a substantially artificial construction for it cuts through a pre-existing system of fields, including a possible small common field north of Pear Tree Farm, and cuts unconformably across the Burnham–Huntspill parish boundary which itself followed a strongly meandering former creek.

[45] Between Lichlake and Mark Bridge, the slightly sinuous line taken by Pilrow suggests that it followed the line of a natural stream which flowed *c*.1 km from the fen-edge.

[46] *Glastonbury Chartulary*, ii, no. 399; see Cole, above p.78.

[47] Williams, *Draining*, 70.

as Meare Pool must have existed by 1091, when, according to William of Malmesbury, St Benignus' bones were carried from Meare to Glastonbury by boat on a river,[48] and in 1294 the Brue was said to be 'embanked and run toward the Mere'.[49] The Brue west of Meare Pool is referred to in 1327 as the watercourse from 'Ferlyngmere' (Meare Pool) to 'Lichelak' (near Burtle) in Tealham Moor.[50] That stretch of the 'new Brue' between Lichlake and the Bristol Channel, formerly known as the 'Fishlake River',[51] must be the channel described in a list of drainage duties on a watercourse running 'from Burtle Pool to the land of Chilton Polden, and from Edington and Chilton to the land of Cossington, and from Cossington to Huntspill, and from Huntspill to the sea'.[52] This document appears to date to the late twelfth century or earlier, for it refers to Robert de Ewyas (who died in 1198) and Ralph de Sancta Barbara, one of Glastonbury Abbey's major tenants in 1189.[53] The use of this watercourse as a navigational canal is demonstrated in 1371, for example, when corn was shipped from Huntspill to Meare.[54]

The date of Pilrow's construction (Figs. 43, 45–7) is also uncertain, though it clearly existed by 1235 when Robert Malerbe was said to be responsible for maintaining Glastonbury Abbey's waterways as far as Mark Bridge.[55] Pilrow may also be the 'Morditch' recorded in 1235–52, whose scouring was a customary service owed by the tenants of Brent, Lympsham, and Berrow, as there is reference to a property adjoining it in Meare manor as it left Meare Pool.[56] The channel now called Pilrow is also recorded in 1315, when there is reference to dykes, sluices, and walls 'for the preservation of the course of water flowing towards the sea, and for the safety of lands in Mark, Mudgley, Wedmore and Biddisham'.[57] It would appear that Pilrow was originally tidal, as in the time of Abbot Sodbury (1322–34) a 'stonework' was constructed at 'Rokusmille' 'to shut out the sea waves',[58] and in 1358/9 the Dean of Wells constructed two sluices at Mark 'to keep the sea-water from ebbing and flowing'.[59]

It is not clear whether Pilrow and the new Brue west of Lichlake were dug at the same time. Today, the Brue is the more major watercourse, but this may not always have been the case. The current junction of Pilrow and the Brue at Lichlake suggests that Pilrow was added later, but earthworks[60] and the line of the Mark–Wedmore parish boundary show that its original course was simply as a continuation of the Brue between Lichlake and Meare. That the Brue west

---

[48] M. Winterbottom and R. M. Thomson (ed.), *William of Malmesbury: Saints' Lives* (Oxford, 2002), 360–5.
[49] *Glastonbury Chartulary*, ii, nos. 478–9.   [50] *Wells I*, 226–8.
[51] Williams, *Draining*, 70 n. 1.   [52] *Glastonbury Chartulary II*, no.1015; *Wells I*, 226–8.
[53] I would like to thank R. Dunning for improving upon the dating given in Rippon, *The Severn Estuary*, 213, and Williams, *Draining*, 70.
[54] *Cal. Inq. Post Mortem* 18, 41–2.   [55] 'Rentalia et Custumaria', 176–8.
[56] Ibid. 38, 41–3, 45, 47, and 50.   [57] *Cal. Pat. R.* 1313–17, 412–13.
[58] *The Chronicle of Glastonbury Abbey*, no. 139.   [59] PWML ii. 134.
[60] These are particularly clear on early air photographs taken by the RAF: CPE/UK 1924 3036.

of Lichlake may have been a later addition has a certain logic, as for the most part Pilrow runs through the low ground west of Wedmore (partly owned by Glastonbury), whereas the current Brue has to cut through increasingly high ground towards the coast (owned by a series of manors in Huntspill). Pilrow also joins the Axe at the port of Rooksbridge, whereas the 'new Brue' runs into what was probably a minor tidal creek at Highbridge, where there is no evidence for a medieval port or landing place.[61]

The original dimensions of these canals are impossible to determine, as their subsequent maintenance may have led to some enlargement (e.g. the River Brue west of Westhay). It is notable, however, that where Pilrow, the Brue, and the Axe survive in their embanked form, and where slight sinuousity in their line suggests that they have not been subject to recent improvements, all three are *c*.4–5 m wide, *c*.3 m deep, with embankments *c*.1 m high and *c*.2–3 m wide

Fig. 45. The embanked Pilrow, looking north from ST 374 503, with the remains of an undated timber revetment retaining the western bank (left) and a stone structure to the east. The embankment survives to the east (right), but has been removed to the west. No proper survey has been carried out of these medieval canals and their associated archaeology. Loxton, on the limestone hills that form the western end of Mendip, can be seen on the skyline.

[61] The sand dunes as mapped in Figs. 42–3 are based on the Soil Survey Sheet 279, with the addition of their extension south-east of Burnham which is based upon observations by local archaeologist Sam Nash: see S. Rippon, 'Roman Settlement and Salt Production on the Somerset Coast: The Work of Samuel Nash—a Somerset Archaeologist and Historian', *Proceedings of the Somerset Archaeological and Natural History Society*, 139 (1995), 99–118. The Parrett estuary is after R. McDonnell, 'Island Evolution in Bridgwater Bay and the Parrett Estuary: An Historical Geography', *Archaeology in the Severn Estuary*, 6 (1995), 71–83.

(the documented canalization of the Sheppey also led to a channel *c*.3.7 m wide: see above). The profiles do vary from a shallow U-shape (e.g. Pilrow north of Mark), to a rather deeper V-shape (e.g. Pilrow in Mark Moor: Fig. 46), though this may be a product of subsequent recutting. In contrast, the Roman canals in Fenland were *c*.10 m wide.[62]

There are no surviving examples of the boats that would have used these rivers and canals, though Redknap has brought together much of the documentary, pictorial, and archaeological evidence for small medieval craft generally.[63] The Magor Pill boat (along with similar vessels from Graveney in Kent, 'Blackfriars 3' in London, and Skuldev in Denmark) appears to have been some 14 m long and 4 m wide,[64] and such vessels might have reached Rooksbridge; other seagoing vessels of this period were up to *c*.20 m long and would have been too large, presumably docking at Uphill.[65] The vessels that would have used the inland waterways of Somerset were probably similar to the flat-bottomed, *c*.2-m-wide Bridgwater barges, Fleet trows, and Parrett flatners.[66]

Fig. 46. Pilrow, looking north from ST 3925 4599. Having been drained, its profile is visible (Fig. 45 shows the same watercourse filled with water looking south from this same spot). The embankments here have been removed. Brent Knoll (left) is visible on the skyline.

---

[62] Crowson et al., *Fenland Management Project*, figs. 44, 58, and 59.

[63] M. Redknap, 'Reconstructing the Magor Pill Boat', in N. Nayling, *The Magor Pill Medieval Wreck*, CBA Research Rep. 115 (York, 1998), 129–42.

[64] P. Marsden, *Ships of the Port of London: Twelfth to Seventeenth Centuries AD*, English Heritage Archaeological Rep. 5 (London, 1996), 55–104; M. Redknap, 'The Historical and Archaeological Significance of the Magor Pill Boat', in Nayling, *The Magor Pill Medieval Wreck*, 143–54.

[65] C. Green, *The Severn Traders* (Lydney, 1999); Mardsen, *Ships of the Port of London*; Redknap, 'The Historical and Archaeological Significance of the Magor Pill Boat', table 20.

[66] Green, *The Severn Traders*, 53, 60–1.

Fig. 47. Pilrow, looking south towards Lichlake from ST 3925 4599 (the same bridge as Fig. 46), here filled with water.

Like Mill Reen and Monksditch on the Gwent Levels, the Axe, Brue, and Pilrow are embanked channels that in some areas are higher than the ground through which they pass. They did not play a part in the drainage of these areas, though by carrying fresh water across them they helped to prevent flooding from upstream. However, the low-lying moors through which they mostly pass can hardly have been of great value, and it is likely that such major feats of engineering were primarily designed to improve Glastonbury's communications, both internally and beyond. The abbey's use of these canals is well documented. In addition to maintaining Pilrow in 1235, Robert Malerbe was also responsible for a watercourse between Clewer and Street (which must be the old course of the Brue through the Axe valley), and the nearby vineyard at Panborough (the 'Northlode' in Theale? see above). As head boatman for the abbey he was also responsible for transporting wine from there, and the vineyards at Pilton and Meare, to Glastonbury.[67] He also had to provide a boat to carry eight men, and acting as its coxswain to convey the abbot to Meare, Brent, Butleigh, Nyland, Godney, and Steanbow (near Pilton), and all the abbot's men and kitchen, including the movable kitchen gear and cooks, and his huntsmen and hounds (Fig. 43).[68] An eighteenth-century antiquary also records that 'The tradition is, that in former times

---

[67] 'Rentalia et Custumaria', 176–8.
[68] R. N. Grenville, 'Somerset Drainage', *Somerset Archaeological and Natural History Society Proceedings*, 72 (1926), 1–13.

the abbots of Glastonbury used to sail by boat in an annual excursion from Glastonbury down the river Brew [Brue], and along Pilrow river to Mark, and to halt several days at this house of rendezvous High-Hall, from which they proceeded by Pilrow river to East Brent, where they had other estates.'[69] Rooksbridge served as a trans-shipment point for goods being transported from Glastonbury to the open sea. In *c*.1400, for example, a seagoing vessel foundered there,[70] while in 1500, St John's church in Glastonbury used two boats to ship some seats from Bristol to Rooksbridge where they were transferred to thirteen smaller vessels which sailed to Glastonbury via Meare (i.e. down Pilrow).[71]

Pilrow, like Mill Reen and Monksditch on the Caldicot Level in Gwent (see above), also supported a watermill, at Rooksbridge, which is a further example of the significance that nodal points in the river/canal system could assume. 'Rokesmulle' is first recorded in 1189, when its value of 40*s* made it Glastonbury Abbey's most valuable mill,[72] and in the early fourteenth century it is recorded that Richard Saint Barbe conveyed to Geoffrey, Abbot of Glastonbury, all his rights in the watercourse called 'Barberewe' beginning at 'Therlemere' (Mark Moor) and extending to a watermill at 'Rokespulle' (on the River Axe at Rooksbridge).[73] The use of canals to power watermills could, however, also lead to a conflict of interest, and in 1358 Abbot Monington's mill at Rooksbridge was accused of stopping up all the water in the watercourse at Meare, leading to a risk of flooding.[74]

Another use to which these canals were put, which conflicted with their navigational role, was fishing. In 1242, for example, the Abbot of Glastonbury's boats were accused of breaking three fisheries in the Axe near Rackley.[75] Wells held a fish-house at 'Rodwere' beside the Pilrow Cut, which in 1378 was the subject of repairs costing 40*s*.[76] The greatest number of weirs appears to have been on the Brue. In 1327, for example, an agreement was reached defining the boundary between the estates of the abbey and the bishop that allowed the abbot access to the weirs of *Hachwere* and *Bordenwere* in Meare and *Pariswere* in Westhay, all on the dean's property.[77] In the early fourteenth century, other weirs are recorded on the Brue at *Coubrigge* just to the north of Westhay;[78] *Brudenwer* (Brue-weir)[79] and 'between *Brudenwere* and *Lichelake* to the west'

---

[69] P. M. Slocombe, *Mark: A Somerset Moorland Village* (Bradford-on-Avon, 1999), 16.

[70] *Cal. Inq. Misc.* vii, no. 163.

[71] W. E. Daniel, 'The Churchwardens Accounts of St. John's Glastonbury', *Somerset and Dorset Notes and Queries*, 4 (1895), 89–96.

[72] R. Holt, 'Whose were the Profits of Milling Corn? An Aspect of the Changing Relationship between the Abbots of Glastonbury and their Tenants 1086–1350', *Past and Present*, 116 (1986), 3–23.

[73] Slocombe, *Mark*, 16–17.   [74] *PWML* ii. 131.   [75] *Somersetshire Pleas*, no. 237.

[76] *Wells I*, 285.   [77] *Glastonbury Chartulary II*, no. 647; *Wells I*, 226–8.

[78] 1311/12 Meare Account Roll: SRO T/PH/Lon 2/16 11216; 1343/4 Meare Court Rolls: SRO T/PH/Lon 2/9 6365. I would like to thank Dave Musgrove for bringing these and the following references to my attention.

[79] 1343/4 Meare Court Rolls: SRO T/PH/Lon 2/9 6365.

both on the Brue west of Meare Pool;[80] and *Northwer* by La Hamme on the Brue east of Meare Island.[81] In addition, the 1515 Beere Survey mentions *Cockeswere juxta Lichelake*, and *Jameswere* in the east part of Meare. The latter is reflected in the name 'James Wear River' which takes the Sheppey across the now reclaimed Meare Pool. The nature of these fish-weirs is not made clear, though in 1301/2 there is reference to moor at *Les Puttes juxta La Shirte*; in 1515 there was pasture, waste, and heath at '*Lerpyttes* in the north part of *La Yoo juxta Swyre*'.[82] 'Puttes' were large wicker baskets used to trap fish in rivers[83] or the intertidal zone.[84]

Fig. 48. A seigniorial landscape of water exploitation and water management. The River Brue, north of Meare Fish House (left), church, and Manor Farm (background, centre left), looking south-west from ST 4490 4175. Meare Pool lay to the north (right), and the Fish House is surrounded by the earthworks of fishponds. The Brue contained numerous fish-weirs, and was used for transporting wine from the vineyard at Meare to Glastonbury.

[80] 1343/4 Meare Court Rolls: SRO T/PH/Lon 2/9 6365.

[81] 1355: Survey of Abbot Monnington: British Library, Egerton 3321.

[82] Meare Account Roll: SRO T/PH/Lon 2/14 11272; Abbot Beere's Survey of Meare (British Library, Egerton 3034).

[83] e.g. the Trent: P. M. Losco–Bradley and C. R. Salisbury, 'A Saxon and a Norman Fish Weir at Colwick, Nottinghamshire', in M. Aston (ed.), *Medieval Fish, Fisheries, and Fishponds in England* (Oxford, 1988); C. R. Salisbury, 'Primitive British Fishweirs', in G. L. Good, R. H. Jones, and M. W. Ponsford (eds.), *Waterfront Archaeology: Proceedings of the Third International Conference on Waterfront Archaeology*, CBA Research Rep. 74 (London, 1991), 76–87.

[84] e.g. the Severn Estuary: J. R. L. Allen and S. J. Rippon, 'Magor Pill (Gwent) Multiperiod Site', 75–97; S. Godbold and R. Turner, 'Medieval Fishtraps in the Severn Estuary', *MA* 38 (1994), 45–6; N. Nayling, 'Medieval and Later Fish Weirs at Magor Pill, Gwent Levels: Coastal Change and Technological Development', *Archaeology in the Severn Estuary*, 10 (1999), 99–113; E. Townley, 'Fieldwork on the Forest Shore: Stroat to Woolaston, Gloucestershire', *Archaeology in the Severn Estuary*, 9 (1998), 82–5.

The construction of these canals would have had a further conflict with the wider landscape. For over 2 km south of the Axe at Rooksbridge,[85] and some 2 km east of Highbridge, both Pilrow and the Brue respectively post-date reclamation, enclosure, and drainage (Fig. 43). In slicing through the field systems in these areas these canals would have divided landholdings and disrupted local communications.

The North Somerset Levels

The North Somerset Levels are less well documented, but do contain a number of major artificial watercourses on the Bishop of Wells's manors of Banwell, Congresbury, and Yatton (Fig. 41). Part of the Congresbury Yeo is clearly canalized, though there is no indication of when this occurred or who was responsible; a medieval date is most likely when the manors of Congresbury and Yatton, through which it passes, were in the hands of the bishops of Wells (these estates were sold and subsequently fragmented in the sixteenth century making a date for the Congresbury Yeo's canalization after that date unlikely). There is no medieval evidence for its use for navigation, but in 1736 William Donne bought iron from Graffin Prankard of Bristol, turning it into nails at Congresbury mill; Prankard's accounts show 232 cwt of iron being 'put on board his own vessel for Congresbury'.[86] On 24 February 1794, the *Sherborne and Yeovil Mercury* included an advertisement for the Congresbury tannery with reference to 'a navigable river running alongside it with the use of a barge of the same'.[87]

The narrower but clearly artificial Banwell River also potentially served as a canal, linking the Bishop of Wells's manor of Banwell with the estuary. Until *c*.1790 it was only intertidal as far south as Banfield Cottages, and possibly Ebden Bridge,[88] though it is heavily embanked as far as the fen-edge at Banwell. There is no evidence as to when the River Banwell was dug, though it existed by 1351 when it is referred to as the *Banewellesyeo*.[89] If the interpretation of the 'Eton' place name, where the Banwell River passes close by the fen-edge at Wolvershill, is correct then it existed by 1325.[90]

---

[85] Between the Axe and Pilrow Cottage (ST 373 510), Pilrow takes a slightly sinuous line with a number of changes in direction in order to skirt around the settlement at Rooksbridge and its associated droveways. This section also clearly post-dates the enclosure and drainage of the areas through which it is passing. At Pilrow Cottage it changes direction again, this time heading almost directly south, in a perfectly straight line to Mark Bridge: this section also appears to post-date some elements of the historic landscape, which may represent rudimentary enclosure of what was otherwise a relatively unimproved area of marsh.

[86] SRO DD/DN 439 f177.

[87] I would like to thank Gill Bedingfield for bringing these references to my attention.

[88] J. R. L. Allen, 'Geological Impacts on Coastal Wetland Landscapes: Sea Level Rise with Illustrations from the River Banwell', *Somerset Archaeology and Natural History*, 141 (1998), 17–34.

[89] SRO DD/SAS C/795: I would like to thank Martin Ecclestone for bringing this reference to my attention.

[90] See Cole, above p.81. I would like to thank Ann Cole for bringing this to my attention.

Conclusions

Though certain wetlands around the Severn estuary were reclaimed during the Roman period, and elsewhere areas such as Fenland were extensively settled without reclamation, these landscapes were largely lost during a widespread period of marine transgression in the late/post-Roman period. The survival of a Roman drainage system on the Wentlooge Level may indicate continuity in drainage technology in this one area, but otherwise the renewed interest in water management seen during the medieval period was built on new foundations. The use of rivers for navigation in coastal wetlands was widespread in the Roman and early medieval periods, and by the eleventh and twelfth centuries, these natural rivers, now usually located within reclaimed landscapes, were increasingly inadequate for the demands placed upon them. This resulted in the construction of a complex network of artificial watercourses that served a wide variety of functions. In a number of cases, notably in Fenland and the Somerset Levels, they had the specific purpose of improving communication by providing easier passage across low-lying backfens, and alternative routes where changes in coastal and/or riverine geomorphology were blocking earlier routes. In linking inland regions with the coast and the world beyond, these canals were of great significance in articulating the local, regional, and even national economy.

Although in many cases artificial watercourses were multi-functional, being used for communication, powering mills, fishing, and carrying upland streams, they did not necessarily drain the areas through which they passed. Watercourses such as Pilrow and the Sheppey were raised above the level of the adjacent reclaimed areas and cannot have taken waters from them: the needs of communication and reclamation were *not* united in single structures and systems. The construction of these watercourses also serves to illustrate the intensity of landscape exploitation during the medieval period, in that where natural rivers were canalized, disputes between the interests of fishing, milling, and transportation were commonplace. Other artificial watercourses, however, served more specialized functions, such as the Rhee Wall on Romney Marsh, which was constructed specifically to flush out a choked estuary, and Mill Reen and Monksditch on the Caldicot Level which were constructed simply to carry freshwater streams across the Levels and to power watermills. Features that at first sight might appear very similar could have had rather different functions. The construction of artificial watercourses was, therefore, a common feature in coastal landscapes, and whilst many did indeed function as canals, improving navigation was not the only reason why nature was tamed.

9

The Water Roads of Somerset

CHARLES AND NANCY HOLLINRAKE

> By 878 Alfred was reduced to hiding in the marshes at Athelney ... at
> this site they built a fort ... [1]

In the spring of 878, King Alfred moved 'under difficulties through woods and
into inaccessible places in marshes'[2] to the island of Athelney, in that part of
Somerset now known as the 'Somerset Levels'—an extensive tract of alluvial
moorland rarely rising higher than 7 m above sea level—and for a period of
about seven weeks organized the resistance to Guthrum and the Danish army
before defeating them at Edington[3] in Wiltshire. Soon afterwards Guthrum
was baptized at Aller, a small 'island' in the moors only 5 km east of Athelney
and attached to the large royal estate of Somerton.

It is often assumed that Athelney, a small 'island' or area of higher ground
measuring approximately 800 m × 200 m, was chosen as a refuge by Alfred
because of its remoteness, situated as it was in the middle of uncharted swamps
and marshes. Contrary to appearances, however, Athelney is not remote and
was far from inaccessible, lying as it does amidst a network of rivers which
would have serviced the large royal estates of Somerton, Curry, and Taunton,
which provided access to the Severn estuary and the sea, and which linked up
with the network of Roman roads running off the Foss Way in the south of
the shire (Fig. 49). Athelney was a focal point of this part of Somerset and
of great strategic importance. All the while that the king and his retinue were
based there, the rivers running by the island—the Parrett and the Cary to the
east, overlooked by the isolated hill of Burrow Mump, and the Tone running
between Athelney and Lyng to the west—must have been bristling with boats
and transports carrying military commanders, messengers, and diplomats, as
well as providing the food and logistical requirements necessary to service
such a large group of people. It is difficult to believe, for instance, that when
Alfred went to Athelney to organize his forces, there was not already some

---

[1] M. Costen, *The Origins of Somerset* (Manchester, 1992), 112–13.
[2] *ASC* s.a. 878 (p. 74).   [3] F. M. Stenton, *Anglo-Saxon England*, (Oxford, 1971) 257.

infrastructure existing on the island which was capable of housing and feeding his retinue.

Alfred also built fortifications at Athelney including a defended *burh* at Lyng (on the edge of the higher ground immediately west of Athelney Island), and possibly on Burrow Mump (Fig. 50), probably not specifically to defend the monastery which he founded (or refounded) there in thanksgiving for his victory, but to protect and control the rivers which flowed through and past his estates and which provided access into the heartland of Somerset.

After defeating Guthrum in Wiltshire, Alfred returned to his Somerset base with his captives (perhaps by water; possibly along the Avon and down the Severn estuary) and had Guthrum baptized at Aller 'island'. The location of Aller is a further indication of the importance, even primacy, of river traffic in this area of Somerset. It is separated from the higher ground of the High Ham plateau and the Somerton estate by a gap of around 400 m through which a river once flowed. This waterway can be traced on maps and on aerial photographs, curving around Aller island. It runs from the Parrett just to the north of Stathe, 2 km west of Aller, curves around Aller, and rejoins the Parrett approximately 2 km to the north-west of Langport and 1 km south-east of Aller. The old watercourse, now called variously Stathemill Rhyne, Oxleazedrove Rhyne, and Midlemoor Rhyne, might originally have been a braided channel of the Parrett rather than a separate river.

After his baptism, Guthrum was taken from Aller to Wedmore, within the Cheddar estate and another of Alfred's manors, to complete the religious ceremony and make the agreement that served to protect southern England by creating the Danelaw. Again, water transport was probably the easiest and quickest method of transporting the large English and Danish retinues. From Aller the party could have travelled, by water, down to the Parrett, and then down that river to the Severn estuary, sailing up the coast to the mouth of the Axe at Brean Down and then along the Axe to Cheddar and Wedmore. An overland route would have been slow, circuitous, and difficult, and the river/sea/river route is the most practical solution.

Rivers are the forgotten highways of early medieval England. The road system constructed during the Roman occupation of Britain, which has functioned to some extent through to the present day, has tended to draw attention away from riverine traffic, which must surely have been the major form of long-range communications in the prehistoric period and whose importance also continued during the Roman centuries. Ilchester, for instance, the largest Roman town in Somerset, successor to an Iron Age *oppidum* and strategically positioned on the Foss Way, is also situated on a major river which flows into the Parrett—the Ilchester Yeo—formerly known as the Ivel: a rare crossing point on this river probably provided the *raison d'être* for the Roman fort there in the later first century AD. Roman Ilchester was not only a town on a major road: it was also a port with large granaries, standing on a river that might

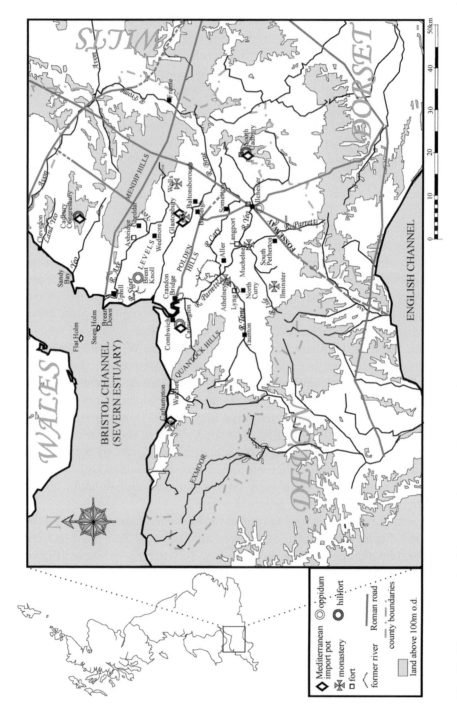

Fig. 49. The water roads of Somerset. Compare Figs. 17 and 43.

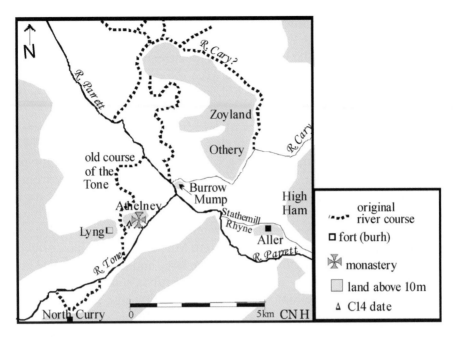

Fig. 50. Athelney and early river courses.

already have undergone some artificial straightening and canalization before the Roman conquest.[4]

In the seventh century, much of the land that would eventually become Somerset straddled the interface between the conflicting worlds of the western British and the West Saxons. In the last quarter of the seventh century the newly converted West Saxon kings apparently founded two monasteries in north-central Somerset: Glastonbury and Muchelney. Glastonbury is well known and is frequently described as 'remote' or 'isolated', whilst Muchelney stands upon a low island amid alluvial moors in an area that still suffers from severe winter inundations. Like Athelney, neither Glastonbury nor Muchelney are isolated places and comparisons between these two sites and Athelney are instructive. Athelney is located adjacent to two major rivers: the Tone which runs through the rich vale of Taunton Deane and originally flowed through the narrow gap between Athelney and Lyng to join the Parrett north of Athelney Island,[5] and the Parrett, flowing north from the hilly country of north Dorset, through central Somerset, and emptying into the Bristol Channel. The Parrett is tidal at least as far as Langport, a late Saxon *burh* and the river port for the royal estate at Somerton.

Glastonbury lies on the Brue, a river whose course was changed by the thirteenth century but which originally flowed north from Glastonbury to join

[4] P. Leach, *Ilchester*, ii (Sheffield, 1994), 6.
[5] M. Williams, *The Draining of the Somerset Levels* (Cambridge, 1970), 59–60.

the River Axe. The Axe flows into the Bristol Channel past Brean Down and a possible Roman port at Uphill, and was the river which serviced the royal estate at Wedmore/Cheddar and on which the defended *burh* at Axbridge was built. The Axe also provided access between the important minster, later the cathedral, at Wells and the sea (the bishops had a palace at Wookey, near to the source of the river).

Muchelney Abbey stands on a low island adjacent to the confluence between the rivers Isle and Parrett which occurs immediately to the south of the monastery, the former river connecting with the important church at Ilminster (minster on the River Isle), a property that was held by Muchelney from the eighth century. Immediately north of the monastery is the confluence of the Parrett and the Ilchester Yeo (Ilchester is approximately 10 km to the south-east).

Muchelney, therefore, is located at the confluence of three major river systems and connects with the sea via the River Parrett. Muchelney might have originally been founded as the royal minster serving the large and important royal estate of Somerton which stands north of the Parrett/Isle junction (Langport is only 1.5 km north of the monastery) and, in addition, the monastery stands immediately north of the large royal estate of South Petherton.

It would be illogical to suppose that this close association between the royal estates, royal minsters, and the river system originated with the West Saxon conquest of Somerset. As suggested previously, river transport must have been important during the Roman period with a river port at Ilchester and, importantly, river ports on the Parrett estuary at Combwich, servicing the country west of the river and at Crandon Bridge servicing the area east of the river (Fig. 49). There was also a possible Roman port at Uphill on the estuary of the Axe, and probably a Roman river port at Cheddar, on the Cheddar Yeo, a tributary of the Axe, and which might have been used for the transportation of Mendip lead (Fig. 51).

In the aceramic, post-Roman centuries in Somerset, the find-spots of imported Mediterranean wares, mainly east Mediterranean amphorae, show a decidedly water-based distribution (although with one notable exception that proves the rule). Within the county imported potsherds have been found at Carhampton on the north Somerset coast,[6] Cannington cemetery on the Parrett estuary (and near to the Roman port of Combwich); Glastonbury Tor and Glastonbury Mound, the latter a small natural clay mound adjacent to the River Brue used as a wharf; and Cadbury Congresbury,[7] a reused hillfort by the Congresbury Yeo, a navigable river which leads into the Bristol Channel at Sandy Bay. The exception is South Cadbury Castle, a refortified hillfort in the south of the county and a possible military base whose 'garrison' may have been stationed

---

    [6] C. and N. Hollinrake, *An Archaeological Evaluation at Eastbury Farm, Carhampton* (unpublished archive Report Number 56 to Somerset County Council, 1994). A paper on the Eastbury Farm Evaluation is awaiting publication in *Medieval Archaeology*.

    [7] P. A. Rahtz, et al., *Cadbury-Congresbury 1968–1973* (Oxford, 1992).

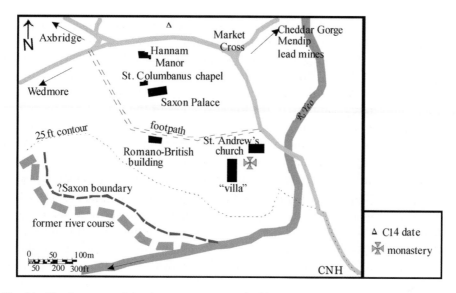

Fig. 51. The Roman and Anglo-Saxon port at Cheddar (Somerset).

there to prevent Saxon incursions into Somerset along the Foss Way corridor (Fig. 49).

Two other sites have also recently provided evidence for occupation during the Dark Age/early Anglo-Saxon period through carbon 14 assessments; Cheddar, which yielded a date centring on about 500 AD from an aceramic ditch,[8] and Athelney Island, which has recently provided sixth, and seventh, century dates from two separate features.[9] In addition, a recent Time Team investigation at Athelney recovered two fragments of imported pottery; a sherd of Phoecian red slipped ware (PRS) and a probable sherd of imported B ware amphorae.[10] Both Cheddar and Athelney should now be assumed to have a 'Dark Age' phase, possibly monastic but also, as with all of these early religious sites, probably also with strategic and economic aspects.[11]

Many other important West Saxon settlements in Somerset were also situated on or named directly from rivers. Taunton, 'settlement on the [River] Tone', was the site of an early fort and minster, and Frome, with an important minster founded by Aldhelm in the early eighth century, is named after the river on which it stands. Other early and important settlements where a river name is combined with 'ton' have been listed by Michael Costen and include, *inter alia*, Williton, Chewton, Petherton, and Bruton, the latter also connected with an

---

[8] C. and N. Hollinrake, 'An Archaeological Evaluation at the Old Showground, Cheddar' (unpublished Report Number 116 to Somerset County Council, 1997).

[9] Somerset County Council Architectural and Historic Heritage Group 1998 and 1999 Annual Reports, p. 25.

[10] Identified by the authors and confirmed by Dr Ewan Campbell.

[11] See also J. Blair, 'Palaces or Minsters? Northampton and Cheddar Reconsidered', *Anglo-Saxon England*, 25, (1996), 97–121 and 114–16, for an alternative view of the dating.

early church founded by Aldhelm.[12] All of these estates were held by the king;
a further indication of the strategic and economic importance of rivers.

Almost all of the river courses in this part of Somerset have been changed
to a greater or lesser degree. Some rivers have been diverted into quite
different artificial channels, the Tone and the Brue for instance, and some have
almost completely disappeared, the lower reaches of the River Cary being one
example, a river which once flowed past Roman sites at Ansford, Charlton
Adam, Littleton, and Weston Zoyland before emptying into the Parrett estuary.
Others, including the rivers described and named in a late seventh-century
boundary clause[13] in Glastonbury's charter for Brent Knoll, have disappeared
so completely that only air photography and fieldwork are capable of plotting
their original channels.[14] Because waterways have been so neglected, much
remains to be recorded, including the wharves and landing stages along their
banks. In the area around Athelney, for instance, *lode* or *load* place names,
signifying 'canal', 'artificial water channel', or 'dyked watercourse' (see Cole,
above pp. 77–8),[15] can be found at Curload (the *lode* for the Curi estate),
Othery, and Weston Zoyland, whilst 'hythe' place names, signifying a wharf
or landing stage, can be used to pinpoint the exact location of old channels
that have moved elsewhere or have been drained out of existence.

With the advent of climate change and with the price of fossil fuels inevitably
rising, water channels and water transport are likely to become ever more
important and the location and study of old river courses and defunct channels
might be of more than just academic interest. If future plans include some
retreat from the drained and managed alluvial landscapes that have been the
norm for the last 200 years in Britain, then the old water courses might once
again have some practical role to play, and their study should perhaps be
considered to be a research priority.

---

[12] Costen, *Origins*, 87–8.

[13] See L. Abrams, *Anglo-Saxon Glastonbury: Church and Endowment*, (Woodbridge, 1996), 70.
The estate is defined only by four rivers; the bounds are in Latin and 'of the most basic sort' and 'The
simplicity of the description has been taken as a confirmation of its age'.

[14] R. Leech, 'Romano-British Settlement in South Somerset and North Dorset', (Ph.D. thesis,
University of Bristol, 1977), 255.

[15] M. Gelling and A. Cole, *The Landscape of Place-Names* (Stamford, 2000) 20–1.

10

Glastonbury's Anglo-Saxon Canal and Dunstan's Dyke

CHARLES AND NANCY HOLLINRAKE

The Anglo-Saxon Canal

In 1986, limited rescue excavations in advance of a Safeway supermarket development on the Old Fairfield at Glastonbury investigated an artificial watercourse which ran from a point adjacent to the River Brue to the putative Anglo-Saxon market place by Glastonbury Abbey, a distance of approximately 1.75 km.[1] The feature was interpreted as a 'closed-end canal', an interpretation that has met with scepticism from some quarters but full acceptance by others. Before discussing the wider background for this feature it is best to describe its physical dimensions and route.

The canal commences at the southern tip of Glastonbury 'island' at Northover (see Fig. 52), in the vicinity of a modern tanning factory which has masked its southern end. This area stands approximately 2 m above the alluvial floodplain of the River Brue and is composed of Lower Lias clays and Upper and Middle Lias siltstones. Northover, a place name meaning 'north of the river',[2] stands on the narrowest part of the Brue floodplain, and a wooden causeway, possibly constructed during the eighth century[3] but still functioning until the twelfth century, once crossed the river valley at this point.[4] The River Brue now runs approximately 80 m south of Northover, but its course was altered in the thirteenth century and the original channel might once have flowed closer to the edge of the solid clay. From Northover, the canal runs

---

[1] C. Hollinrake and N. Hollinrake, 'A Late Saxon Monastic Enclosure Ditch and Canal, Glastonbury, Somerset', *Antiquity*, 65 (1991), 117–18; C. Hollinrake and N. Hollinrake, 'the Abbey Enclosure Ditch and a Late-Saxon Canal: rescue Excavations at Glastonbury, 1984–1988', *Proc. Somerset Archaeol. and Nat. Hist. Soc.* 136 (1992), 73–94.

[2] M. Gelling and A. Cole, *The Landscape of Place-Names* (Stamford, 2000), 200, 203.

[3] Pers. com. Richard Brunning, Somerset County Council, who obtained radiocarbon dates during a Ph.D.-linked investigation of the causeway.

[4] The wooden causeway was replaced by a stone causeway, situated *c.*40 m to the west, in about the 12th century; the latter lies below the modern road.

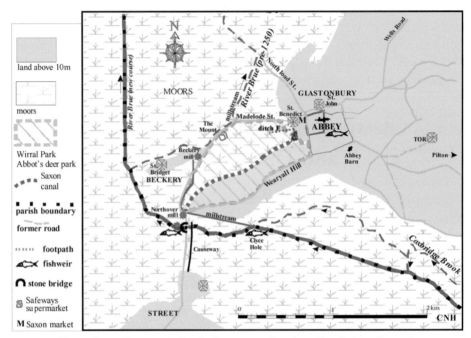

Fig. 52. Glastonbury (Somerset): Anglo-Saxon canal and medieval river channels.

along the northern edge of Wearyall Hill, on a level course approximating to the 10-metre contour line. This part of the canal was filled in during the earlier part of the twentieth century[5] but is still visible over a length of about 500 m to the north-east of the factory as a slight depression at the base of the hill. This stretch is followed by a deep, linear, hollow that still fills with water in winter months and which was used as withy-beds in the nineteenth century; beyond this is a further slight depression, most of which has been backfilled by modern rubble within the last fifteen years. Although the canal follows the 10-m contour, it also follows the edge of the hard clay of the hill; beyond the hill, to the north, are the soft alluvial silts and peat deposits of Wirral Park. Most of the northern part of the canal, from the main road onwards, has now been destroyed by modern development, although a tiny area still survives immediately north of the Safeway supermarket building.[6]

The canal had been noted previously: In the early nineteenth century the Revd John Skinner sketched its course across the Fairfield and suggested that it was a 'camp of the Belgic Britons', and it was still a notable feature until the early part of the twentieth century. In 1886, the local antiquary and archaeologist Mr John Morland asked, 'what was the meaning of the old *canal* under Wearyall?' (his italics),[7] and in 1893 the *Central Somerset Gazette* reported a meeting of the Glastonbury Antiquarian Society which discussed the

---

[5] Information from the late Mr Stephen Morland.   [6] Now Morrisons supermarket.
[7] *Proceedings of the Glastonbury Antiquarian Society*, 1 (1886), 46.

canal; the members suggested that it probably dated to the thirteenth century and was used for transporting produce from Glastonbury's nearby manors of Butleigh and Baltonsborough into the abbey, and Mr J. G. Bullied[8] suggested that there was a mill at the Glastonbury end.[9] Its bank was used as the base for the modern road which was constructed as a turnpike in 1821, and sections through the bank confirming its structure were recorded during development works in 1989.[10] This road replaced the medieval road around the perimeter of Wirral Park, formerly a deer park belonging to the abbot, founded in the twelfth or thirteenth century and encompassing Wearyall Hill (originally Wirral) and the low-lying land immediately north of the hill and the canal.

The canal followed the edge of the clay as far as the area of the modern supermarket building, after which it ran through soft alluvial silts. Construction would obviously have been easier, and the canal would have been more structurally sound, if its builders had continued to follow the heavy clay. That option would probably have brought the watercourse towards the south-west corner of the abbey precinct, but the northern terminus of the canal must have been considered an important enough location for the more difficult course to be taken through the alluvial deposits.

The supermarket development took place before the inception of PPG16 and the developers only allowed very limited investigation, all of which was funded by local bodies, principally the Glastonbury Antiquarian Society. The excavation details have been published elsewhere and will not be repeated here.[11]

The canal was sectioned in two places, in Trenches I and IV, approximately 170 m apart. It was slightly over 1 m deep and around 5 m wide, with a flat base and 45-degree sides. The base levels of the canal were identical in both sections.

In Trench IV, near to the northern terminus of the canal, the watercourse had been cut through soft, silty clays and the sides were revetted by a number of sharpened oak stakes. A carbon 14 determination from one of these provided a secure pre-Conquest date (HAR-9207: 690-1030AD at 2sigma and 830-990AD at 1sigma).

The remains of more wooden stakes and timbers rammed into the base at this point indicated the probable location of wharves or landing stages, and twelfth- to fourteenth-century pottery sherds in the upper silts of the feature (the lowest fills contained no pottery) also suggested that it was built in the pre-Conquest period.

The northern terminus of the canal must have been in the vicinity of St Benedict's church. This church, originally dedicated to St Benignus (St Benedict's is a medieval corruption), was formerly a chapel attached to the parish church of St John the Baptist, and abbey documents describe its founding

---

[8] Father of Arthur Bullied, discoverer and excavator of the Glastonbury Lake Village.

[9] Pers. comm. Mr Sandy Buchanan.

[10] C. and N. Hollinrake, 'An Archaeological Watching Brief at Wirral Park' (unpublished report to Somerset County Council, 1989).

[11] Hollinrake and Hollinrake, 'The Abbey Enclosure Ditch'.

and consecration in 1100.[12] The church shares exactly the same alignment as the late Saxon churches within the monastic precinct, and is situated at the western tip of a triangular area which has been postulated as the late Anglo-Saxon market place.[13] William of Malmesbury's 'Life' of Benignus describes the translation of St Benignus from Meare - a small Glastonbury manor only about 5 km away - to Glastonbury in 1091, an event which appears to suggest that the canal was functioning at that period. After stating that the body was carried from Meare to Glastonbury by water, part of the text reads:

When the relics came to the shore and had been carried to an appropriate place on land, perhaps half-way between river and monastery (*media fere via inter flumen et monasterium*), a sermon was preached from a raised place (*ex eminentioris stationis*) to the people standing around. ... At the place where the holy body then rested was built forthwith a church in the saint's name ... .[14]

The importance of that text is that St Benedict's church is not at all half-way between the river and the monastery. The course of the pre-thirteenth-century river is known, and it flowed along the same line as the medieval mill-stream. The distance from the nearest point of that river to St Benedict's is about 750 m, whilst the distance between St Benedict's and the Lady Chapel in the monastery is 200 m and between St Benedict's and the position of the abbey enclosure ditch about 100 m, the same distance as between St Benedict's and the location of the canal wharves noted above.

We believe that the abbacy of Dunstan and of his immediate successors, broadly the mid-tenth century, is the most likely period for the construction of the canal, possible reasons for its construction being the rebuilding of the abbey by Dunstan, which would have required much building stone, and the political importance of Glastonbury at this period (for a time it housed the national treasure and was chosen as the burial place for three English kings). The watercourse might, therefore, have had a dual function, as a practical working canal and as a processional way into the town and abbey from the river, this last possibly for royal funerals and the translation of relics. The radiocarbon assessment given above does indicate, however, that the canal could be earlier, ninth century for instance.

The Background

Glastonbury is sometimes known as the Isle of Avalon and the abbey claimed that its British name had been *Inniswytrin*, both names indicating an island. Glastonbury is not an island: it has dry-land access across a narrow neck of land to the east, but large expanses of alluvial moors extend to the west and

---

[12] James P. Carley (ed.), *The Chronicle of Glastonbury Abbey* (Woodbridge, 1985), 161–2.

[13] M. A. Aston, 'The Towns of Somerset', in J. Haslam, *Anglo-Saxon Towns in Southern England* (Chichester, 1984), 178.

[14] *William of Malmesbury: Saints' Lives*, ed. M. Winterbottom and R. M. Thomson (Oxford, 2002), 360–5.

north, and the floodplain of the River Brue forms the southern boundary. These low-lying lands still regularly flood in the winter and spring months, and flooding would have been even more extensive before the moors began to be drained in the medieval period. As many of Glastonbury's manors were dispersed throughout these moors, Glastonbury must always have relied heavily on water transport, and an account survives detailing the services due from a mid-thirteenth-century tenant that includes his main duties as an abbey waterman and boatman.[15] Within Glastonbury itself there are a number of place names which suggest this familiarity with waterways: *Northover*, at the southern terminus of the canal, appears to mean 'north of the river bank', whilst two medieval street names recorded in abbey deeds, *Northload* and *Madelode* (meaning Middle Lode), suggest either artificial waterways leading from the river towards the settlement area[16] or landing stages. Both of those streets are earlier than, and aligned differently to, the medieval street pattern which was set out after the Norman Conquest. The position and name of 'Southlode' (implied by the name of *Madelode*) has not survived (although a case could possibly be made for the 'canal' itself to be the 'Southlode' as the spacing between these three places is fairly equidistant). *Madelode* was the medieval name of the road now known as St Benedict's Street, the street containing the church mentioned above, and Madelode Bridge spanned the medieval mill-stream (part of which utilized the original, redundant course of the River Brue). During the 1986 Fairfield excavations a smaller (and possibly earlier) ditch than the canal, Ditch F, was recorded in Trench IV. This feature was aligned at about 90 degrees to the canal and contained twelfth-century potsherds in its upper fill plus some large, unworked blocks of yellow limestone.[17] The line of this ditch was traced for some distance to the west and this line, if projected, would have reached the river in the vicinity of Glastonbury Mound, or the Mount, a small natural clay island adjacent to the River Brue which once functioned as a wharf. Rescue excavations there in 1971 prior to its destruction recovered Romano-British material, sixth-century imported Mediterranean amphora sherds (B ware), and late Saxon pottery, together with large quantities of iron slag from the tenth to twelfth centuries when it was used for smithing.[18] It is possible, therefore, that Ditch F might also be the 'Madelode'.

Baltonsborough and Dunstan's Dyke

There is, in Baltonsborough, a watercourse known as 'Dunstan's Dyke'. Tradition asserts that Dunstan was born in this small village which is situated approximately 6 km south-east of Glastonbury and which was a Glastonbury

---

[15] S. Morland, 'Glaston Twelve Hides', *PSANHS* 128 (1984), 38; see also Rippon, above pp. 223–5.
[16] Gelling and Cole, *The Landscape of Place-Names*, 20, 200.
[17] Hollinrake and Hollinrake, 'The Abbey Enclosure Ditch', figs. 5 and 6C.
[18] J. Carr, 'Excavations on The Mound, Glastonbury, Somerset, 1971', in *PSANHS* 129 (1985), 37–62.

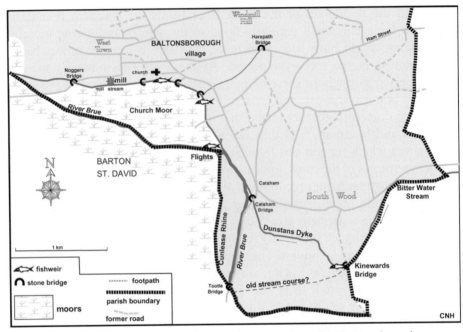

Fig. 53. Baltonsborough (Somerset): Dunstan's Dyke and medieval river channels.

demesne manor from the mid-eighth century until the Dissolution. The Ordnance Survey shows Dunstan's Dyke as a 1.25-km length of the River Brue
running north from Tootle Bridge to a large, artificial, stone-lined pool known
as 'The Flights' (see Fig. 53). This identification is probably incorrect. The
evidence relating to Dunstan's Dyke was reviewed around eighty years ago
by local antiquary and archaeologist John Morland who used two medieval
descriptions of the bounds of Glastonbury's Twelve Hides[19]—the earliest dating from the thirteenth century and interpolated into the writings of William
of Malmesbury[20] and the other and more detailed account from a perambu-
lation by Abbot Bere in 1503—to locate the boundary points listed in those
documents.[21] All discussions of the bounds agree that Dunstan's Dyke was an
artificial watercourse which ran to the north-west from 'Kinewards Bridge' and
provided the power for Baltonsborough Mill (Fig. 53).

This mill, recorded in Domesday Book,[22] is situated west of the parish
church on an ancient road; this is called the *Harepath* east of the church,
Ham Street further to the east, and then continues westward of the mill
across Wallyers Bridge (?Welshman's Bridge) also mentioned in the boundary

---

[19] Glaston XII Hides - an area around Glastonbury over which the abbey had sole control and which
expanded from *c.*600 hectares in 1086 to around 16,000 hectares by the later medieval period. See
Carley, *Chronicle* 274 n. 17, and Morland, 'Glaston Twelve Hides'.

[20] *The Early History of Glastonbury* ed. J. Scott (Woodbindge 1981), 48–150.

[21] J. Moorland, 'The Brue at Glastonbury', *PSANHS*, 68 (1922), 78–85, and Morland, 'Glaston
Twelve Hides'.

[22] DB i. 90[v].

description, along a causeway spanning alluvial ground and on to the villages of Butleigh and Street. The surrounding land is too flat and the local streams too small to provide sufficient water to power a mill-wheel, and the River Brue probably meandered too slowly through its floodplain and at too low a level to be considered. Instead, the mill-leat is fed by a stream, formerly known as the 'Bitter Water', that rises on higher ground to the south-east and provided a strong and constant flow.

The mill-leat, from Kinewards Bridge to Baltonsborough Mill and beyond where it joins the Brue, has a length of approximately 4 km. Like the Glastonbury canal it follows the edge of the hard clay adjacent to the interface between the clay and the alluvial moors, is *c*.5 m wide, is embanked, has many straight cuts, and contains a number of small weirs as it gently runs down the slope.

That part of the River Brue shown incorrectly on Ordnance Survey maps as 'Dunstan's Dyke' starts at Tootle Bridge (probably a thirteenth-century structure) and joins the mill-leat just north of Catsham Bridge. This stretch of the river is a wide, deep, artificial channel, that was probably constructed by the thirteenth century to supply extra water to the mill-leat.[23]

That new river channel diverted the Brue from its original course (now just a field ditch known as the Cunlease Rhyne which formed, and still forms, the boundary between the parishes of Baltonsborough and Barton St David) into the mill-leat. An overflow sluice was constructed on the mill-leat north of the junction with the new river, so that when the water flow became too great for the narrow leat it could empty into a purpose-made pool—the Flights—which connected back into the older course of the river via a short channel. This was a major engineering project, probably undertaken during (?or before) the thirteenth century when Glastonbury was engaged in extensive modifications and changes to its many river systems, including the complete diversion of the Brue from its old course north of Glastonbury. The confidence and expertise shown on the Baltonsborough project, and the other major river diversions undertaken by Glastonbury during the medieval period, is surely a reflection of a long acquaintance with and understanding of water management techniques.

The Cunlease Rhine is one of many abandoned river channels within the Glastonbury moors, including one that formed the north boundary of the adjacent manor of Butleigh (before that parish acquired extra land, reclaimed from moors drained in the medieval period). Now no more than a small stream called Old Rhyne, this was formerly known as the *olde yo*[24]—the old river—and flowed from higher ground at Butleigh joining the Brue just north-east of Street. The late Saxon boundary clause for Butleigh mentions four hythes along this 3-km stretch of water,[25] testimony to the importance of local river

[23] Morland, 'Brue at Glastonbury'.

[24] Dom A. Watkin, *The Great Chartulary of Glastonbury*, ii, Somerset Record Soc. 63 (Frome, 1952), 426.

[25] S270a boundary point numbers: 16 - Hoctan*yth*; 17 - Self*ith*; 18 - Wele*syth* (Welsh hythe) and 19 - the *yith* on Bregedswer (possibly named after Bregored, recorded as a supposedly pre-AD 670

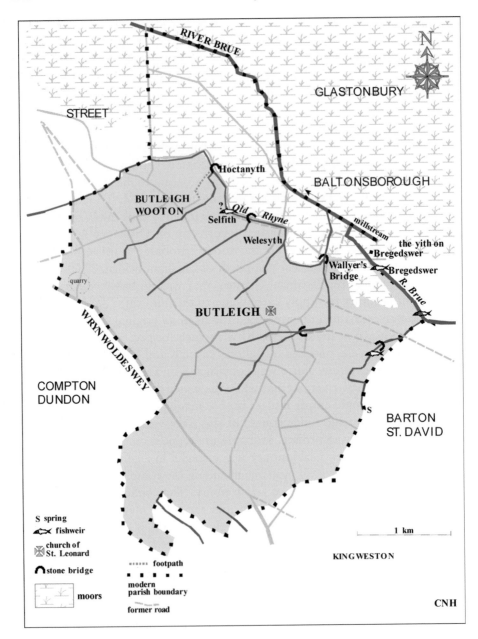

Fig. 54. Butleigh (Somerset): tenth-century charter-bounds.

traffic in the late Anglo-Saxon period (See Fig. 54). Sometimes, in winter months when the ground is sodden, the *olde yo* reappears; a great, wide expanse of waterlogged ground marking its old course from Butleigh to Street.

(pre-Anglo-Saxon) abbot of Glastonbury). The boundary clause is clearly a late Anglo-Saxon addition to a probably genuine charter of 802.

There is ample evidence from Glastonbury's estates scattered throughout the Somerset moors that the monastery was skilled in water management, and although the documentary evidence available for the changes brought about by the abbey relates to the later medieval period,[26] that skill cannot have risen out of a vacuum. Common sense dictates that a monastery which had depended on water transport from its original foundation would, over the centuries, have also learnt to manage and, when necessary, change or divert those rivers.

[26] See M. Williams, *The Draining of the Somerset Levels* (Cambridge, 1970).

11

Early Water Management on the Lower River Itchen in Hampshire

CHRISTOPHER K. CURRIE[†]

In an earlier article, published in 1997, the author discussed the possibility of Anglo-Saxon navigation on the lower reaches of the River Itchen between Southampton and Winchester.[1] The so-called 'medieval canal' up the Itchen to Winchester, which some authorities have extended as far as Alresford, has long been a matter of academic debate. This watercourse, supposedly built by Bishop Godfrey de Lucy (1189–1204), has had much support in the past by such distinguished authorities as the *Victoria County History*[2] and Maurice Beresford;[3] some more recent commentators, such as Keene,[4] have been more sceptical. It was thought that the matter had finally been brought to rest by Roberts,[5] who showed convincingly that the tradition of the canal to Alresford was based on an uncritical acceptance of antiquarian writings. It therefore appeared that the whole story of early medieval navigation on the river was a fabrication, until fieldwork by this author discovered the remains of old channels in the vicinity of Gater's Mill on the lower reaches of the river. These seemed to be associated with a charter-boundary of 1045 for South Stoneham

The author is grateful to George Watts, Kevin White, Alan Morton, and Prof. Barbara Yorke for their suggestions on drafts of the text in various early forms. Kevin White, Alan Morton, Bob Thompson, Edward Roberts, and George Watts are further thanked for visiting the site with the author, and for making their own keen observations. Jack Sturgess kindly provided the photograph used in Fig. 56. Ann Cole is thanked for her comment on *stæð* place names, and John Blair for his support and encouragement. Finally, thanks go to Harold Barstow, who offered the author much useful primary information about the history of the area. The conclusions are the author's own.

[1] C. K. Currie, 'A Possible Ancient Water Channel around Woodmill and Gater's Mill in the Historic Manor of South Stoneham', *Proc. Hants. Field Club and Archaeol. Soc.* 52 (1997), 89–106. [*Editor's note*: Very sadly, Christopher Currie died shortly before this book went to press, and inevitably the process of final revision falls to me. I have naturally confined myself to stylistic changes, except for one necessary reference to a further publication, which is added below as n. 9.]

[2] *VCH Hants* (1900–14).

[3] M. Beresford, 'Six New Towns of the Bishop of Winchester', *MA* 3 (1959), 187–214.

[4] D. Keene (ed.), *A Survey of Medieval Winchester*, i (Oxford, 1985).

[5] E. Roberts, 'Alresford Pond, a Medieval Canal Reservoir: A Tradition Assessed', *Proc. Hants Field Club and Archaeol. Soc.* 41 (1985), 127–38.

referring to a 'new' and 'old' river: the phrase 'new river' ('niwan ea') in this context suggests a man-made channel.

The Charter Evidence

A full analysis of the charter-boundaries relevant to this study is given in my article of 1995, to which readers are referred.[6] This present work follows this study, but confines discussion to the passages in the charters directly related to the presence of a 'new river'.

The earliest charter for South Stoneham (S 944) dates from 990 × 992, and records a grant of land to an unnamed party by King Æthelred II. The bounds of this estate appear to cover roughly the same area as the later charter of 1045 (S 1012). In the earlier one the bounds start on the Itchen, move along the king's boundary to the 'bitch's pole', and thence to 'Wadda's stake'. In the 1045 charter the same apparent land is granted by King Edward to the Old Minster at Winchester. The bounds here start at Swaythling, and probably move down the contemporary equivalent of the Mansbridge Road to the 'old Itchen' ('ealden Icenan'). From here, they move along the top of an orchard to the 'new river' ('niwan ea'), then along the boundary to the 'claypits' ('lampyttas'), and along the boundary again to 'Wadda's stake'. The 'boundary' referred to is probably the undefined 'king's boundary' of the first charter, and need not itself have changed significantly; however, three extra points have sprung up between the original 'Itchen' and 'Wadda's stake'.

Although the second charter could be elaborating on the first by giving extra points, it is also possible that the additions have been made because the landscape between the original points had changed. The local topography is such that a change in the boundary seems the less likely option. The favoured interpretation is therefore that the 'new river' came into being between 990 and 1045.

The same argument can be made for the appearance of the 'claypits' in the second charter, and the disappearance of the 'bitch's pole' mentioned in the first. Possibly the claypits were dug in relation to the making of the new river, to provide clay for banking, or some other functional task. In digging them, the 'bitch's pole' may have been removed. It is noteworthy that the copse to the immediate south (the direction in which the charter-bounds are moving) of the recently identified channel is known as Marlhill Copse, 'marl' being a term used for earth dug out of the ground as a fertilizer. This name was probably given to explain the existence of pits in the area, and they may have subsequently been used for agricultural purposes. The hill was known as 'Malhull' or Marlhill as early as 1333.[7]

---

[6] C. K. Currie, 'Saxon Charters and Landscape Evolution in the South-Central Hampshire Basin', *Proc. Hants. Field Club and Archaeol. Soc.* 50 (1995), 103–25.

[7] S. Himsworth (ed.), *Winchester College Muniments* (3 vols., Chichester, 1981–4), ii, no. 1592.

The 1045 charter gives a list of possibly relevant features after the bounds, including 'the millstead at Mansbridge' ('se mylnstede æt Mannæs Bricge'). The identification of this feature is not clear-cut. The initial impression is that this 'Mansbridge' mill may be Gater's Mill, although an earlier charter for North Stoneham (S 418), dated 932, mentions a mill at 'North Mansbridge'. Like the 1045 charter, this mill is mentioned after the bounds. In my 1995 article, it was argued that the mill now known as Gater's Mill may have once belonged to the estate of Stoneham that was subsequently divided into North and South Stoneham.[8] It is possible that the mills mentioned here are one and the same. It is equally possible that the 1045 charter refers to a newly erected mill at Woodmill. On this unresolved point, the reader is referred to my 1997 article for a fuller discussion of the history of these two mills.

The Present Remains[9] (Fig. 55)

As late as 1940, the 25-inch Ordnance Survey map[10] showed a substantial channel heading east-south-east from Woodmill to the southern corner of Riverside Park. This channel was parallel to the substantial levee bank that then followed the course of the river up to its upper tidal limits. These banks could have been a medieval or earlier creation to prevent the flooding of adjoining fields. The fields of the lower Itchen have probably always been highly prone to this, especially when high spring tides coincide with periods of heavy flow. This has remained a problem in the area until recent memory. That a levee bank should have reached some 250 m beyond Woodmill to the junction of Woodmill Lane with Manor Road indicates the substantial size of the channel it once followed.[11]

This channel may have been a remnant of the 'new river' mentioned in 1045. Its physical remains demonstrate that it was equal in width to the main river. It has been gradually backfilled over the period 1940–75, but is still marked today by a substantial hollow along the edge of Woodmill Lane, with a bank to the south, up to 1.2 m high, representing the former levee. Near its junction with Manor Road, Woodmill Lane seems to cut across the line of the channel and levee bank. This lane has all the appearance of having been an old routeway from South Stoneham to Bitterne, and may have medieval or earlier

---

[8] Currie, 'Saxon Charters', 110.

[9] [*Editor's note:* The topographical argument originally set out in ibid. was attacked by A. D. Russel, 'Some Comments on the Southampton Evidence in C. K. Currie's "Saxon Charters and Landscape Evolution in the South-Central Hampshire Basin"', *Hants Field Club and Archaeol. Soc. Newsletter*, NS 25 (spring 1996), 21–2. Russel argued that the 15-m-wide channel only assumed its present width and alignment when the adjoining land was landscaped and raised for construction of the golf course; the earlier ditch was much slighter, and its northwards continuation could be assumed to be a silted-up natural channel. Nonetheless, CKC restated his original position in Currie, 'A Possible Ancient Water Channel', and here. The dispute can perhaps only be resolved by excavation. However, the inference which CKC drew from the two charter-bounds does not in itself seem to be weakened by Russel's point.]

[10] OS 25-inch sheet LXV.3, 1940 edn.   [11] OS 25-inch sheet LXV.3, 1865, 1897, 1910 edns.

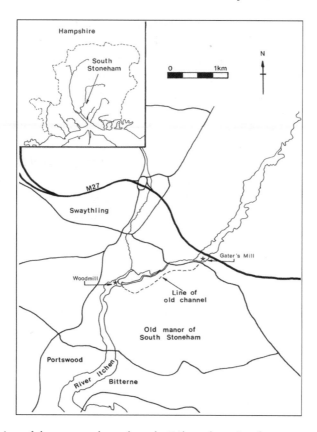

Fig. 55. Location of the proposed canal on the Itchen above Southampton.

origins. Its present course seems to date from the period after the 'new river' had fallen into disuse.

On the east side of Woodmill Lane (SU 4421 1505), the present line of the channel is continued by a broad hollow up to 15 m wide. At the bottom of this hollow is a small stream, representing the local catchment of water flowing off Town Hill. Ordnance Survey 25-inch OS maps appear to show that this stream was cut between 1910 and 1933, although the physical remains of what seems to be the edge of an earlier, wider, silted channel are visible to the south-east of this stream. It is possible that the scarp visible here represents a continuation of the levee bank clearly shown further west on early 25-inch maps. This situation continues until the conjectural alignment is crossed by a bridge leading into Riverside Park at SU 4456 1529. Stonework in the side of the bank adjacent to this structure has tool markings characteristic of *c.*1840 on it.[12]

The situation here is complicated by the fact that the original surveyor's 2-inch drawing for the first edition Ordnance Survey one-inch map (Fig. 56) seems to show the channel taking a more direct route to join up with the

[12] Bob Thompson, formerly of Southampton Museums, pers. comm.

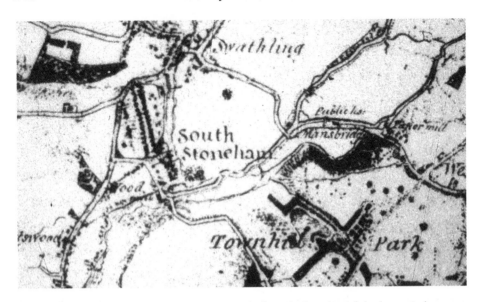

Fig. 56. The Ordnance Survey surveyor's two-inch scale drawing of the lower Itchen area, showing the water channel discussed in this chapter.

channel on the other side of Woodmill Lane.[13] If so, this channel would have been buried beneath recent dumping for the present golf course. However, its form had almost certainly been changed by the time of the South Stoneham tithe map (*c*.1845), which shows the channel paralleling the large bend in the main river (Fig. 57): it does not do this on the 1808 drawing. By the first edition of the 25-inch map in 1865–6, this situation had changed further, the original channel being replaced by a series of narrower ditches. Although it might be argued that the small scale could result in error, the 1808 map shows the decided widening of the channel at exactly the same spot as does the tithe map. As the 1808 drawing corresponds in accuracy with many other features shown on the tithe map, there is no reason to doubt it in this case. It would seem therefore that the original channel may have been lost between Woodmill Lane and this bridge.

Continuing north-east from the bridge, the channel becomes increasingly overgrown and stagnant as it follows the base of the slope of the steep-sided hill along the edge of Marlhill Copse. The channel here is up to 15–20 m wide, depending on the extent of silting and other natural factors. There are many fallen trees and alder and willow scrub in the channel as far as SU 4576 1545, about 750 m from the above-mentioned bridge, when the channel turns north towards Gater's Mill, leaving the edge of Marlhill Copse.

At approximately SU 4475 1542, on the south side of the channel, is an earthwork bank, up to 1.5 m high and about 10 m across its base. Cut into the hillside between the bank and the hill is a ditch, up to 1.5 m deep. This is

<hr />

[13] HRO, Original Surveyor's Drawing, sheet 12, 1808.

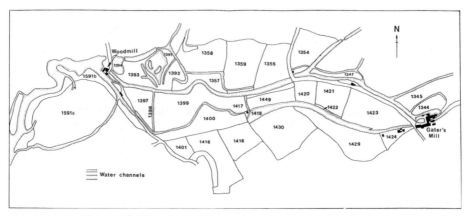

Fig. 57. A detail (traced) from the South Stoneham tithe map of 1845, showing part of the water channel marked as no. 1418, 'lake'.

about 100 m in length, although its extent has not been accurately measured. It enters the main channel by cutting across the bank at right angles, then follows parallel to the main channel, and terminates abruptly in a dead end. There is currently no drainage flowing into it from the hill, and no immediate explanation for its existence. It may be contemporary with the main channel, or a subsequent feature.

After the channel has left Riverside Park to continue eastwards through a scrubby piece of former meadow, the remnants of an old hedgeline follow the line of the channel on its north side. This stands on a very degraded bank, but there is no trace of a ditch. This hedge appears to stand some 5 m north of the conjectured line of the north bank of the channel, as if leaving a deliberate gap between itself and the channel. Whether this was a walk alongside an ornamental pool, a possible towpath, or a feature unknown has yet to be tested. The tithe map marks a short section of it as 'pathway',[14] but it is completely isolated at both ends, and seems to have been the remnants of a relict feature even then. The present footpath does not respect the hedgeline, and cuts across it on a number of occasions.

This stretch of the channel, where it is probably in its best condition, is 15 m or more wide in places. It is marked on the tithe map as a long thin 'Lake' of about 1.3 hectares, and owned by Edward Gater, the lessee of Gater's Mill. Although it is possible that this explains the origin of the watercourse as an ornamental feature or fishpond, there is reasonably good documentary and physical evidence to suggest that an earlier channel had existed on this alignment.

Just before the main channel reaches the main Mansbridge Road near Gater's Mill, it narrows suddenly, and is crossed by a trackway crossing a concrete pipe (SU 4524 1552). This feature seems to mark a short section of modern

---

[14] HRO, 21M65 F7/217/1-2, tithe map and award for South Stoneham.

infill between the track and the modern road just below Gater's Mill mill-pool. A reasonably substantial ditch was shown here when the Ordnance Survey last mapped this area in 1967,[15] suggesting this infilling has probably occurred since that date. In 1845 the channel extended right up to the old road.

Today there are two river channels at Gater's Mill. Currently that on the west or left takes the main River Itchen around Gater's Mill, whilst a second, eastern, channel passes through the mill. As will be shown below, this situation was radically different in the medieval period.

Anglo-Saxon River Engineering

This volume describes several examples of Anglo-Saxon river engineering. From the number of mills recorded in Domesday Book, the pre-Conquest English must have undertaken it often. At nearby Titchfield there is a long artificial leat, nearly 800 m long, feeding the existing village mill. If this mill can be assumed to be on the site of that mentioned in Domesday, then this artificial watercourse is one of a number of suspected major river alterations dating from before the Conquest. This suggestion is supported by a charter of 948 (S 535) for the estate of Segensworth (later in Titchfield) which refers to a meadow between 'the Meon and the mill ditch'.[16]

Mills do not necessarily need such leats on larger streams and rivers as a matter of course. They are usually constructed because the siting of a mill across the main stream would be a major obstacle to access up and down stream. On larger rivers, this would include access for boats undertaking local trade, but even on smaller rivers the blockage of the main stream was a frequent cause of litigation, as it prevented salmon and other migratory fish from getting to the upper reaches to spawn. Salmon have always been important in the economy of any river system, and rights to fish were jealously guarded. Therefore, the importance of making parallel leats to prevent mills blocking rivers cannot be overstressed, and it is assumed that the Anglo-Saxons would have constructed them. It is notable that many of the existing artifical leats associated with mills in England are close to the width of the original river course. To consider that historic societies would have thought a narrow side-ditch sufficient passage for migrating fish (as we often do today) fails to appreciate the importance placed on this resource. It is in the light of this that the references to 'old' and 'new' rivers in the vicinity of Woodmill and Gater's Mill in the 1045 charter are significant, suggesting as they do that river engineering associated with both mills, almost 1.5 km apart, had been undertaken by this date.

Many earlier writers have seriously underplayed the fish-passage argument, preferring to concentrate their efforts on the question of navigation. Navigation could nevertheless have taken place, if only as a secondary consequence. Biddle

[15] OS 25-inch 1967 edn., plan SU 4515.
[16] M. Hare, 'Investigations at the Anglo-Saxon Church of St. Peter, Titchfield, 1982–1989', *Proc. Hants. Field Club and Archaeol. Soc.* 47 (1991), 117–44, at 119.

and Keene[17] quote the 1045 charter as evidence for alterations to the Itchen in the early medieval period, and note that although this may have been carried out to facilitate navigation, there is no further mention of such possibilities until the episcopate of Godfrey de Lucy. Although Roberts[18] has disproved de Lucy's supposed canal to Alresford, it seems possible that boat traffic could have passed upriver by a channel bypassing Woodmill and Gater's Mill. Although the small scale of the 1808/1810 maps may have exaggerated its size, it was clearly still seen then as a continuous feature. The present physical remains of this channel show that, where it has survived unmodified, it was apparently large enough to allow small flat-bottomed boats to pass along it. This does not necessarily mean that boats did use this particular channel, but it shows that small boats *could* potentially have passed around Gater's Mill, either by this channel or by the equivalent of the present river channel.

In a recent synthesis of the reasons for the decline of *Hamwic*, Morton[19] has suggested that many of its functions had migrated upstream to Winchester by the early tenth century at the latest. If the construction of the artificial river was to facilitate the moving of supplies into that town, it might be expected that it would have been undertaken by that date. The evidence suggests, however, that the work was carried out between 992 and 1045, and there is no clear evidence that boats could have reached Winchester until the building of the Itchen Navigation in the post-medieval period.

That the 'new river' was designed, at least partly, to allow boat traffic to pass around obstructions in the river seems a possibility, but the destination of that traffic must remain conjectural. A mill already existed at 'North Mansbridge' in 932, probably on the site of Gater's Mill. By 1086 there were two mills and two fisheries in the two Stonehams,[20] all of which would probably have caused obstructions on the Itchen. A mill and a 'sea weir' are first mentioned at South Stoneham in 1045, but not in 992. Could the need for the artificial river have been the building of a substantial new mill, with an important fishery, at Woodmill between 992 and 1045? Or was it simply that the problem had existed for much longer, but had been tolerated until some unknown factor came into play, forcing the hand of the authorities to carry out what would have been a substantial undertaking?

In the earlier article,[21] this author threw doubt on the old arguments against de Lucy's supposed canal to Winchester. Although he is inclined to support Roberts's[22] arguments against navigation to Winchester in the thirteenth century, the evidence for earlier navigation is not clear-cut. My

[17] M. Biddle and D. Keene, 'Winchester in the Eleventh and Twelfth Centuries', in M. Biddle (ed.), *Winchester in the Early Middle Ages* (Oxford, 1976), 241–448, 270.

[18] Roberts, 'Alresford Pond'.

[19] A. D. Morton (ed.), *Excavations at Hamwic*, i CBA Research Rep. 94 (London, 1992), 75.

[20] DB i. 41ᵛ, 43ᵛ.

[21] Currie, 'A Possible Ancient Water Channel'.   [22] Roberts, 'Alresford Pond'.

1997 article puts forward a case to suggest that an inquisition into the canal in 1275 was fed with evidence that was misleading, to prevent a revival of the Bishop of Winchester's right to levy tolls on navigation on the river. Although the thirteenth-century revival of the conjectured navigation seems to have been unsuccessful, the need to prevent it can be given as an argument that navigation of some sort may once have existed. It is possible that this navigation was of a limited kind, and may not have involved laden boats of any size reaching Winchester. However, there is now evidence that navigation as far as the bishop's manor of Bishopstoke, from which the cargoes could have been moved on to Winchester by road, may once have been possible. Conceivably a distorted memory of this passage caused the bishops to try to revive their rights in the thirteenth century.[23]

Since the writing of the 1997 article, another piece of evidence has come to light. In researching the earlier article, the author came across a reference to a *stæð* on the Itchen in the bounds of a charter of 960 (S 683) for Bishopstoke. This *stæð* appears about 2 km upstream from Gater's Mill. Grundy[24] had suggested that it meant 'landing place', but the author was advised that such an interpretation was now outdated, and that the meaning 'shore' was more correct. More recently, Gelling and Cole[25] have offered reasons for reconsidering this advice. It is their opinion that the term *stathe* may have been used to imply a place where boats were unloaded. Above (pp. 74–5) Ann Cole considers the difficulties of interpreting this word, suggesting that in charter-bounds it often implied a simple river bank. However, this is never certain, and the above evidence that small boats may have been able to pass Woodmill and Gater's Mill in the eleventh century (if not earlier) is supported by the proximity of a Roman villa close to the *stathe* with access to the Roman road from Roman Bitterne to Winchester.

It is uncertain whether the channel discovered by fieldwork was made before the date of the Bishopstoke charter, or in an attempt to keep navigation open when it was later obstructed, but this new evidence does further support the idea that boats may have navigated at least the lower reaches of the river, and thus strengthen the proposal that there was indeed navigation as far as Bishopstoke in the late Saxon period.

Conclusion

Although archaeological excavation may be the only means of dating conclusively the waterchannel discussed here, it does seem to be a man-made feature. Whilst not ruling out the possibility that parts of the feature are later, there

---

[23] For details of the arguments relating to the complex evidence on this matter, see Currie, 'A Possible Ancient Water Channel'.

[24] G. B. Grundy, 'The Saxon Land Charters of Hampshire with Notes on Place and Field Names', *Arch J*. 2nd ser. 28 (1921), 55–173, at 114.

[25] M. Gelling and A. Cole, *The Landscape of Place-Names* (Stamford, 2000), 91–2.

is evidence to suggest that it may be associated with the 'new river' recorded in the 1045 South Stoneham charter. This charter does not explain why a 'new river' was needed, but it is most likely to be related to obstruction to fish migration, navigation, or a combination of both factors. This obstruction seems to have resulted from the creation of Woodmill and Gater's Mill. The importance of the salmon and eel fisheries on this part of the river is well documented, and the important industrial complexes that grew up at both mill sites would have had significant impact on them.

Although it is not possible to prove conclusively that the earthwork was connected with river navigation, the evidence presented suggests that there is a case to be answered. Regardless of how this question is finally resolved, the navigation discussed here was unlikely to have extended further that Bishopstoke. Edward Roberts's 1985 argument against the existence of the de Lucy canal to Winchester and Alresford remains convincing.

12

Transport and Canal-Building on the Upper Thames, 1000–1300

JOHN BLAIR

> So farwell Cricklad, come off that ground,
> We'el sail in Boats, towards London Town,
> For this now is the highest station,
> Up famous Tems for Navigation,
> But when th'tis joyn'd with Bath Avon,
> Then row your Wherrys father on,
>
> For by power of Lockes, Rains, and Fountains,
> They'l make Boats to dance, upon the mountains.
>
> Thomas Baskerville, 1692[1]

Baskerville's optimistic doggerel was written in an age of improvement. Two or three centuries earlier, the Thames had been easily navigable upriver to Henley, with difficulty to Oxford, and not significantly beyond.[2] So it is hardly surprising, given the great expansion in written sources after 1250, that historians of the medieval Thames highway have looked mainly at navigation from Oxford downwards. In this well-documented era the uppermost stretch of the river, from Cricklade to Oxford, was of slight importance, and it has seemed wasted effort to hunt for clues that things had once been different.[3] In 1973, R. H. C. Davis suggested that the building of mill-weirs with flash-locks

---

For comments on an earlier draft, I am very grateful to Ann Cole, Kanerva Heikkinen, John Langdon, John Maddicott, Robert Peberdy, and Christopher Whittick.

[1] BL, Harl MS. 4716, fo. 4ᵛ. For this vision of linking the Thames and Severn via the Bristol Avon, see T. S. Willan, *River Navigation in England* 1600–1750, 2nd impression (London, 1964), 9–10; J. Chandler (ed.), *Travels through Stuart Britain: The Adventures of John Taylor, the Water Poet* (Stroud, 1999), 192–3.

[2] R. Peberdy, 'Navigation on the River Thames between London and Oxford in the Late Middle Ages: A Reconsideration', *Oxoniensia*, 61(1996), 311–40 for the chronology and causes of the late medieval decline of navigation above Henley. Cf. E. Jones, 'River Navigation in Medieval England', *JHG* 26 (2000), 66.

[3] A case for navigation on the Radcot—Oxford stretch is made by J. F. Edwards, 'The Transport System of Medieval England and Wales: A Geographical Synthesis' (University of Salford Ph.D. thesis, 1987), 265–84, to which I am indebted for some references used below, and which (notwithstanding

(figs. 3, 70) from *c*.950 onwards had the accidental effect of facilitating passage between Oxford and Henley by breaking the river up into a series of deep, slow-flowing linear pools, until, after the 1190s, an excess of weirs proved obstructive rather than useful and the process went into reverse.[4] Davis's theory has not had the influence it deserves, and even he did not seriously consider the possibility of navigation above Oxford. But he did make the crucial point that earlier medieval conditions on the upper Thames could have been very different from late medieval ones, and could have been affected by specific natural and man-made forces which are hard to predict on general grounds. Might there have been a former time when somewhere towards Cricklade was 'the highest station | Up famous Tems for Navigation'?

The processes of silting, geomorphological change, and human intervention, described in other chapters, are extremely complex. Did accelerating alluviation, caused by runoff from more intensively farmed arable during the tenth to thirteenth centuries,[5] clog the river up, or help the major channel to flow more strongly by blocking minor ones? Did mill-dams and bridge-causeways obstruct river traffic, or facilitate it? Did the natural process of meander migration (Rhodes, above) make the river gradually less usable through time? But it may be that on the upper Thames, as in north-west England (Edmonds, above pp. 33–6), we should avoid geographical determinism and expect human effort to overcome physical obstacles so long as the incentives for doing so were strong enough. There is a good deal of circumstantial evidence that during the seventh to twelfth centuries—maybe far earlier—economic activity and cultural contacts focused on the line of the Thames from source to estuary (Fig. 58).[6] This evidence includes the loss patterns of eighth-century coinage (especially continental issues and the probably London-minted 'series L'), which show that the west midlands had commercial links with the south-east as well as with East Anglia and the south coast ports, and the network of salt-roads which ran south-eastwards from Droitwich in Worcestershire to the upper Thames (Fig. 5). So long as the river corridor was worth using, labour and resources would have been invested to keep it usable.

---

the criticisms of Jones, 'River Navigation') seems to be vindicated by the additional evidence offered here.

[4] R. H. C. Davis, 'The Ford, the River and the City', *Oxoniensia*, 38 (1973), 258–67 (reprinted in his *From Alfred the Great to Stephen* (London, 1991), 281–91).

[5] M. Robinson and G. Lambrick, 'Holocene Alluviation and Hydrology in the Upper Thames Basin', *Nature*, 308 (1984), 809–14; M. Robinson, 'Environment, Archaeology and Alluvium on the River Gravels of the South Midlands', in S. Needham and M. G. Macklin (eds.), *Alluvial Archaeology in Britain* (Oxford, 1992), 197–208.

[6] J. Blair, *Anglo-Saxon Oxfordshire* (Stroud, 1994), 8, 14, 80–7; J. Blair, 'The Minsters of the Thames', in J. Blair and B. Golding (eds.), *The Cloister and the World: Essays in Medieval History in Honour of Barbara Harvey* (Oxford 1996); M. Metcalf, 'Variations in the Composition of the Currency', in T. Pestell and K. Ulmschneider (eds.), *Markets in Early Medieval Europe: Trading and 'Productive' Sites, 650-850* (Macclesfield, 2003), 37–47, especially Fig.4.2; J.R. Maddicott, 'London and Droitwich, c.650–750', *Anglo-Saxon England*, 34 (2005), 7–58. For a wider perspective on the Thames

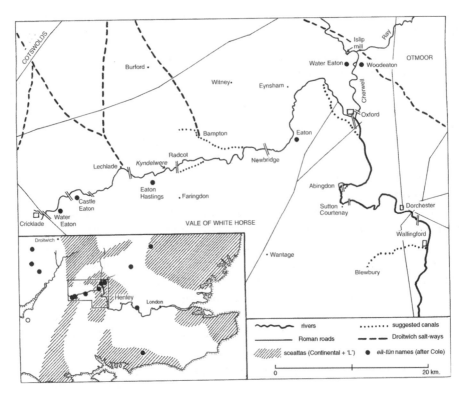

Fig. 58. The upper Thames region: general map locating the places discussed in this chapter.

Thus the extent to which the river remained open for navigation may have been less to do with hydrological change than with changing transportation needs. One potential factor is a gradual improvement in roads and bridges which, after the twelfth century, could have made negotiating difficult watercourses less worth the trouble. Another is the emergence of a more clearly defined hinterland supplying the growing city of London with grain. By 1300, as a recent study shows, regular commercial contacts with the city extended upriver to Henley and Abingdon, but no further.[7] However,

obligations to provide fixed quantities of grain to households in London or to carry it to the city for sale or consumption were imposed on some manors which were up to 40 miles distant over land, much further from London than the markets which normally supplied the city by road around 1300. ... Such obligations probably originated at a much earlier date and thus preserved a shadow of a system in which direct transfer of produce to the metropolitan ... centre was the best strategy for supplying the household or for selling the surplus. Subsequently, with the development of intermediate markets,

as a long-term 'gateway' see A. Sherratt, 'Why Wessex? The Avon Route and River Transport in Later British Prehistory', *Oxford Journal of Archaeology*, 15 (1996), 211–34.

[7] B. M. S. Campbell, J. A. Galloway, D. Keene, and M. Murphy, *A Medieval Capital and its Grain Supply: Agrarian Production and Distribution in the London Region c.1300* ([London], 1993), 46–54, 194–5.

a more flexible and efficient system of urban supply evolved, which in the metropolitan hinterland included several market towns specializing as grain entrepôts.[8]

Although this comment refers to road transport, the emergence of a more focused and intensive supply zone would clearly have had an adverse effect on any regular downriver transport from the south midlands to London. It is indeed precisely through the assertion of obsolescent vested interests—carriage services by boat from Dorchester to London, the right to carry merchandise along the river from Radcot to London—that we will recognize 'the shadow of a system' of long-distance transport. At the same time, we should be alive to the possibility that local river traffic networks—operating within the upper Thames region rather than extending out from it—were also more active in the earlier than the later middle ages, coexisting with the long-distance networks and perhaps stimulated by them.

In the absence of manorial accounts, trade and transport on the upper Thames before the mid-thirteenth century can be recognized only by piecing together small, scattered clues. The following survey of the few written sources that have come to light, organized in the form of an upriver progress, cannot be exhaustive. It nonetheless serves its purpose, which is to prove beyond doubt the existence of traffic over long stretches and sometimes substantial distances.

Navigation on the Upper Stretches of the Thames before 1300: Some Written Evidence

It has been the practice of recent historiography—essentially a reasonable one—to envisage the Thames as divided into three main lengths: from Cricklade to Oxford, from Oxford to Henley, and from Henley to the sea. The navigability of the lowest stretch is not in doubt, while that of the highest has not been seriously considered. It is on the middle stretch, swift flowing and attractive to mill-builders, that the number and effect of weirs and locks has been most debated by late medievalists. Whether the sources would allow that debate to be extended back before the fourteenth century is unclear, but they certainly give some indications that navigability was expected and, on occasion, facilitated. At Wallingford, tenants in the 1060s 'did the king's service with horses or by water as far as Blewbury, Reading, Sutton [Courtenay] and Benson'.[9] The last three places would have involved river journeys of 18 km upstream and 24 downstream; Blewbury, by contrast, is 6 km from the Thames, and

---

[8] B. M. S. Campbell, J. A. Galloway, D. Keene, and M. Murphy, 174 (and cf. 55). In D. M. Palliser (ed.), *The Cambridge Urban History of Britain: I: 1600–1540* (Cambridge, 2000), 191, Derek Keene expresses the view that 'for much of the tenth century ... London's main traffic was inland along the Thames, and its contacts with the Oxford region were strong'. Pamela Nightingale and Robert Peberdy point out to me that, since bulk goods for foreign export went down the Thames and continental luxuries came up it, the feeding of London would not have been the only dynamic involved.

[9] DB i. 56. Cf. J. Fletcher, *Sutton Courtenay: The History of a Thames-Side Village* (Sutton Courtenay, 1990), 14–20, who argues on topographical grounds for the late Anglo-Saxon construction of wharfage at this royal manor.

raises an interesting possibility explored further below (pp. 264–6). Nearby, at Dorchester-on-Thames, a custumal of *c.*1220 reveals a system of obligations to carry the Bishop of Lincoln's grain downriver.[10] A Chiselhampton tenant takes it to Oxford and Wallingford, 'and there it is put in a ship (*navis*) and carried to London if the lord bishop so wishes'; a Dorchester tenant has the duty of carrying the grain to the bishop's ship, while another goes with it to London to steer and assist. These custumals record other practices which were well established (going back in at least one case to the Conquest),[11] and the arrangements for transporting grain could well have been of long standing. Like others which we shall encounter, they refer specifically to *down*river traffic.

Just above Sutton Courtenay, Abingdon and its abbey (reformed *c.*954) provide explicit evidence that the Thames was canalized to facilitate downriver traffic from Oxford. The abbey's twelfth-century Chronicle relates an episode in the time of Abbot Orderic (1052–66):

The River Thames, on which shipping (*navigium*) goes to and fro, flows past the monastery of Abingdon on its south side. In Abbot Orderic's time the course of that river which stretched out (*porrectus*) beyond the church's plot of land called Barton by the people there, next to the hamlet called Thrupp, caused great problems for rowers. For the lower ground was raised a long way above the upper,[12] so that that channel often contained too little water. So the citizens of Oxford—for their shipping made the passage most often—petitioned for the course of the river to be diverted (*fluvii cursus … dirivetur*) through the church's meadow, which lies below it to the south, on the understanding that a hundred herrings should be paid from each of their ships (*navis*) to the monks' cellarer by custom forever after.[13]

The abiding importance of the new cut is underlined by the attempts of early twelfth-century Oxford boatmen—vigorously resisted by the abbey—to use it

---

[10] Oxford, Queen's College, MS 366, fos. 26, 27, 27^v. (This is an early 14th-century MS, but it transcribes surveys of about a century earlier.)

[11] J. Blair, 'Estate Memoranda of *c.*1070 from the See of Dorchester-on-Thames', *English Historical Review*, 116 (2001), 114–23, at 119–20.

[12] 'Nam tellus inferius longe quam superius altior subrecta.' Presumably the chronicler did not mean to say that the Thames flowed uphill. Some possible solutions are: (a) taking 'altus' to mean 'lower' (and perhaps reading *subiecta* for 'subrecta'), that the downstream ground fell away so steeply from the upstream that the water drained away; (b) reading *subsecta* for 'subrecta', that the downstream river bed was cut away much deeper than the upstream; (c), reading *subvecta* for 'subrecta' and taking 'tellus' to mean river-borne earth, that the phrase should be translated 'for the earth, carried much deeper downstream than upstream …'. In fact the terrain looks completely flat today. I am very grateful to James Binns and John Hudson for advice on this problem.

[13] J. Stevenson (ed.),*Chronicon Monasterii de Abingdon*, i Rolls Ser. 2.i (London, 1858), 480–1; J. Hudson (ed.), *Historia Ecclesie Abbendonensis*, i (Oxford, 2007), 218–19. Cf. the abbey's 'De abbatibus' (*Chron. Mon. de Ab.* ii, Rolls Ser. 2.ii (London, 1858), 282) where the citizens ask the monks 'to allow them to make a channel (*meatus*) through their meadow, which is on the south side of the church, more easy for ships (*naves*) than by the other channel (*alveus*)': this version may be slightly earlier than the other (J. Hudson, (ed.) *Historia Ecclesie Abbendonensis*, ii (Oxford, 2002), pp. xxii–xxiii). For this episode, and the earlier mill-leats constructed by Æthelwold, see G. Lambrick in M. Biddle et al., 'The Early History of Abingdon, Berkshire, and its Abbey', *Medieval Archaeology*, 12 (1968), 47; C. J. Bond, 'The Reconstruction of the Medieval Landscape: The Estates of Abingdon Abbey', *Landscape History*, 1(1979), 59–75; Bond above, pp. 179–80, 189–90.

without paying the toll.[14] The Chronicle implies that some of these cargoes were destined for London,[15] and indeed the herrings (salted or smoked) must have come via the Thames estuary, presumably in the emptied cargo boats returning upriver.[16] In 1205 King John gave a man freedom from toll and hindrance for one ship (*navis*) on the Thames between Oxford and London.[17]

Oxford marks the confluence of the Cherwell, the one upper Thames tributary where written sources point to at least some river traffic. In 1241 it was complained that the Prior of St Frideswide's, Oxford, had 'obstructed part of the River Cherwell by a weir which he enlarged so that boats could not cross it as they used to'.[18] Three royal commissions concern a tributary of the Cherwell, the Ray (Fig. 62). The first (1271) complains that wattle barriers and nets across the water of Otmoor have narrowed it, causing flooding and impeding vessels carrying 'divers necessaries or small or other fish'; the second (1294) orders the removal of weirs from the Thames and 'the river of the moor of Otmoor' between Otmoor and London, because vessels cannot pass as formerly; while the third (1375) orders a survey of 'the hythe called la Ree' of Otmoor, which had been blocked by trees and sluices.[19] Improbably small though the Ray looks for navigation, it could have had a special role in the trans-freighting of goods between the Thames and Wash river systems via the Buckinghamshire Ouse; a market for Droitwich salt[20] near the source of the Ray hints at the preparation of foodstuffs to be sent in either direction.

On the Thames above Oxford, we enter a different geological terrain (below, p. 271) and unexplored territory in the historiography of early river traffic. In twelfth-century legend, the royal abbess Frideswide and two of her nuns were miraculously transported upriver from Oxford to Bampton in no more than an hour.[21] This feat of an angelic boatman would have taken much longer by normal human means. However, regular *down*river traffic along the same stretch is indicated by the service, owed in 1317 by each pair of yardlanders at

---

[14] J. Hudson (ed.), *Historia Ecclesie Abbendonensis*, ii (Oxford, 2002), 138, 174–6, 312.

[15] In the version in Stevenson (ed.), *Chron. Mon. de Ab.* ii. 282, the citizens of London and Oxford jointly petition the abbey for the canal.

[16] Cf. R. Peberdy, 'Navigation on the River Thames between London and Oxford in the Late Middle Ages: A Reconsideration', *Oxoniensia*, 61(1996), 324–5 for a much later instance of this symbiotic river-traffic between Oxford and London.

[17] *Rot. Lit. Pat.* i.i. 52.

[18] J. Cooper (ed.), *The Oxfordshire Eyre, 1241*, Oxford Record Soc. 56 (Oxford, 1989), 131.

[19] *Cal. Pat. R. 1266–72*, 597–8; *Cal. Pat. R. 1292–1301*, 114; *Cal. Pat. R. 1374–7*, 157. The point is made by Edwards, 'The Transport System', 284. For the geography of the Ray across Otmoor see C. J. Bond, 'Otmoor', in [R. T. Rowley (ed.)], *The Evolution of Marshland Landscapes* (Oxford, 1981), 113–35.

[20] At or near Piddington, where in the 1360s each virgater 'will give the lord a penny for *saltesilver* yearly at Martinmas, or they will carry the lord's salt from the market where it will be bought (*de foro ubi emptus fuerit*) to the lord's larder': S. R. Wigram (ed.), *Cartulary of the Monastery of St. Frideswide*, ii, Oxford Hist. Soc. 21 (Oxford, 1896), 113. As Fig. 62 shows, Piddington was on one of the saltways from Droitwich.

[21] J. Blair, 'St. Frideswide Reconsidered', *Oxoniensia*, 52 (1987), 71–127, at 83–5.

Chimney, of taking one boat (*batellum*) laden with the lord's corn to Oxford.[22]
Chimney was a Thames-side outlier of Bampton Deanery manor (Fig. 64),[23]
and the service implies that grain from the considerable rectory demesne had
regularly been sent downriver for sale in Oxford, either for local use or because
the outermost ripples of the London market secured a better price.[24] Among
the rectorial tenants, the Chimney men presumably had the most to do with
the river and its traffic; the probable early medieval canal at Bampton, to be
described below (pp. 272–8), entered the Thames nearby.

Moving yet further upriver, to Radcot near the Gloucestershire boundary, we
encounter references to the regular loading of grain onto boats for downriver
transport. In 1205 King John granted to the monks of Beaulieu, whose grange
at Faringdon (on the Berkshire side, 3 km south of the Radcot crossing) had
just been founded, that 'the said monks' ships (*naves*) shall have free passage
along the Thames from Faringdon to the sea free of all levy and custom'.[25]
A century or so later, deeds of the lords of Radcot recite this grant—in the
slightly expanded form that the monks have 'free passage along the whole
Thames with their ships and merchandise from their manor of Faringdon as far
as the town of London' (Appendix, text F, pp. 290–1 below)—and promise
not to impede the monks' passage around *Kyndelwere* mill with their ships and
boats ('cum navibus et navicellis suis'), nor to prevent 'merchants of their corn
or other goods from loading or unloading their ships, halting in or adjoining
the course of the Thames, or mooring their ships against the monks' land as
they see fit' (text E, pp. 289–90 below). At Faringdon the Beaulieu monks
had taken over a royal manor which was already incipiently urban by 1086,[26]
and was a major producer of grain surpluses in the thirteenth century. The
message of these texts is that some of the grain was shipped downriver in
commercial vessels (*naves*), potentially as far as London, to an extent which
caused friction with the monks' neighbours at Radcot. Commissions to clear
the Thames up to Radcot of weirs and other obstructions imply at least some
interest in navigation to this point as late as the 1360s.[27]

---

[22] Exeter Cathedral, Dean and Chapter MS 2931: each unfree yardlander 'navigabit bladum domini
usque Oxon' cum uno homine et dimidio batello, ita quod duo virgatarii faciant duos homines et unum
batellum integrum'.

[23] *VCH Oxon.* xiii. 80–6.   [24] Cf. Campbell et al., *A Medieval Capital*, 67–8.

[25] Printed L. Landon (ed.), *The Cartae Antiquae Rolls 11–20*, Pipe Roll Soc. ns 17 (London, 1939),
109–11, and S. F. Hockey (ed.), *The Beaulieu Cartulary*, Southampton Record Ser. 17 (Southampton,
1974), 3–5: 'et quod naves eorundem monachorum liberum habeant transitum per Tamisiam a
Farendone usque ad mare absque omni exactione et consuetudine.' This is a riverine counterpart to
the letters patent by which, in 1209, John allowed free passage for the monks' carts to carry corn
and other goods from Faringdon to Beaulieu: *Rot. Lit. Pat.* i.i. 90. Beaulieu's Faringdon grange was
at Wyke, just north of the town, where cropmarks of a huge barn complex were observed in 1990:
Royal Commission on the Historical Monuments of England: Newsletter, 6 (spring 1992); M. Aston,
Monasteries (London, 1993), fig. 105.

[26] DB i. 57ᵛ (the entry includes nine *hagae* in the *villa*); R. Fleming, 'Rural Elites and Urban
Communities in Late-Saxon England', *Past and Present*, 141 (1993), 3–37, at 17.

[27] *Cal. Pat. R. 1350–4*, 204 (1351); *Cal. Pat. R. 1367–70*, 346–7 (1369); *Cal. Close R. 1369–74*,
11 (1369).

Above Radcot, such explicit sources fail. There are, however, two early and important Thames crossings—Cricklade and Lechlade—which deserve attention as possible transfer points from road freight to water freight on the uppermost reaches (cf. Hooke, above p. 38). Both names include the element *gelād*, 'river crossing liable to flooding'; at Cricklade the crossing was on the important Roman road from Cirencester to Winchester and London, while at Lechlade the 'piece of ground called Lade', at the head of the bridge, is mentioned in 1246.[28] Cricklade was a Burghal Hidage borough and a mint from *c*.980 onwards, and proprietors of major estates further downriver had houses there in the eleventh century;[29] in the early thirteenth century it was the home of a family of Jewish financiers with links to Oxford.[30] At Lechlade salt routes from Droitwich converged on the Thames, and the early importance of the river crossing is underlined by the boundary kink which kept it just inside Gloucestershire and thus, probably, the Hwiccian kingdom; it was later a notable freight point and market, its fair considered 'one [of] the most eminent in England' in 1692.[31] These facts do not prove that the uppermost river *was* used for transport in the tenth to thirteenth centuries, but they point to groups of people who may have had some strong incentives for using it.

One further point, derived not from documents but from place names, is worth making at this stage. As Ann Cole shows (above, pp. 78–82), the element *ēa-tūn* ('river settlement') can persuasively be associated with special functions or duties to keep watercourses clear of obstructions. The concentration of four such names on the uppermost Thames (Water Eaton and Castle Eaton below Cricklade, Eaton Hastings below Lechlade, Eaton below Newbridge), and two more on the lower Cherwell (Woodeaton and Water Eaton below Islip), therefore suggests a local concern for what she calls 'keeping the narrower reaches of rivers open for navigation, and for maintaining fords in a fit state to allow both road and river traffic to pass' (Figs. 18, 58).[32]

Turning now from the written to the physical and topographical evidence, we may begin with two kinds of man-made obstacle across the path of the

[28] Cole, above, p. 77; M. Gelling and A. Cole, *The Landscape of Place-Names* (Stamford, 2000), 81; *Cal. Charter R.* 1226–57, 296.

[29] D. Hill and A. R. Rumble (eds.), *The Defence of Wessex* (Manchester, 1996), 199–201; D. Hill, *An Atlas of Anglo-Saxon England* (Oxford, 1981), maps 222–4. For houses in Cricklade owned by Abingdon and Westminster Abbeys, see S. Kelly (ed.), *Charters of Abingdon Abbey*, ii.2, Anglo-Saxon Charters 8 (Oxford, 2001), 526–31, and DB i. 67.

[30] C. Roth, *The Jews of Medieval Oxford*, Oxford Hist. Soc. NS 9 (Oxford, 1951), 64–6: the family of Lumbard of Cricklade (d. 1277), who moved to Bristol, and then to Oxford in the 1240s. In 1204, apparently as part of a series of killings involving Oxford Jews, a Jew was killed by one Sampson of Berkshire, who took sanctuary in Cricklade church; Sampson probably came from Cricklade (the church there is dedicated to St Sampson), and he had a brother called Robert of Duxford, a place on the south bank of the Thames just downstream of Bampton (fig. 64) (*Curia Regis Rolls*, iii. 145).

[31] *VCH Glos.* vii. 106–16; Maddicott, 'London and Droitwich', 43–4; Thomas Baskerville's travels, BL, Harl MS. 4716, fo. 8.

[32] As Cole observes on p. 80 above, the concentration of Eaton names on the upper but not the lower Thames supports a particular association with river routes which—being narrow and liable to obstruction by silt, weeds, and debris—required regular maintenance.

river which are likely—whether positively or negatively—to have affected its configuration and flow.

Bridge-Causeways, Mill-Dams, and Locks

Most later medieval bridges across the broad floodplain of the upper Thames, as across that of the Trent, were approached from both directions by long causeways of revetted earth, punctuated at intervals by small stone flood-arches.[33] The actual bridges, mostly first mentioned in the thirteenth century, are now of various dates from c.1350 onwards. The causeways could be older (on the assumption that they originally served timber bridges later replaced by main-span arches), but the technology of vaulted flood-arches is unattested before the late eleventh century and became widespread during the twelfth and thirteenth; the causeways at Abingdon were built as late as 1416–22.[34] Possibly the first and certainly the most magnificent in the series is Grandpont at Oxford, datable to c.1070–1100, and there archaeology has shown that the masonry causeway replaced a late Anglo-Saxon trestle bridge and cobbled ford.[35] Probably this exemplifies an earlier category of upper Thames crossings which, whether timber bridges, brushwood causeways, or fords,[36] were altogether lighter structures than their successors.

From c.1100, therefore, the river was traversed by a multiplying series of earthwork and masonry barriers which (notwithstanding the flood-arches, inherently prone to blockage and now often buried) must have held back surplus water and formed static pools, encouraging silting in a period when the silt content of the water was in any case high. Again this is illustrated by Oxford, where the construction of Grandpont stimulated a rapid build-up of the ground level, and narrowing of channels, during the twelfth and thirteenth centuries.[37] At all the bridges the river will now have funnelled through relatively narrow openings, increasing in velocity at those points and scouring out deeper pockets on the bed.[38] This process could have been detrimental to river use—by narrowing the navigable corridor—or alternatively beneficial—by concentrating water in a series of deeper lengths in the main channel.

---

[33] D. Harrison, *The Bridges of Medieval England* (Oxford, 2004), 105–6, 113, 142–3, 182. For individual upper Thames bridges see G. Phillips, *Thames Crossings: Bridges, Tunnels and Ferries* (Newton Abbot, 1981), 16–76; Thomas Baskerville's valuable descriptions of their state in 1692 are in British Library, Harl MS. 4716.

[34] Harrison, *The Bridges*, 45, 110–13, 134–5. A. Cooper, *Bridges, Law and Power in Medieval England, 700–1400* (Woodbridge, 2006), 8–15, concludes that bridges were rare in England before 900, and that the eleventh and twelfth centuries were the high point of bridge-building activity. Note that, except for *stan bricge* at Shifford in 958 (S 654), no 'bridge' place name on the upper Thames (in contrast to its tributaries) is recorded in a pre-1100 source.

[35] A. Dodd (ed.), *Oxford before the University* (Oxford, 2003), 13–16, 32–5, 53–6, 65–82.

[36] Harrison, *The Bridges*, 102–5 for such structures elsewhere; J. Blair and A. Millard, 'An Anglo-Saxon Landmark Rediscovered', *Oxoniensia*, 57 (1992), 342–8, for a paved ford on a minor stream in west Oxfordshire.

[37] Dodd (ed.), *Oxford before the University* 54–6, 67–9, 81–7.

[38] Harrison, *The Bridges*, 80–1.

These points are very similar to the ones made by Davis about mill-dams, a category of structures which—together with fish-traps, with which they were often combined—increased markedly in both number and sophistication throughout England during the eleventh to thirteenth centuries.[39] Some light is thrown on the after-effects by lawsuits and dispute settlements, products of the ceaseless war between mill proprietors eager to raise their leats or ponds as high as possible, and upstream landowners irate at the flooding of their property (especially during the hay harvest, when it prevented both mowing and drying).[40] A typical example is the quarrel over the mill on the Shill Brook at Bampton between two generations of millers and the proprietors of the Deanery manor, which broke out around 1230 and resurfaced in 1268 (Appendix, texts **G** and **H**, pp. 291–2 below). The miller was allowed to raise the water and the sluices to the level of markers fixed in the pool, but no higher; he was to remove the sluices for six weeks in midsummer, and also—as a punitive measure—if he ever allowed the water to exceed the agreed level and thereby flooded the Deanery grounds.

Other disputes concerned dams that were on navigable rivers and apparently involved locks. At *Kyndelwere*, an important mill and fishery on the Thames above Radcot Bridge, it was specified in 1261 that the dam should contain a weir at least 10 feet wide and 7 feet deep (Fig. 70; below, p. 282, and Appendix, text **C**, p. 288). The previous year, the operator of Islip mill had provoked violence by narrowing the weir in his dam across the Cherwell from 25 feet to 4 feet (according to the men who demolished it) or 22 feet (according to the jurors who heard the case), flooding fields and meadows for miles around (Appendix, text **I**, pp. 292–4 below).

Civil pleas of this kind were directed against flooding rather than obstruction, but this is in the nature of legal processes designed to redress injuries to property. Long-distance users of the river would not have found easy remedy at law against obstructions in places where they had no tenurial stake, and their grievances will escape us unless occasionally reflected in hundred jurors' presentments of encroachments (as with the blockage to Cherwell navigation by a weir in 1241).[41] Just a hint that actions over flooding may sometimes have carried other concerns in their wake can be detected in the 1260 Islip case. The

---

[39] J. Langdon, *Mills in the Medieval Economy: England, 1300-1540* (Oxford, 2004), 74–6 for mills; M. Aston (ed.),*Medieval Fish, Fisheries and Fishponds in England*, Bar British Ser. 182 (2 vols., Oxford, 1988), D. J. Pannett, 'Fish Weirs of the River Severn with Particular Reference to Shropshire', in Aston (ed.), *Medieval Fish, Fisheries and Fishponds*, and Currie above, p.251, for fish-weirs.

[40] This kind of lawsuit was extremely common: see for instance *Curia Regis Rolls*, i–xiv (1922–61), covering 1196–1232, subject indexes under Actions (ix) ('de stagno levato', 'de fossato levato','de stagno molendini', 'de placito excluse molendini', 'de cursu aque obstructo', etc.), and several cases in *PWML*. For the opening of sluices during the hay harvest see Langdon, *Mills*, 287. In the texts printed below it is an explicit condition at *Kyndelwere* and Bampton (Appendix, texts **D, G, H**) and is implicit in the outbreak of violence at Islip, which occurred in late July (text **I**); at Rushey weir a boat was provided in the same season 'ad cariandum prostratores gurgitum' (H. E. Salter (ed.), *Cartulary of Oseney Abbey*, iv, Oxford Hist. Soc. 97 (Oxford, 1934), 519).

[41] Above, n 18. Note in this context that, in the instruction to remove weirs from the Ray in 1271, the injury to proprietors of flooded land is coupled with a general complaint about obstruction of

indictment lists nearly a hundred alleged attackers of the mill-dam, of whom about fifteen have identifiable toponyms (below, pp. 292–3). These mostly refer to places not—as one might expect—in a radius of Islip, but extending in a line north-eastwards along the Ray valley (Fig. 62), from Charlton-on-Otmoor via Merton, Arncott, Blackthorn (the reeve), Piddington (the bailiff and two others), to [Steeple?] Claydon between the Ouse and Ray watersheds (three men). This is a distance of 21 km: a dam at Islip cannot have caused flooding in the Claydons. Could these men have been prompted to join the attack by some perception among communities along the Ray that the dam was hindering their use of the river?

One important message of these dispute records is that the weirs (*gurgites*) in the dams—10 feet (*c*.3.05 m) at *Kyndelwere*, 25 feet (*c*.7.62 m) at Islip—were much wider than was needed simply for water control: they must be interpreted as flash-locks, such as R. H. C. Davis and, more recently, Robert Peberdy have associated with the improvement and eventual decline of river transport below Oxford.[42] The existence at this early date of navigable flash-locks *above* Oxford has not previously been noted: in the light of the evidence reviewed above (and, intriguingly, the proximity of both *Kyndelwere* and Islip to *eā–tūn* place names), it contributes to a strengthening case for regular and long-term management of the river which accommodated the interests of navigation. While the natural geomorphological effects of causeways and dams remain ambiguous, these signs of dynamic human intervention demand our attention.

Artificial Channels: Some Topographical Evidence

Starting again in the Henley to Oxford stretch and proceeding upriver, topographical and archaeological traces of improvement seem to concentrate in the upper reaches. Apart from Reading, where twelfth-century revetments and a rubble 'hard' near the Kennet–Thames confluence are presumably associated with the nearby abbey,[43] the first persuasive evidence comes from the burghal town of Wallingford. This is a place where we might expect some investment in communications, and possibly we can see it in the manipulation of a network of streamlets, flowing into the Thames from the west, described by the very minor watercourse terms *broc* and *lacu* in late Anglo-Saxon charterbounds (Fig. 59).[44] The channel, now called Bradford's Brook, by which this water catchment enters the Thames south of Wallingford is manifestly

---

shipping on the Thames and its tributaries: *Cal. Pat. R. 1266–72*, 597–8. See above, p. 9 n. 29, for the difficulty of claiming individual commercial interest on public waterways.

[42] Davis, 'The Ford'; Peberdy, 'Navigation', 311–13. For the problematic word *gurges* see Langdon, *Mills*, 81–2. At *Kyndelwere* and Islip it seems fair, given the width of the gaps, to translate it as 'lock'. Cf. the injunction in 1235 that *gurgites* on the lower Thames are to be 'in eadem latitudine et eodem statu et apertura inter virgas cleiarum' as they had been in Henry II's reign: *Cal. Close R. 1234–7*, 187.

[43] J. W. Hawkes and P. J. Fasham, *Excavations on Reading Waterfront Sites, 1979–1988* (Salisbury, 1997). The Holy Brook, which runs parallel to the Kennet at this point, has a very artificial appearance; could it be a canal associated with the abbey or the earlier minster?

[44] *Haccanbroc* in S 496 (?942, probably genuine); *Tibbælde lace* and *sand lace* in S 354 and S 517 (both forgeries, though preserving genuine Old English descriptions).

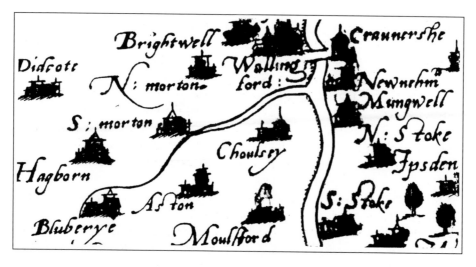

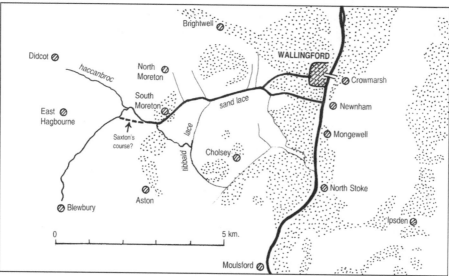

Fig. 59. Watercourses to the west of Wallingford. *Above*: Saxton's map of *c*.1578, showing what he perceived as a single main course from Blewbury via South Moreton to the Thames near Wallingford. *Below*: The area in the nineteenth century, showing natural watercourses (thin line) and new and canalized ones designed to capture water flow (thick line); some data after Grayson, 'Bradford's Brook'. Stream names from late Anglo-Saxon charter-bounds are in italic. Areas of gravel terrace are indicated by stipple.

not natural: it slices cleanly through the First Gravel Terrace. Here—not for the last time—Christopher Saxton's map of Oxfordshire, Berkshire, and Buckinghamshire (*c*.1578) proves remarkably useful: it shows that this config-uration already existed then, and also that Saxton perceived a single substantial stream which rose in Blewbury and flowed to Wallingford via South Moreton. This stream is evidently the product of one or more episodes of intervention, the effect of which was to gather the catchments of the late Anglo-Saxon rivulets

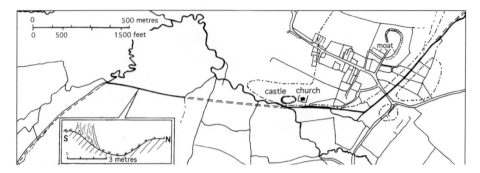

Fig. 60. South Moreton (Berks.): remains of the possible canal.

into a single substantial course.[45] Where it passes through South Moreton village it is of a regular and clearly artificial character (Fig. 60), at one point following the interface between the First and Second Gravel Terraces; this course can be projected westwards, partly along relict ditches, to suggest a lost channel which would conform better to Saxton's schematic plot than does the present meandering brook.[46]

In the absence of excavation, the interpretation of this as a relict canal must be tentative. But the possibility is there, and is strengthened by the appearance of Blewbury in the Domesday list of places to which Wallingford owed carrying services 'with horses or by water'. It is also worth raising because of three implications that can be remembered when examining other upper Thames tributaries: that artificial channels were likely, after long periods of neglect, to silt up and disappear with the reassertion of natural drainage systems; that they may nonethless survive better where they cut through the gravel terraces than where they cut through alluvial deposits; and that Saxton's maps may accurately portray a stage in this geomorphological evolution mid-way in time between the thirteenth century and the first Ordnance Survey maps.

At Abingdon we are spared such uncertainties thanks to the chronicle passage cited above (p. 258), one of only three explicit references to the building of a late Anglo-Saxon canal.[47] Notwithstanding some oddities of the language, and different interpretations of what it describes,[48] there seems to be only one convincing way of fitting it to the topography (Fig. 61). The difficult stretch of the Thames extended from Barton Court, immediately north-east of the abbey, to Thrupp a mile or so upstream. This must therefore be identified

---

[45] See A. J. Grayson, 'Bradford's Brook, Wallingford', *Oxoniensia*, 69 (2004), 29–44. Mr Grayson puts more emphasis on drainage, defence, and milling than I do, and less on transport, but I am most grateful to him for bringing this example to my attention.

[46] Where it runs along the south side of South Moreton castle and church, and then north-eastwards along the edge the village, the channel is straight, deeply cut, and *c.*6.5 m. wide. West of the church it is lost completely, but can be picked up further west as a shallow relict ditch of similar width, lined by very large old trees. The stretch further upstream, running north-eastwards from Blewbury, is now ploughed out but is marked as a ditch or hedge-bank on the earliest large-scale Ordnance Survey maps.

[47] See above, p. 5, for the others.   [48] Above, nn. 12–13.

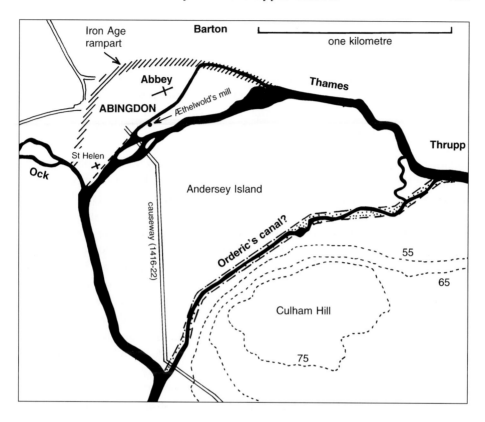

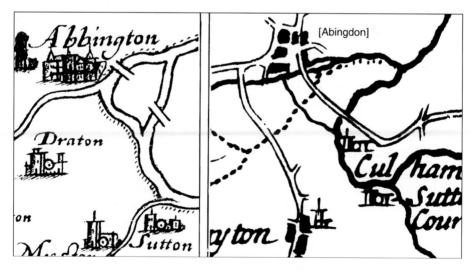

Fig. 61. Abingdon (Berks.): interpretation of the configuration of watercourses in the tenth to eleventh centuries, compared with (*bottom left*) Saxton's map of *c*.1578, and (*bottom right*) Morden's map of 1695.

with the modern main channel, sweeping in a large bend around the south edge of the abbey precinct. If Orderic's diversion went 'through the church's meadow … below it to the south', it is unlikely (in the absence of any relevant earthwork or cropmark[49]) to have bisected the flat gravel island of Andersey, and certainly did not traverse Culham Hill. The only possible candidate is the rivulet, now called Swift Ditch, which runs along a strip of alluvium between them (Fig. 35). It is again interesting that Saxton's map (Fig. 61, bottom left) shows the two channels as of equal width.[50] Both courses are in fact natural versions of the Thames at different stages in prehistory. What Abbot Orderic did in the 1050s or early 1060s was to enlarge the lesser but more direct course, diverting river traffic from the north to the south side of Andersey Island: his workmen laboured through silt, not gravel. Over the centuries since then, the main flow has reverted once again to its 'natural' Anglo-Saxon course beside the late Iron Age valley-fort, and the abbey and town which succeeded it.

Oxford, the next major place upriver, would have had compelling reasons to improve navigation in its commercial heyday, the eleventh to thirteenth centuries. Here the Thames divides into two branches (shown as such on Saxton's map): the present main course, more easterly and nearer the town, and the Seacourt/Hinksey Streams to its west. Ed Rhodes has pointed out (above, pp. 147–9 and Fig. 30) that the two main stretches of the latter, comprising the Wytham Stream and Hinksey Stream, are straight and perfectly in line; he suggests the possibility that they formed part of a single dead straight canal cutting across Osney Island. In the absence of any evidence for such a cut, this can only be conjecture. It is, however, worth noting that in about 1184 the modern navigation channel, on the east side of Osney Island, was called *Eldee*, i.e. *eald-ea*, 'old river'.[51] The implication is presumably that it was 'old' because it had been superseded by some more recent 'new river', which can only be the Seacourt/Hinksey line to the west: whether this cut across Osney Island or (as suggested on Figs. 58 and 62) skirted it, it must have been a good deal more important than later and was evidently perceived as artificial. As at Abingdon, and possibly on the Blewbury to South Moreton stream, we can again recognize a natural watercourse which was bypassed in the early or central middle ages by an artificial one, but which has reasserted itself since.

On the Cherwell-Ray river system, the *eald-ea* term is attested no less than four times (Fig. 62). The earliest, in charter-bounds of 1005 × 1012 ('of myclan dic into þere ealdan ea'), was somewhere on the edge of Otmoor but cannot be located.[52] The other three occur during the 1220s to 1260s in the context of

---

[49] The meandering relict channels shown from cropmarks in Bond, 'Reconstruction', Fig. 6 are clearly natural.

[50] Morden's map of 1695 (Fig. 61, bottom right) actually makes Swift Ditch the larger, but that was after the improvements carried out in the early 17th century by the Oxford-Burcot Commission.

[51] *Cartulary of Oseney*, iv. 60; *VCH Oxon.* iv. 23. For the changing watercourses around Oxford, see also Blair, *Anglo-Saxon Oxfordshire*, 104.

[52] S 943 (Beckley and Horton-cum-Studley).

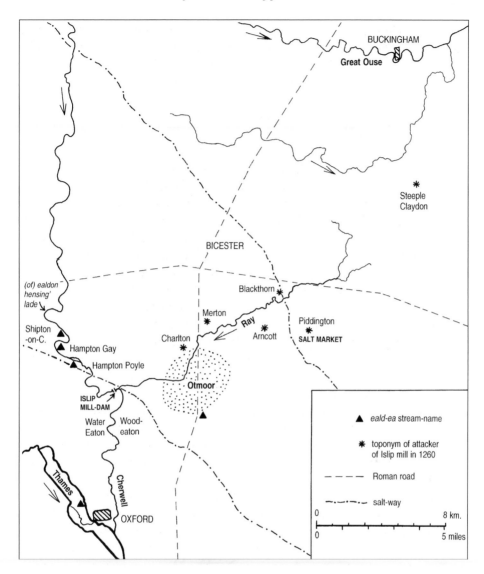

Fig. 62. The Cherwell and Ray river system, illustrating possible overland links to the Great Ouse and Wash systems. The asterisks, marking places from which some of the attackers of Islip mill in 1260 derived toponyms, are included to support the suggestion that that episode was prompted by concerns running the length of the Ray.

Oseney Abbey's dealings over mills, weirs, and leats in the parishes of Shipton-on-Cherwell, Hampton Gay, and Hampton Poyle, all of which bordered the Cherwell.[53] These deeds mention the *Sippeme holdhe* (i.e. *Sciphæma-ealdea*);

[53] H. E. Salter (ed.), *Cartulary of Oseney Abbey*, vi, Oxford Hist. Soc. 101 (Oxford, 1936), 77–8, 102–4. These texts include agreements over the mill-weirs and mill-leats of the three parishes, including the 'law hurdles' ('lawe hurdel') in the weirs, and give the impression that this stretch of the Cherwell was intensively occupied by mills.

the confluence below Hampton Gay mill called *Oldhe*; and—beside Hampton
Poyle manor house and mill a short way downstream—the meadow 'between
Cherwell and *le Eldehe*'. Still extant at Hampton Poyle is the back-stream
of the Cherwell which enclosed this meadow: of artificial appearance, and
resembling the 'barge-gutters' which, as D. J. Pannett has shown, enabled craft
to avoid fish-weirs on the Severn (Fig. 63).[54] Perhaps *eald-ea* was commonly
used to distinguish the original river course, encumbered by a weir, from a
bypass channel around it. At least one of these terms must have been archaic
when written down in 1261: *Sippeme holdhe*, 'the old river of the people of
Shipton', uses the Old English construction *-haema-* 'people of'.[55]

    The inference that there were 'old rivers'—and therefore 'new rivers'?—at
several points along this stretch of the Cherwell is reinforced at Whitehill,
the next estate upstream: its charter-bounds of 1004 include the 'old Hensing
lād', apparently as an offshoot or branch of the Cherwell.[56] The term *lād* is

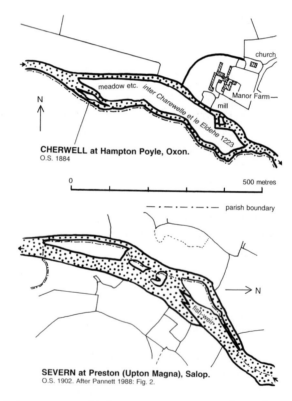

Fig. 63. Natural river courses with barge-gutters: an example on the Cherwell compared
with one on the Severn.

    [54] Pannett, 'Fish Weirs'.
    [55] I am grateful to Matti Kilpiö and Joy Jenkyns for the view that this construction is likely to
pre-date the Old English to Middle English transition.
    [56] S 909; J. Cooper, 'Four Oxfordshire Anglo-Saxon Charter-Boundaries', *Oxoniensia*, 50 (1985),
15–23, at 16.

discussed above (p. 77) by Ann Cole, who defines it as 'either a road or track, or an artificial waterway or canal', and notes its later widespread use (as -*lode*) to mean canal. At Whitehill the context points to this aquatic sense: it may well be that the *lād* here (even though itself 'old') is to be contrasted—as an artificial rather than a natural channel—with *eald-ea* in the other texts.

Upriver from Oxford we encounter different terrain, and different opportunities for human intervention, as has been vividly described by Mary Prior:

Above Oxford the Thames is slow and sluggish with a very slight gradient, but down river the gradient is sharper. Above Oxford the tributaries of the Thames run briskly down the sides of the vale of the upper Thames into the slow-moving river, rather like the drainage from a corrugated iron roof into a gutter which has itself only the slightest incline. If the upper Thames was like a gutter, the Thames from Oxford downstream was more like the drainpipe into which the gutter debouched. The gradient affected the distribution of mills, as mills can only work where there is an appreciable gradient. Therefore, above Oxford, mills were associated with the tributaries of the Thames rather than the mainstream, whilst from Wolvercote, on the outskirts of Oxford, and through Oxford downstream to London mills were also associated with the main river.[57]

Here, then, we should not envisage a weir-encumbered waterway like the late medieval Thames below Oxford. Silted and weed-choked shallows, and the general narrowness of the river, would have posed their own problems, but ones that might be overcome by deepening the natural course or digging parallel artificial ones, followed by regular maintenance (as reflected in the *ēa-tūn* names?). Overall, navigation should have been more straightforward on the uppermost than on the middle stretch of the Thames: a fact that could have encouraged the formation of a river-using community which, while not cut off from long-distance transport, was also to an extent localized and self-contained. In the rare cases where mill-dams did traverse the main channel, as at *Kyndelwere* and Rushey, the need to build up a big head of water would have resulted in larger engineering enterprises, and a correspondingly greater impact on traffic. Probably the biggest single cause of hydrological change between the eleventh and thirteenth centuries would have been the multiplying bridge-causeways, at Cricklade, Castle Eaton, Hannington, Lechlade, Radcot, Newbridge, and Oxford.

With these thoughts in mind, we can turn to two well-attested cases of artificial waterways feeding into the upper Thames. One is a potentially late Anglo-Saxon canal of impressive length, linking one of the royal and ecclesiastical centres of west Oxfordshire to the river. The second, a shorter and perhaps later canal on the south bank, was associated with a mill-dam on the main channel which by the mid-thirteenth century may have been the effective head of navigation for large craft.

---

[57] M. Prior, *Fisher Row* (Oxford, 1982), 107–8.

The Bampton Canal

The first case involved the artificial manipulation of two natural rivulets, the Shill Brook and Highmoor Brook, which feed into the upper Thames floodplain from the limestone and clay to its north (Figs. 64–7). Both run in their own alluvial channels and are certainly natural, but they are linked by a section of stream which runs east from Black Bourton to the north-western edge of Bampton town. There are strong grounds for concluding that this link is artificial, and was cut to form a navigable watercourse from Black Bourton via Bampton to the Thames at Shifford.

By the 1580s the configuration of these features was much as it is today, as is clearly described by Raphael Holinshed:

There are two fals of water into Isis beneath Radcote bridge, whereof the one commeth from Shilton ... by Arescote [Alvescot], blacke Burton and Clanefield. The other also riseth in the same peece, and runneth by Brisenorton unto Bampton, and there receiving an armelet from the first that breake off at blacke Burton, it is not long yer they fall into Isis, and leave a pretie Iland.[58]

Exactly this is shown on Saxton's near-contemporary map (Fig. 65), where the two streams and the 'armelet' form an H-plan.[59] During the last three centuries, however, the natural water flows have reasserted themselves, leaving the 'armelet' between them as in places little more than a muddy ditch.

Crucially, it is impossible for Holinshed's 'armelet' to be a natural channel. Although it traverses various soils, including a patch of alluvium, a 600-m section between Lower Farm and the Highmoor Brook cuts cleanly through the Second Gravel Terrace in a way that could not result from normal geomorphological processes[60] (Fig. 66). Today, and on maps since the eighteenth century, it is very sinuous, but close inspection shows that an initially straight channel has been modified over time through the meander migration effect described above by Ed Rhodes. This is especially clear at one point, where interfaces between natural gravel and alluvium in the vertical, still-eroding banks define a former channel some 7.0 m wide, across the course of which the present ditch snakes from side to side (Fig. 66, inset, section A). Although the exact configuration of the original cut must be conjectural, the impression is that it was dug in a series of straight sections.

Further south, where the Shill Brook runs in winding channels through the Deanery grounds to the west of Bampton church, geophysical survey and two machine-dug sections have yielded important results.[61] The magnetometer plot (Fig. 67) shows that the southernmost stretch of the canalized stream

[58] R. Holinshed, *The First and Second Volumes of Chronicles* (London, 1587), 48.

[59] Morden, in 1695, depicts the link as part of one continuous watercourse flowing from north-west to south-east, with the lower stretch of the Shill Brook as a separate stream rising in Clanfield.

[60] Ed Rhodes visited the site with me and confirmed this.

[61] I am extremely grateful to Arnold Aspinall for carrying out the survey, and to George Dudley for help with digging the sections.

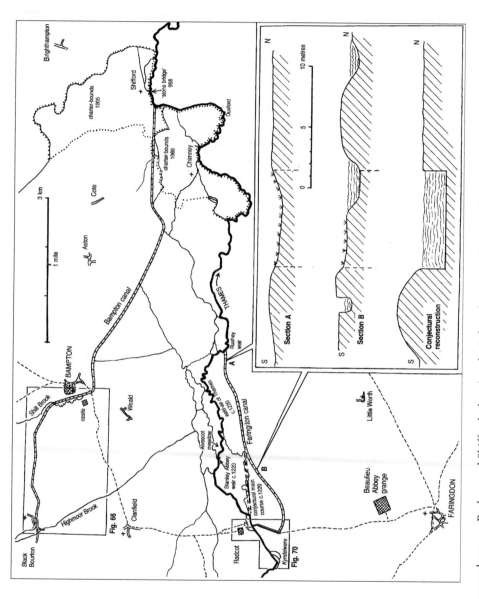

Fig. 64. The Thames between Radcot and Shifford, showing the inferred courses of the canals at Faringdon and Bampton. The inset sections show the silted-up modern profiles of the Faringdon canal at two points, and a hypothetical reconstruction.

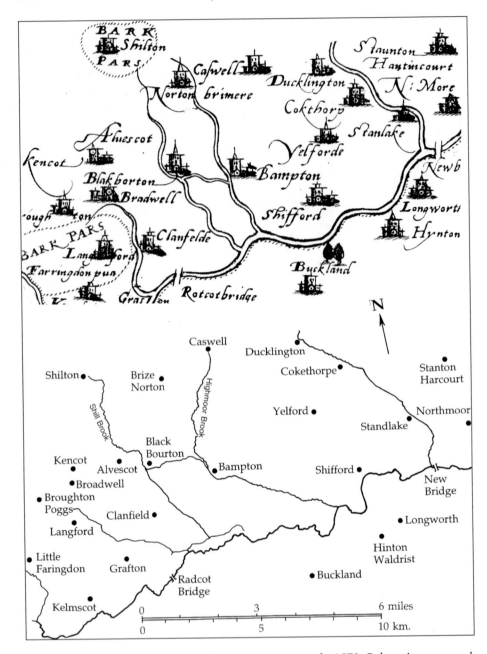

Fig. 65. Bampton and its environs. *Above*: Saxton's map of *c*.1578. *Below*: A true-to-scale interpretation.

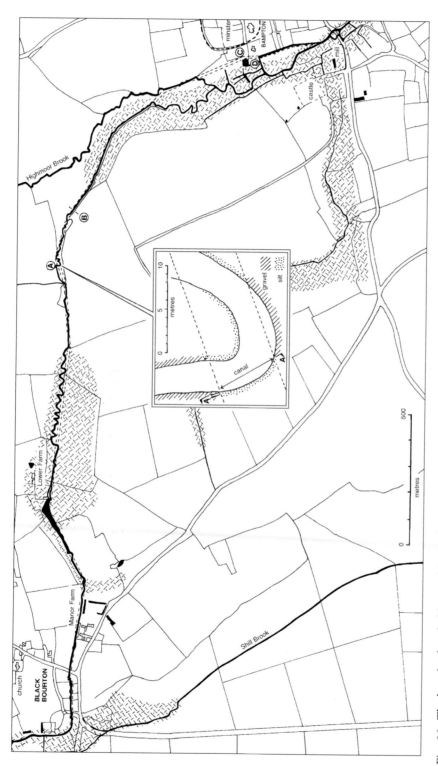

Fig. 66. The length of probable canal from Black Bourton to Bampton (see Fig. 64 for location), as existing in 1884 (OS 25-inch 1st edn.). Alluvial silt is indicated by broken cross-hatching; the subsoil is otherwise Oxford clay and gravel terrace. The suggested primary course of the canal, before modification by meander processes and silting, is indicated by double broken lines. The inset detail shows the evidence visible at one point for a meander cutting across the original straight course (natural gravel shown as hatching, silt as stipple). For the sections indicated as A, B, C, and D, see Fig. 68.

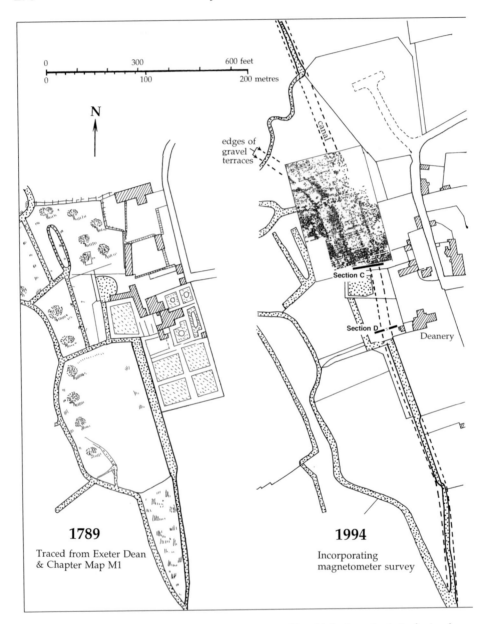

Fig. 67. The watercourses at Bampton Deanery (see Fig. 66 for location). *Left*: As shown on an estate map of 1789. *Right*: The modern layout, showing the geophysical evidence for a lost length of the canal, and the location of machine sections **C** and **D**.

once continued further south, passing close to the Deanery house and lining up with a linear pond. The machine sections **C** and **D** across this filled-in stretch prove that the stream and pond did indeed connect as a continuous and clearly artificial watercourse, dug into the natural gravel just inside the

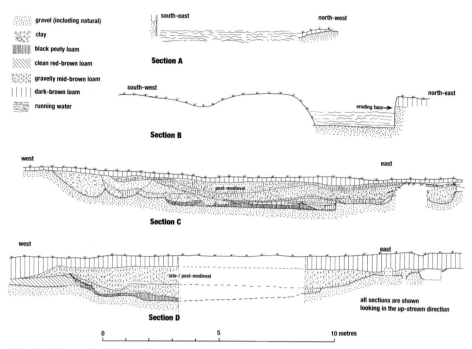

Fig. 68. The probable canal at Bampton: sections.

edge of the Second Gravel Terrace. As seen in these sections (Fig. 68) the channel had been part-filled and recut several times, resulting in a broad, shallow depression some 10–12 m wide which was finally levelled over by post-medieval dumping. The surviving traces of the primary cut suggest that it was *c.*7.4 m wide, had a roughly flat bottom and vertical sides, and was dug at least 1.5 m into the natural gravel (so probably rather more than 2 m below original ground surface); its first fill was a dark humic soil, interleaved with gravel collapses from the sides. Intermediate recuts were associated with shallow-sloping revetments of rubble embedded in clay on both banks, and perhaps with smaller ditches flanking the main channel on either side.

Although the earlier phases produced no dating evidence, it seems likely that this sequence extends well back into the middle ages. It is hard to interpret the primary cut as anything other than a further stretch of the canal (to which it corresponds closely in width), positioned carefully along the edge of the gravel terrace to provide a watercourse with firm, dry banks outside the alluvial floodplain. A factor in the decline of the canal may have been Bampton mill, on the Shill Brook just below the Deanery, where the dam was causing flooding by the 1220s (below, Appendix, texts **G** and **H**, pp. 291–2). The conversion of the more southerly stretch into a linear pond bypassed by the main stream can perhaps be dated to before 1317, when fishponds in the Deanery garden are mentioned.[62]

Downstream of this point the Shill Brook skirts the southern edge of Bampton and enters the alluvial floodplain of the Thames. Eighteenth-century maps[63] show that it turned eastwards to the south of Aston and Cote in the form of the gently sinuous Great Brook, entering the Thames between Shifford and Chimney (Figs. 29, 64). Here a vital clue is provided by the Shifford charter-bounds of 1005, which mention 'two weirs, one above the *lād*, the other below' ('ii weras, oþer bufan þære lade, oþer beneoðan').[64] The aquatic context suggests that this *lād*, like the one on the Cherwell in 1004 (above, pp. 270–1), was a 'lode' or canal: the most straightforward reading of the phrase is that the weirs were on the Thames, one just upstream of the point where a *lād* entered the river and the other just downstream. This *lād* was surely the Great Brook, which meets the Thames immediately south of the manorial centre at Shifford.

Putting the evidence together, we can recognize an artificial, 7-m-wide watercourse between Black Bourton and Bampton which underwent major natural and human modification before the late middle ages; and, further downstream at Shifford, an explicit reference to the same watercourse as a 'canal' (*lād*) in 1005. Maybe it was this that the Chimney yardlanders had once used to carry grain from the rectorial demesne—much of which was around Bampton itself—to the Thames and thence to Oxford.

The Passage of the Thames below Kyndelwere and Radcot Bridge

The references from 1205 to ships (*naves*) passing downriver from the Beaulieu Abbey grange at Faringdon to the sea (above, p. 260) have important implications for management of the Thames, for the stretch from Radcot to Oxford would scarcely have carried anything more than small craft without human modification. The documentation for Beaulieu's estates is good, and shows the care and concern of the early thirteenth-century Cistercians—those pre-eminent high medieval water managers—to control the river and its banks.

Beaulieu's first known riverine dispute here was with the monks of Stanley, who held Worth, immediately east of Faringdon on the Berkshire bank.[65] It was settled by complicated agreements of 1219 and 1222 (Appendix, texts **A** and **B**, pp. 287–8 below) which allowed Stanley to have a weir $7\frac{1}{2}$ feet wide 'in the old course of the Thames called *Eldee*', maintained the water level in that channel at six inches, and imposed sanctions and safeguards to avoid flooding

---

[62] Exeter Cathedral, Dean and Chapter MS 2931: 'et sunt in gardino duo vivarii.'

[63] *VCH Oxon.* xiii. 32, 63, 100.

[64] H. E. Salter (ed.), *Eynsham Cartulary*, i Oxford Hist. Soc. 49 (Oxford, 1906–7), 23 (= S 911). Possibly the downstream weir was associated with the 'stone bridge' mentioned at this point in the Longworth charter-bounds of 958 (S 654).

[65] *VCH Berks.* iv. 494.

of Beaulieu's land. The weir comprised two posts set into a sill-beam flush with the stream-bed, and was closed by a strong hurdle or boards. Luckily we can locate it (Fig. 64), at the crossing now called Old Man's Bridge: a meadow outlier of Alvescot was stated in about 1210 to 'extend by that branch of the Thames which divides Oxfordshire from Berkshire, and it lies to the east of the monks of Stanley's weir and to the north of the Thames'.[66] Notable here is the use yet again of the expression *eald-ea* 'old river', underlined by the Stanley monks' negotiations a few decades later with a north-bank landowner concerning the 'dead-stream' of the Thames.[67] But the river at Old Man's Bridge is anything but dead today: it is the main course of the Thames, and a weir $7\frac{1}{2}$ feet (*c*.2.3 m) wide and a water level of six inches (*c*.0.15 m) are equally inconceivable. The width would have been more restricting than the 10-foot (*c*.3 m) lock upstream at *Kyndelwere*, while the six-inch draught would have been inadequate for any kind of traffic except manhandling the flattest of punts. It can be concluded that in the early thirteenth century the present main course was a subsidiary one, but that it was perceived to have been the main course at some earlier date, and was already (as it still is) the county boundary: in other words, the river had been diverted, but has reverted since then to its natural channel.

Where then was the main course in the 1220s? South of the present Thames, between Radcot Bridge and Rushey weir, the earliest detailed Ordnance Survey maps show sections of a very substantial artificial channel aligned with the Thames, up to 8 m wide, and in places with a parallel flood-relief channel to its north. It is now broken into discontinuous lengths, but the intervening sections can be traced along field ditches as a linear depression some 7 m wide with a slight bank to the south (Fig. 64, inset). Although the meadows here are certainly liable to flooding, an earthwork on this scale seems excessive for a drain, and it is one potential candidate for whatever had replaced the *Eldee* at some date before *c*.1220.

This channel would have entered the natural course of the Thames at Rushey weir, due south of Bampton, but the point upstream where it left it is not so obvious. As currently visible, it starts at the south end of the straight, embanked causeway leading southwards from Radcot Bridge (Figs. 69–70),[68] and is fed

[66] *Cartulary of Oseney*, iv. 499. This was the western end of Burroway: the extra-parochial portion indicated in *VCH Oxon*. xiii. 2, parcel M on map. It was complained in 1261 that the Abbot of Stanley had blocked up a path between Radcot and Buckland in a place called 'le Olde Ehe' next to the said abbot's mill, obstructing public passage (TNA, PRO, JUST 1/701, m. 23), presumably at Old Man's Bridge where there is still a footpath.

[67] A list of now-lost Stanley deeds includes 'Domini Willelmi de Valence de la Dedelake Tamisie: quietaclamatio' (W. de G. Birch, 'Collections towards the History of the Cistercian Abbey of Stanley', *Wilts. Arch. Mag.* 15 (1875), 239–307, at 274); Valence was lord of Bampton 1248–96 (*VCH Oxon*. xiii. 22). Cf. the 'Deadlake' of the Plym: *Place-Names of Devon*, i, EPNS 8 (Cambridge, 1931), 5.

[68] Litigation in 1387 concerned 'the royal road called "cawsey" in Faringdon which extends to Radcot bridge' (*PWML* i. 5). The bridge is first mentioned in 1208, when King John gave protection to one 'Brother Aylwyn' (a monk of Beaulieu?) who had undertaken its repair (*Rot. Litt. Pat.* i.i. 87–8). In the 13th century it could be called the 'great bridge of Radcot' (Bodleian Library, MS Barlow

Fig. 69. The fourteenth-century Radcot Bridge in 1799, seen from the north (Oxfordshire) bank and the east (downstream) side. Two original arches flank the rebuilt navigation arch, broken during the battle in 1387. Visible on the opposite bank is the north end of the Faringdon causeway as it bends towards the bridge. (S. Ireland, *Picturesque Views of the River Thames* (London, 1801), i, opp. p. 39.)

from the broad (17-m-wide) linear pond which flanks that causeway on its west (upstream) side. This pond could have served as a dock[69] (and would have been very convenient for the loading of bulk goods such as grain), but its configuration makes it a most implausible conduit for the main waters of the Thames. This arrangement suggests that the channel was a 'blind lode', open only at the downstream end and terminating in a hythe at the other.[70] Water flow into it could have been controlled by a sluice at one or other end of the long pond.

Convincing though this canal looks on topographical grounds, there is an insuperable objection to identifying all of it as the main navigation channel: the

49, fo. 100 (pencil foliation), a terrier of Wyke Grange including meadow east of the royal road leading 'a magnam pontem de Roddecot'). The existing 14th-century structure is the survivor—and the largest—of a sequence of three medieval bridges over the branches of the Thames here, all broken in the battle of 1387 (J. N. L. Myres, 'The Campaign of Radcot Bridge in December 1387', *English Historical Review*, 42 (1927), 20–33: I take 'pavimentum pontis interruperunt in tribus locis' to mean that a breach was made in each of the three bridges); Baskerville describes their state in 1692 (BL, Harl MS. 4716, fo. 16). See E. A. Pocock, *Radcot and its Bridge* (Clanfield, 1966).

[69] Presumably it had provided the material for the causeway; Mr Haskins (below, n. 71) believes it to be a withy-bed. See above, p. 236, for a relict canal used as a withy-bed.

[70] Compare the late Anglo-Saxon canal at Glastonbury and some later ones in the Fens: Gardiner and Hollinrake, above, pp. 94, 238; J. R. Ravensdale, *Liable to Floods: Village Landscape on the Edge of the Fens, AD 450-1850* (Cambridge, 1974), 25.

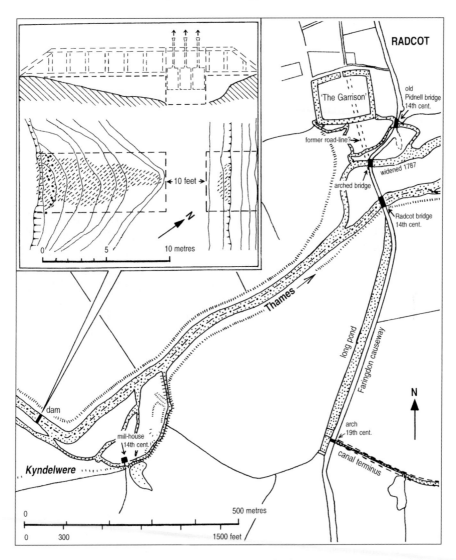

Fig. 70. The Radcot and *Kyndelwere* area (see Fig. 64 for location). Topographical detail as in 1883 (OS 25-inch 1st edn.); post-medieval buildings are omitted. The line of alternate dots and dashes is the county boundary, also the boundary between the lordships of Faringdon and Radcot (see Appendix, texts **C, E, F**). *Inset*: plan and section of the *Kyndelwere* dam, after survey in 1997 by Paul Haskins. Contours are at 25-cm intervals; rubble is shown as broken hatching, stakes as black dots. The superstructure shown in broken line is imaginary, and intended merely to illustrate how the recorded remains would be consistent with a flash-lock and an aperture 10 feet (*c*.3 m) wide, as stated in 1261 (text **C**).

thirteenth-century references to vessels passing upstream to *Kyndelwere*, which is immediately *above* Radcot Bridge, are incompatible with a constriction at this point. The conclusion must be that there was another and now completely lost watercourse, maybe running in a straight line eastwards from Radcot Bridge and meeting the canal at a half-way point (as suggested in Fig. 64), which carried through traffic. The south-western arm of the canal would then have been specifically for the use of carts and boats trans-freighting on the Faringdon causeway. If this interpretation involves some frustrating uncertainties, it at least emphasizes in broad terms the intensity with which this stretch of the river was used and managed.

A short way upstream, on the south bank of the river's natural course, is a distinctive D-shaped side channel. This can be identified as the main leat of *Kyndelwere*, Beaulieu Abbey's much-documented mill (Fig. 70). Thanks to the excellent fieldwork of Mr Paul Haskins we now know a good deal about its configuration in the later middle ages, including the plan of the fourteenth-century mill-house.[71] Most remarkable is the discovery, on the bed of the present Thames, of the mill-dam itself: some 4 m thick, constructed of rubble reinforced with stakes, and with a break towards the north (left-bank) end which would accommodate the 10-foot (3-m) gap stipulated in 1261 (Appendix, text C).[72] The special long-term importance of this mill is underlined by its prominence in the thirteenth-century texts, as also by the exceptionally high value—35s.—which Domesday Book had placed upon the mill and fishery of Faringdon in 1086.[73] A surviving account for 1269/70 shows that *Kyndelwere* produced grain liveries not greatly less than some of Beaulieu's entire granges, and more than twice as much multure as any other mill belonging to the abbey.[74] There must have been a great deal of coming and going around the mill with grain-laden boats, belonging both to tenants obliged to grind here and to other landowners who found it a convenient mill to patronize.

It may be no coincidence that the highest Thames-side canals yet recognized, at Bampton and Radcot Bridge, flanked the river just downstream of this major economic focus, and major check point in the river's flow. Intriguingly, a nearby 'hard' or landing place on the Radcot bank of the Thames had a strange

---

[71] P. Haskins, 'Kindlewere: A Medieval Mill and Fishery on the Thames' (Oxford University Department for Continuing Education dissertation, 1997; copy deposited in the Oxfordshire Sites and Monuments Record). He shows that the name 'Kindlewere' survived until 1787, after which the site was incorporated into Camden Farm. I am deeply grateful to Mr Haskins for making his discoveries so freely available, for giving me a tour of the site, and for sharing with me his long experience of the local topography.

[72] This very rare discovery can be compared with the 12th-century dam excavated at Castle Donington, though unlike that it is not set slantwise to the flow of the river: P. Clay and C. R. Salisbury, 'A Norman Mill Dam and Other Sites at Hemington Fields, Castle Donington, Leicestershire', *ArchJ* 147 (1990), 276–307 (and cf. Langdon, *Mills*, 80).

[73] DB i. 57ᵛ.

[74] S. F. Hockey (ed.), *The Account-Book of Beaulieu Abbey*, Camden 4th ser. 16 (London, 1975), 24, 28, 73–5. As well as the texts edited below, there are references to two murders at *Kyndelwere* in the Berkshire eyre of 1284 (TNA, PRO, JUST 1/44 mm. 3d, 15). In 1220 the monks acquired half an acre of meadow in Radcot: 1,005 feet long, between the Thames and a ditch bordering the meadow of Radcot, and extending from the old weir towards the west; the ditch itself was acquired in a subsequent deed (*Beaulieu Cartulary*, 16–18; the initial grant is dated to 1220 by the accompanying fine, H. E. Salter (ed.), *The Feet of Fines for Oxfordshire* (Oxfordshire Record Soc. 12 (Oxford, 1930), 64). This looks like the creation of a 'barge-gutter' and 'bylet' (cf. Pannett, 'Fish Weirs', 374–5 for Shrewsbury Abbey), and can be interpreted as a modification of *Kyndelwere*.

name, 'the Jury's Hard',[75] which seems to indicate the site of a special court concerned with riverine disputes.[76] The context of Faringdon's commercial prominence, and of Beaulieu's anxiety to control the land around Radcot Bridge, could be that *Kyndelwere* was a break point between different scales of traffic, perhaps even the head of navigation for large-scale freight.

Some Conclusions

It is now possible to see something of the physical background to the trade and transport suggested by our exiguous texts. Passage along difficult stretches of the Thames and Cherwell, and to the Thames from places up to 10 km away, was made possible by the digging of bypass cuts or 'barge-gutters' in some cases, canals using the waters of natural tributary streams in others. While these could run through gravel, clay, or alluvium, the South Moreton, Black Bourton, and Bampton cases suggest a preference for clean cuts through the gravel near manorial centres (and presumably therefore mooring and loading sites). They seem characteristically to have been around 7 m wide, and to have been cut in straight or gently sinuous lengths: it is reasonable to call them canals. As the present case studies show, later change—whether through neglect, human modification, meander migration, or reversion to natural drainage patterns—has often disguised them to the point of near-invisibility: it is currently impossible to know how common they may once have been.

When were these canals created? Most of the cases discussed above are visible to us by the early to mid-thirteenth century. The *eald-ea* place names are probably older than this—certainly so in the case of the *Sciphæma-ealdea*. The two occurrences of *lād* in charter-bounds are at Whitehill in 1004 (when it was 'old') and at Bampton/Shifford in 1005. On the other hand, the Blewbury

---

[75] Appendix, text C, p. 288 below. The only other known reference to this place is in 1271, when Matthew de Besilles was given as deodand 'batellum illud in quo Gilbertus filius Walteri le Messer nuper in aqua Thamisie ad locum qui dicitur La Juresherd, qui est infra libertatem predicti Mathie de Radcote, per infortunium submersus est, una cum quinque quarteriis frumenti et dimidio, quatam [*sic*] catena ferrea, quadam sera, et undecim saccis in eodem batello inventis' (TNA, PRO, C66/90 m. 29 (= *Cal. Pat. R. 1266–72*, 610); also in JUST 1/705 m.21 (assize roll, 1285)). Matthew's jurisdictional liberty was recorded in 1279 (E. Stone and P. Hyde (eds.),*Oxfordshire Hundred Rolls of 1279*, Oxfordshire Record Soc. 46 (Oxford, 1968), 56).

[76] Text C indicates that in 1261 the jurys of three riparian hundreds (Shrivenham and Wantage/Gainfield in Berkshire, Bampton in Oxfordshire) were assembled at *Jureyscherd* to determine on which side of the county boundary *Kyndelwere* lay, and how wide and deep its weir should be. This procedure recalls disputes over land near Bridport in 1122 determined by 'sixteen men, three from Bridport and three from Burton Bradstock and ten from the neighbourhood', and over Sandwich harbour in 1127 determined by 'an assembly of wise men living near the sea' of whom twelve were from Sandwich and twelve from Dover (R.C. Van Caenegem (ed.), *English Lawsuits from William I to Richard I*, i, Selden Soc. 106 (London, 1990), 197, 217). Cf. the 'Shepway' and 'Brodhull' courts of the Cinq Ports (K. M. E. Murray, *The Constitutional History of the Cinq Ports* (Manchester, 1935)); or the East Anglian *hethewarmoot* attended by masters of fishing vessels (Gardiner, above, p. 99).

to Wallingford canal must, if it existed at all, have post-dated the tenth-century charter-bounds here with their *brocs* and *lacus*. The main fixed point is of course the Abingdon canal, which we know to have been dug between 1052 and 1066. These indications converge on the last century or so of the Anglo-Saxon period, though the general economic context makes it inherently likely that the system was not merely maintained but developed into at least the twelfth century.

What kinds of craft could have plied on the upper Thames? Our texts refer to 'ships' ('naves') at Abingdon and Dorchester, 'ships and little ships' ('cum navibus et navicellis') at *Kyndelwere*; the Abingdon text adds the interesting detail—if it is more than a literary flourish—that the unmodified Thames at Thrupp had been difficult for rowers ('remigantibus'). At Bampton, on the other hand, each pair of Chimney tenants used a 'boat' ('batellum') to carry grain to Oxford. These references provide an exiguous prelude, earlier in time and further upriver, to the much fuller evidence for the 1290s to 1340s adduced by John Langdon from purveyance accounts (above, pp. 114–26). We can probably assume that a *batellum* and a *navicella* were comparable in size to each other, and both smaller than a *navis*. As Langdon shows, vessels operating between Henley and London in the later period were pre-eminently the flat-bottomed barges then called 'shouts', exemplified by the wreck of *c.*1400 called 'Blackfriars 3' (14.64 m long, 4.3 m in the beam) but sometimes apparently much bigger than that.[77] Shouts are first mentioned by name in 1204,[78] and no earlier wrecks of precisely this kind have yet been found. But the basic kind of coastal and estuarine ship attested in southern Britain between the tenth and twelfth centuries, only slightly more rounded in profile and with a similarly flat keel, could have negotiated flash-locks almost as easily as a shout. Recorded examples (Fig. 3), the Graveney boat of *c.*920 (4.0 m) and the two London Custom-House boats of c.1160–90 (both at least 3.5 m), were slightly narrower in the beam than Blackfriars 3.[79]

Were the *naves* to be seen on the Thames at Abingdon and Dorchester, and even at Radcot, on this scale? Some clue may be given by the recorded dimensions of flash-locks, since these must have accommodated the biggest craft in normal use. On the Severn below Gloucester, for instance, it was ancient practice by the 1380s to leave 18-foot (*c.*5.5-m) gaps in all weirs 'for the passage of all men, boats and ships going through there': an arrangement

---

[77] Langdon, above pp. 115–16, 122–6; P. Marsden, *Ships of the Port of London: Twelfth to Seventeenth Centuries AD*, English Heritage Archaeological Rep. 5(London, 1996), 55–104.

[78] *Rot. Litt. Pat.* i.i. 42b; the term is of Dutch origin. I owe this reference to Robert Peberdy.

[79] V. Fenwick, *The Graveney Boat*, BAR British Ser. 53 (Oxford, 1978); Marsden, *Ships of the Port of London*, 41–54; above, p. 7. The 13th-century Magor Pill wreck has a similar profile to Graveney but also some important differences, notably a deeper keel: N. Nayling, *The Magor Pill Medieval Wreck*, CBA Research Rep. 115 (York, 1998) 48, 146–9. D. M. Goodburn, 'Do we have Evidence of a Continuing Late Saxon Boat Building Tradition?', *Internat. Jnl. of Nautical Archaeology*, 15 (1986), 39–47, argues for a distinctive English tradition, extending from Sutton Hoo via Graveney to Blackfriars 3, which should be contrasted with the Nordic tradition.

that would have allowed the thirteenth-century Magor Pill boat (3.7 m in the beam) to carry Forest of Dean iron ore to the Severn mouth.[80] If the 7.62-m weir which the Islip mill-dam contained up to 1260 was for the same purpose, it implies that ships on this scale could have been taken up the Cherwell, and maybe to 'the hythe called la Ree' on Otmoor. By comparison, *Kyndelwere* was markedly smaller at 10 feet (c.3.04 m) wide and 7 feet (c.2.14 m) deep, even though the depth could have accommodated a substantial draught.[81] Passage through this lock is demonstrated by the Beaulieu monks' right, vindicated in about the 1280s, to go back and forth around ('circa') the mill with their *naves* and *navicellae* (Appendix, text E, p. 290), but it was surely only the *navicellae*, ranging from logboats, punts, and small rafts up to the smallest kind of shout, that could use the Thames from this point upwards. The one aquatic fatality recorded near *Kyndelwere*, in 1271, involved a *batellum* carrying just under a ton of wheat, a chain and padlock, and eleven sacks.[82] Local people negotiating side streams would presumably have used craft of this kind, even if—like the pairs of Chimney men with their *batella*—they sometimes made longer journeys. But it remains completely imponderable whether canals such as the one at Bampton carried craft of a size that their width of around 7 m might, on a maximal interpretation, imply.

The proposition that vessels on the scale of Graveney or Blackfriars 3 could get well above Oxford in this period—on the Thames as far as Radcot, on the Cherwell and Ray as far as Otmoor—is therefore not an absurd one, though one must agree with Langdon (above, pp. 12, 125) that this was almost certainly a one-way traffic, the ships being returned upriver empty, or with only light cargoes such as salted herring. It must also have been the case that once those craft had negotiated the hazards of the middle Thames, and reached the sluggish stretch above Oxford, they would then have had to contend with blockages, poor maintenance, and vagaries in the water level, even if with fewer weirs.

In the absence of purveyance accounts we cannot know whether the vessels in regular use below Henley were already larger still: in other words, whether Henley was already the 'break point' that it would later be.[83] But if Langdon is right in seeing a shift of mill-dams from lesser to major rivers during c.1086–1300 (above, p. 9n), this must mainly have affected the middle Thames, and would in turn have made the upper and lower stretches more separate from each other than they had previously been. It does on the whole seem likely that

[80] *PWML*, i. 157–8; Nayling, *The Magor Pill Medieval Wreck*, 112–13, 150–2 (though the ore actually found on this wreck seems to have come from Glamorgan).

[81] The existing piers of the main arch in Radcot Bridge (above, n 68) are marginally further apart at 12 feet (3.66 m); the main arch in Newbridge, the next major bridge downriver, is 18 feet (5.50 m) wide.

[82] Above, n 75. In the 1240s a John the Punter fell off a *batellum* on the Thames near Abingdon: M.T. Clanchy (ed.), *Roll and Writ File of the Berkshire Eyre of 1248*, Selden Soc. 90 (London, 1972–3), 346.

[83] Above, Langdon, pp. 120–1; Peberdy, 'Navigation'.

the discrepancy in volume of traffic between the lower Thames on the one hand, and the middle and upper Thames on the other, was less marked in 1000 than in 1300. On the other hand, much of this upper Thames traffic need not have been on a long-distance scale. The men who steered logboats and punts between *Kyndelwere* and Oxford may have been happiest and most confident on their home ground, and the weirs below Oxford may have encouraged the formation of a self-contained—if highly professional—little community of carriers and water managers above it. This picture of two Thames-using worlds, a long-distance and a local one, is not inconsistent. Rather, it underlines how both the social unity and the economic success of the early English state were founded on a multiplicity of localized, but dynamic and interconnecting, systems.

Appendix

Radcot and Kyndelwere

A *Concord between the monks of Stanley and those of Beaulieu, settling a dispute concerning the course of the Thames. The monks of Stanley can build a lock (described) in the old course [at old Man's Bridge], maintaining a water level of half a foot and opening the lock in case of flood. 1217. (BL, Cotton MS Nero A.xii, fos. 40ᵛ–41ᵛ)*

Hec est finalis concordia facta inter domum de Stanlee et domum Belli Loci Regis super controversia dudum habita inter dictas domos de cursu aque Thamisie: scilicet, quod monachi de Stanle firmabunt duo magna ligna in veteri cursu Thamisie qui vocatur Eldee, habentia inter se spatia vij pedum regalium et dimidii, et claudent aperturam illam una forti claya loco excluse; hoc precipue providentes, ut aqua Thamis' ibidem contineatur infra canalem [ad?] mensura[m] dimidii pedis. Inundatione vero superveniente, ammovebunt dictam clayam donec aqua minuatur usque ad dictam mensuram, scilicet dimidii pedis infra canalem. Licebit etiam dictis monachis de Stanle claudere predictam aperturam asseribus si voluerint, ita tamen quod si monachi Belli Loci R[egis] aliquod inde dampnum incurrerint, monachi de Stanle dampnum eorum integre resarcient et eorum indempnitatem[84] per omnia providebunt. Si quis autem monachus vel conversus de Stanle vel de Bello Loco Regis hanc formam pacis imposterum turbaverit, a domo propria eliminetur, non reversurus nisi de consensu utriusque domus. Facta est autem hec concordia coram abbatibus de Quarrer', de Thama, de Kyngeswode et de Brueria, anno incarnationis dominice mᵒ ccᵒ xvijᵒ etc.

B *Further agreement between the monks of Stanley and those of Beaulieu concerning modification (described) of the same lock to avoid flooding of Beaulieu's land, to be carried out before mowing, or at latest before autumn, unless prevented by flood or secular force. 1222. (BL, Cotton MS Nero A.xii, fos. 41ᵛ–43)*

Hec est conventio facta inter domum de Stanle et domum Belli Loci Regis, anno gratie mᵒ ccᵒ xxijᵒ in crastino Sancti Martini, apud Farendon coram A. abbate de Waverle, a domino Cist' et capitulo generali iudice delegato, et domino R. abbate de Bruere, gerente vices abbatis de Borele coniudicis dicte abbatis de Waverle: videlicet, quod abbas de Stanle, quam citius poterit, solviam illam quam ponitur in fundo Thamis' ad gurgitem de Ealdeya, in qua firmantur duo ligna que ponuntur ex utraque parte gurgitis, locari faciet et firmari, ita quod equabitur fundo recti canalis ipsius aque ut per predictam solviam aqua non impediatur, quin habeat liberum cursum suum. Et positio ipsius solvie fiet in presentia abbatum et fratrum utriusque domus et in presentia legalium hominum ex consensu utriusque partis electorum si partes voluerint. Perficientur autem hec ante secationem et collectionem fenorum, vel ad ultimum ante autumpnum, nisi inundatio aquarum vel maior vis secularium impedierit. Providere etiam debent abbas et monachi de Stanle ut aqua Thamis' in predicto loco contineatur intra canalem [ad?] mensuram dimidii pedis. Inundatione autem superveniente, ammovebunt predicti abbas et monachi clayas et firmaturas prenominatas in ore gurgitis donec aqua minuatur usque ad predictam mensuram. Promiserunt vero abbas et monachi de Stanle quod, quam citius poterunt, facient quamdam aperturam prope molendinum suum quatinus cum aqua superexcreverit, si per negligentiam dicte claye non fuerint ammote, possit aquarum superfluitas per prefatam aperturam transire et transfluere. Memorati vero abbas et monachi de Stanle promiserunt coram dictis iudicibus, in verbo veritatis, superscriptam conventionem in omnibus et per omnia pro posse suo firmiter et fideliter

---

[84] indempnitati *MS*.

esse se observaturos. Et si dicta conventio sicut prescriptum est non fuerit observata, abbas et monachi de Stanle stabunt arbitrio predictorum iudicum super tota causa eisdem antea commissa.

C *Verdict of jurors 'for Jureyscherd', of Shrivenham, Wantage and Gainfield, and Bampton hundreds, that Kyndelwere is in Berkshire, and has been from of old; that the water of the weir should be 7 feet deep, and should have free passage over if it rises higher; and that the width of the weir should be 10 feet and a barleycorn. 1261. (BL, Cotton MS Nero A.xii, fos. 74ᵛ–75; another transcript, in a 14th-century hand, is in Bodleian Library, MS Barlow 49, fo. 110ᵛ (pencil foliation)). It seems likely that this verdict was in some way connected with the eyres held in both Oxfordshire and Berkshire in 1261, even though it has not been located in their (incomplete) extant rolls.*

[*rubric*] Veredictum iuratorum pro Jureyscherd, de Kyndelwer' et de nocumento stagni et gurgitis, anno regni Regis Henrici xlv.

Nomina iuratorum hundredi de Schryvenham: Johannes de Gopeshulle, Paganus de[85] Kni3ttinton', Jacobus[86] de Offinton', Bricius de eadem, Matthias de Ordynston', Nigellus de Colleshull, Ricardus de eadem, Willelmus Lumbard', Ricardus Sturmy,[87] Robertus le Gras, Robertus le[88] Frankelayn de Offinton', Andreas Alewy, Thomas de Faulore.

Nomina iuratorum hundredi de Wanetyng et Gamenefeld: Galfridus de Pusye, Thomas Hamond, Willelmus de Nywynton', Henricus Saumon, Robertus de Offynton', Ricardus le Wyde Were, Rogerus de Comba, Robertus de Aula, Willelmus de Marisco, et Willelmus Serle.

Nomina iuratorum hundredi de Bampton: Willelmus Patric, Robertus de Chadlyngton', Thomas le Portir, Hugo Frankeleyn, Petrus de la Pyre, Willelmus de Aula, Gerardus le Bussh', Robertus de Eggerche,[89] Robertus de Ulmo, Robertus Stokes, Johannes Hagge, et Adam le Bacheler.

Qui iurati dicunt per sacramentum suum quod gurgis de Kyndelwere est in Berk', et bene patet[90] esse sicut solet esse antiquo tempore. Et aqua gurgitis debet esse de profundo vij pedum.[91] Et si contigerit quod plus fuerit aqua, tunc aqua liberime habeat cursum suum ultra. Et gurgis debet esse in latitudine decem pedum et unius grani ordei.

D *Presentments of attacks on Beaulieu's mills at Eaton Hastings and Kyndelwere by Thomas de Giltham, bailiff of Benedict de Blakenham [lord of Eaton Hastings[92]]. The abbot accuses Thomas and others of attacking the Eaton Hastings mill and removing boards and posts adjoining the sluices, to the value of 20 shillings, on 28 October 1283; and of attacking Kyndelwere and removing four sluices, to the value of 100 shillings, on 10 June 1284. Thomas says that Benedict has meadows next to both mills, and is entitled to expect the abbot to remove the sluices during the hay-mowing season, or to remove them himself if the abbot defaults; he removed the boards, posts, and sluices because the abbot and his bailiffs had refused to do so and the meadows had flooded. 1283–4. (Berkshire eyre of 1284, TNA, PRO, JUST 1/48 m. 4, JUST 1/43 m. 2, JUST 1/45 m. 4d; the basic text here is from 1/48, with non-trivial variants from 1/43 and 1/45 noted in footnotes.)*

Thomas de Giltham attachiatus[93] fuit ad respondendum Abbati de Bello Loco Regis de placito quare ipse simul cum Osberto de Waltham,[94] Roberto Wiryng',[95] Ricardo le Fraunkeleyn de Eton', Hugone le Clauer et Ricardo[96] le Schulle Lokere, bona et catalla ipsius Abbatis ad valenciam viginti solidorum apud Eton' inventa vi et armis ceperunt et asportaverunt, et alia enormia ei intulerunt, ad grave dampnum ipsius Abbatis et contra pacem etc. Et unde predictus Abbas queritur quod predictus Thomas simul cum predicto Osberto et aliis, in vigilia

---

[85] *Barlow; om. Nero.*
[86] *Christian names only in Barlow from this point on; Nero has initials.*
[87] *Barlow*; Sturny *Nero.*   [88] *Barlow*; de *Nero.*   [89] Eggereche *Barlow.*
[90] potest *Barlow.*
[91] vij pedum *written over erasure Nero*; septem pedum *as part of original text Barlow.*
[92] *VCH Berks.* iv. 529.   [93] Gyleham summonitus *1/45.*   [94] Wautham *1/45.*
[95] Wyrring *1/43*, Wyryng *1/45.*   [96] Ricardo *omitted 1/43, 1/45.*

Apostolorum Simonis et Iude anno domini Regis nunc undecimo,[97] venerunt ad molendinum ipsius Abbatis in predicta villa, et borda et postes iuxta exclusas eiusdem molendini in
stagno existentes, ad valenciam xx s., ceperunt et vi et armis asportaverunt, et alia enormia
ei intulerunt ad grave dampnum ipsius Abbatis et contra pacem domini Regis, unde dicit
quod deterioratus est et dampnum habet ad valenciam xx s. Et inde producit sectam.

Et Thomas venit, et defendit vim et iniuriam quando etc. Et dicit quod quidem Benedictus
de Blakeham, cuius ballivus ipse fuit, habet quoddam tenementum in predicta villa de Etone,
et quoddam pratum adiacens predicto molendino, et dicit quod ratione predicti tenementi
sui, tempore quo prata sua falcare debent, exclusas molendini predicti Abbatis amovere
debet si predictus Abbas vel ballivi sui ad hoc faciendum fuerint negligentes. Et dicit quod
tempore quo fuit ballivus predicti Benedicti, pratum ipsius Benedicti domini sui inundabatur
tempore quo falcari debuerant ratione exclusarum predicti Abbatis, quas ballivi sui amovere
noluerunt, per quod ipse Thomas, tanquam ballivus predicti Benedicti, borda et postes fixos
in aqua ad aquam illam includendam amoveri fecit, sicut ei bene licuit, eo quod pratum
predicti domini sui inundatum fuit per inclusionem aque predicte, et quod predictas exclusas
non asportavit, nec alia enormia ei fecit, ponit se super patriam, et Abbas similiter. Ideo fiat
inde iurata. Postea predictus Abbas non est presens. Ideo ipse et plegii sui de prosequendo
in misericordia. Queruntur nomina plegiorum inter brevia finita etc.

Idem Thomas attachiatus fuit ad respondendum predicto Abbati de placito quare ipse
simul cum Johanne de Waltham,[98] Johanne de Hoghton',[99] Benedicto de Blakeham, Willelmo le Perer, Galfrido le Whreyte, Thome le Wrghte,[100] Roberto le Blake, Ricardo Maheu,
et Ricardo le Clerk, bona et catalla ipsius Abbatis ad valenciam centum solidorum in
molendino ipsius Abbatis apud Kyndelwere inventa vi et armis ceperunt et asportaverunt et
alia enormia ei intulerunt, ad grave dampnum ipsius Abbatis et contra pacem etc. Et unde
predictus Abbas queritur quod predictus Thomas simul cum predictis Johanne et aliis, die
Sabbati proxima ante festum Sancti Barnabe Apostoli hoc anno, venit ad molendinum ipsius
Abbatis in predicta villa et quatuor exclusas ipsius Abbatis precii centum solidorum cepit et
asportavit et alia enormia ei intulit, unde dicit quod deterioratus est et dampnum habet ad
valenciam centum solidorum. Et inde producit sectam etc.

Et Thomas venit et defendit vim et iniuriam quando etc. Et dicit quod quidem Benedictus
de Blakeham, cuius ballivus ipse Thomas aliquando fuit, habet quoddam tenementum in
predicta villa, et quoddam pratum eiusdem tenementi sui adiacens predicto molendino
predicti Abbatis. Et dicit quod tempore quo predictus Benedictus vel ballivi sui prata
sua falcare voluerit, predicti Abbas vel ballivi sui predictas exclusas amovere debent; et
si predictus Abbas vel ballivi sui in hoc remissi fuerint, bene licet predicto Benedicto
vel ballivo suo exclusas illas amovere. Et quia in quindena Sancti Johannis Baptiste
proxima preterita, postquam predictus Thomas prata predicti domini sui predicto molendino
adiacentia falcari[101] fecerat, predictus Abbas nec ballivi sui exclusas predictas amovere
noluerunt, per quod pratum domini sui inundabatur; ipse Thomas ipsas exclusas amovit
prout predictus Benedictus et ballivi sui hoc de iure facere possunt, et semper hucusque
facere consueverunt. Et quod nullam aliam transgressionem ei fecit, ponit se super patriam,
et Abbas similiter. Ideo fiat inde iurata. Postea predictus Abbas non est presens. Ideo ipse et
plegii sui de prosequendo in misericordia. Queruntur nomina plegiorum inter brevia finita
etc.

E *Sir Matthew de Besilles and his wife Elizabeth oblige themselves and their heirs not to
impede the monks of Beaulieu in their free passage around the mill of* Kyndelwere *with
their ships and boats (reserving the fishery); nor to impede merchants of their corn or other
things from loading or unloading their ships on or near the Thames, or mooring their ships
against the monks' land.* [c.1275×95[102]] (BL, Cotton MS Nero A.xii, fos. 47ᵛ–48ᵛ)

---

[97] in vigilia … undecimo *1/43, 1/48 (in both cases written over an erasure)*; die Sabbati proxima
ante festum Sancti Barnabi hoc anno *1/45.*

[98] Wonesham *1/45.*   [99] Hotgton' *1/43,* Hogeton' *1/45.*

[100] Wreyhte *1/43,* Wreycte *1/45.*   [101] falcare *1/43, 1/48,* falcar' *1/45.*

[102] The dating of **E** and **F** is based on the following information (from *Cal. Inq. Post Mortem,* i. 68,
110; iii. 197; v. 284–5; viii. 161): Elizabeth, co-heiress of John d'Avranches from whom she inherited

Omnibus Christi fidelibus presens scriptum visuris vel audituris, Mathias de Besilles miles, dominus de Redcote, et Elizabet uxor sua, eternam in Domino salutem. Noveritis nos unanimi assensu concessisse et per presens scriptum nos obligasse, pro nobis et heredibus nostris vel assignatis, quod non inquietabimus, contradicemus nec ullo modo impediemus, per nos vel nostrorum servientium seu tenentium aliquos, abbatem et conventum Belli Loci Regis vel servientes suos in molendino de Kyndelwere quam libere possint ire et redire cum navibus et navicellis suis circa dictum molendinum, tam ante quam retro, ad negotium et commodum suum faciendum, quandocumque et ubicumque sibi viderint expedire, sine calumpnia nostri, heredum nostrorum vel assignatorum imperpetuum; ita tamen quod famuli dictorum religiosorum de dicto molendino nullam nobis fraudem in piscariam nostram conferant vel iacturam. Concessimus etiam et obligavimus nos quod nullo modo impediemus mercatores bladi sui vel alterius rei quam possint onerare et exonerare naves suas, pausare in filo vel iuxta filum Thamis', et naves suas attachiare versus terram dictorum religiosorum ubi melius elegerint, sine contradictione imperpetuum. In cuius rei testimonium sigilla nostra presenti scripto sunt appensa. Hiis testibus etc.

F *Geoffrey de Besilles recites King John's charter [1205][103] giving the monks of Beaulieu free passage along the Thames with their ships and merchandise from Faringdon to London; recites and confirms* E; *and further allows the monks to wash their sheep in the river at Kyndelwere and to strengthen and embank the riverbank there (reserving the fishery), and confirms all his ancestors' gifts to the monks in the demesne of Radcot.* [1315 × 1339] (BL, Cotton MS Nero A.xii, fos. 48ᵛ–50)

Universis Christi fidelibus presens scriptum visuris vel audituris, Galfridus de Besilles, dominus de Redcote, salutem in Domino sempiternam. Noveritis quod—licet celebris memorie dominus Johannes quondam rex Angl' dedisset et concessisset abbati et conventui de Bello Loco Regis per cartam suam quod ipsi et eorum successores liberum transitum habeant per totam Tamisiam cum navibus et mercandisis suis a manerio suo de Farendon' usque ad villam London'—verumptamen processu temporis, super huiusmodi transitu et quibusdam aliis locorum circumstantiis iuxta molendina [sic] de Kyndelwere dictorum monachorum, controversia extitit mota inter dominum Mathiam de Besilles, dominum de Redcote, patrem meum, ac Elisabeth uxorem eius matrem meam, et predictos religiosos abbatem et conventum de Bello Loco Regis supradictos; tandem, intervenientibus pacis amatoribus, dicta controversia conquievit[ur] in hunc modum: videlicet, quod dictus dominus Mathias pater meus et Elizabeth mater mea, unanimi voluntate et assensu ac per scriptum suum, pro se et heredibus suis, predictis abbati et conventui et eorum successoribus quod ipsi et homines sui ac mercatores et servientes eorum libere possint ire et redire cum navibus et navicellis suis in aqua Thamisie circa molendinum eorum religiosorum de Kyndelwere, tam ante molendinum quam retro, ad negotium et comodum suum faciendum quandocumque et ubicumque sibi viderint expedire, absque calumpnia seu contradictione prefati domini Mathie patris mei seu Elizabeth matris mee vel heredum suorum concesserunt. Ego vero predictus Galfridus, concessionem predictam patris mei ac matris mee ratam habens et gratam imperpetuum, volo et concedo pro me et heredibus meis seu meis assignatis predictis abbati et conventui et eorum successoribus, pro salute anime mee, antecessorum et heredum meorum, et per presens scriptum confirmo, quod predictus abbas et conventus et eorum successores, homines ac servientes eorum licite possint, in prefata aqua Thamisie ubi competentius viderint faciendum circa prefatum molendinum suum de Kyndelwere, oves suas lavare, lotas ad terram deportare, et terram suam prope ripam dicte aque Thamisie undique firmare et wallare et contra fluxum et alluvionem dicte aque absque consumptione custodire modo quo viderint et sciverint meliori, absque omni impedimento, calumpnia seu contradictione mea, heredum vel assignatorum meorum seu hominum aut servientium nostrorum imperpetuum; iure meo et heredum meorum in predicte aque piscaria sepe salvo. Omnes etiam donationes et concessiones antecessorum meorum de quibuscumque possessionibus

---

Radcot, was born *c*.1254. Geoffrey, her son by Matthew, was born *c*.1276. Matthew died in 1295; Elizabeth died, still holding Radcot, in 1315, after which Geoffrey held it until his death in 1339.

[103] Above, n. 25.

seu libertatibus eisdem religiosis in dominio meo de Redcote, quoscumque tempore factas, ratifico et confirmo pro me et heredibus meis, sicut carte quas inde habent testantur. In cuius rei testimonium presens scriptum sigilli mei impressione roboravi. Hiis testibus etc.

The Bampton Mill-Pond Dispute, 1268

G *Litigation between the Dean and Chapter of Exeter [proprietors of Bampton rectory] and Thomas the Miller of Bampton. Whereas there was litigation [in 1229 × 1231[104]] between Serlo former Dean of Exeter and the Chapter, and Thomas's father Adam, concerning the mill-pond at Bampton and its sluices, which Adam had raised to the nuisance of the Dean and Chapter, and it was agreed that the sluices would not be raised above markers placed in the pond; and that the sluices would be removed yearly to the Dean and Chapter's court in Bampton between 3 June and 15 July, in recognition of which they would pay Adam half a quarter of wheat yearly: Thomas has now raised the pond and sluices higher, and has not allowed the Dean and Chapter to take the sluices to their court. And whereas it was also agreed with Adam that if he allowed the water to rise above the markers, the sluices would be taken out and placed for a day and a night [in the Dean and Chapter's court] for that infringement: Thomas has not allowed the sluices to be taken out as agreed. The Dean and Chapter say that they have thereby suffered damage to the value of £40. The parties then came to agreement, for which Thomas gave half a mark; and it was agreed ... [see H]. 30 June 1268. (Oxfordshire eyre of 1268, TNA, PRO, JUST 1/702A m.6d.)*

Thomas le Mouner de Bampton' summonitus fuit ad respondendum Decano et Capitulo Exon' de placito quod teneat eis concordiam factam inter Serlonem quondam Decanum Exon' et predictum Capitulum, et Adam le Mouner patrem predicti Thome cuius heres ipse est, de quodam stangno [*sic*] cum pertinentiis in Bampton' et exclusis eiusdem stagni. Et unde predicti Decanus et Capitulum queruntur quod cum predictus Serlo quondam Decanus Exon' et Capitulum tulissent assisam nove disseisinis coram Williemo de Raleg' justiciario ad hoc assignato versus Adam le Mouner, patrem predicti Thome, cuius heres ipse est, de quodam stagno quod idem Adam levavit ad nocumentum liberi tenementi ipsorum Decani et Capituli in Bampton'; et quadam concordia facta est inter eos coram prefato Willelmo justiciario, per quam predictus Adam [convenit[105]] quod pessones et signa apponerentur in predicto stagno, et quod excluse excedere non deberent summitatem pessonum et signorum predictorum, ita quod inumdatio [*sic*] aque transcendere non deberent [*sic*] predicta signa; et etiam cum convenisset inter eos per predictam concordiam quod predicte excluse trahi deberent et deferri annuatim in curiam predictorum Decani et Capituli et ibidem morari per tres septimanas ante festum Sancti Johannis Baptiste et per tres septimanas post, et propter quamdam amotionem predicti Decanus et Capitulum dare deberent predicto Ade et heredibus suis dimidium quarterium frumenti per annum: predictus Thomas exaltavit predictum stagnum ultra pessones et signa predicta, et similiter exclusas predictas, et similiter non permittit predictum Decanum et Capitulum deferre predictas exclusas ad curiam ipsorum Decani et Capituli prout convenit inter eos. Et etiam cum convenisset inter eos quod si supradictus Adam vel heredes sui permitterent aquam predicti stagni in tantum inundare quod predicta signa[106] posita in predicto stagno aliquatenus per inundationem aque coopirerentur, quod excluse trahi deberent et deponi [in curiam predictorum Decani et Capituli[107]] per unum diem et unam noctem pro transgressione predicta: predictus Thomas, licet aqua per inundationem transcendat metas et signa predicta, non permittit ipsos predicta[s] exclusas trahere, prout inter eosdem Serlonem Decanum et Capitulum et Adam patrem predicti Thome convenit, unde dicunt quod deteriorati sunt et dampnum habent ad valenciam xl li; et producunt sectam.

---

[104] Serlo was dean 1225–31; the case was heard before William Ralegh, who was active as a judge 1229–39. There is apparently no fine among the feet of fines for either Oxfordshire or Devon.

[105] *Conjectural restoration of omission.*   [106] stagna *MS.*

[107] *Conjectural restoration of omission.*

Postea concordati sunt, et predictus Thomas dat dimidiam marcam pro licencia concordandi, per plegium Johannis de Exon' etc. Etiam habent cyr[ographum]. Et convenit inter eos quod predictus Thomas concessit ... [*continues with summary of* H].

H *Agreement between the Dean and Chapter of Exeter and Thomas the Miller of Bampton, reaffirming the earlier agreement as set out in* **G**. *15 July 1268.* (Cartulary transcript, Exeter Cathedral, Dean and Chapter MS 3672, p. 12.)

Hec est concordia facta apud Exon', a die Sancti Johannis in tres septimanas anno regni Regis Henrici filii Johannis Regis Anglie quinquagesimo secundo, inter Willelmum Decanum Exon' ecclesie et eiusdem ecclesie Capitulum (per Johannem de Exon' concanonicum attornatum suum) ex una parte, et Thomam le Mouner de Bampton ex altera, videlicet: quod predictus Thomas concessit et recognovit pro se, heredibus et assignatis suis coram Ricardo de Middelton', Ade de Gremuil, Rogero de Messindene et Thoma Trevet, justiciariis domini Regis tunc ibi presentibus, quod non licebit sibi decetero exaltare stagnum molendini sui in Bemptun' ita quod inundatio aque per predictam exaltationem excedat vel transcendat summitatem pelsonum vel signorum in predicto stagno positorum. Excluse vero eiusdem stagni summitatem predictorum pelsonum vel signorum nullatenus debent excedere, nec a loco debito amoveri. Preterea predicte excluse trahi debent et deferri annuatim in curiam predictorum Decani et Capituli in Bemptun' et morari ibidem per tres septimanas ante festum Nativitatis Beati Johannis Baptiste [et] per tres septimanas post dictum festum, propter quam ammotionem dabunt dicti Decanus et Capitulum memorato Thome et heredibus vel assignatis suis dimidium quarterium frumenti circa Nativitatem Sancti Johannis Baptiste. Et si forte dictus Thomas vel heredes sive assignati sui, vel aliquis alius per ipsos, tenuerit aquam predicti stagni ita quod predicta signa aliquatenus per inundationem cooperiantur, trahi debent excluse et deponi per unum diem et unam noctem in curia predictorum Decani et Capituli pro huius transgressione. Hec omnia ex utraque parte recognita fuerunt et reconcessa coram prefatis justiciariis, et in rotulis eorum redacta apud Oxon'. Et presenti scripto in modum cyrographi confecto sigilla nostra alternatim in robur et testimonium huius concordie apposuerunt.

The Islip Mill-Dam Dispute, 1261

I *Litigation between Richard Overeya and a group of local men. Richard accuses 97 men (named) of coming in force and against the peace to his weir on the Cherwell at Islip on 28 July 1260, breaking the weir, and pulling out and burning the stakes and hurdles. They answer that Richard has a weir there in which there used to be a gap in mid-stream, 25 feet wide without any obstruction of stakes or hurdles, allowing floodwaters to pass through; that a year ago Richard raised the weir and obstructed the course of the river with stakes and hurdles, reducing the aperture to 4 feet; that around 25 July this barrier caused flooding of fields and meadows for four leagues around; and that the defendants took out the stakes and hurdles which had been newly put there and laid them on dry ground, but did not burn them; and that they did not come in force or against the peace. The jury finds that the aperture of the weir has long been 25 feet wide; that a year ago Richard narrowed the aperture by three feet and raised the weir by three feet, and so much strengthened the weir with hurdles that the water flooded, causing great damage; and that the defendants dismantled both the old and the new stakes and hurdles, but did not burn them. 1261.* (Oxfordshire eyre of 1261, TNA, PRO, JUST 1/701 m. 10.)

Johannes le Baillif de Pidinton', Johannes le Provost, Robertus le Long, Ricardus le Mercer, Hugo le Gray, Thomas Bonde, Willelmus de Cleydon', Radulfus de Pullitot', Radulfus de Cleydon', Willelmus de Hedindon', Johannes filius Thome de Pidinton', Willelmus Dure de Pidinton', Hugo filius Gervasii, Thomas de Cleydon', Martinus le Provost de Blakethurn', Hugo Caleman, Rogerus filius Roberti Red, Ricardus filius Roberti de Chiltre, Ricardus le Sweyn, Hugo le Sweyn, Rogerus le Sweyn, Alanus le Sweyn, Andreas filius Gunnild', Hugonis de Grete, Rogerus filius Hugonis de Grete, Rogerus le Pipere, Johannes Jordan, Johannes Cok, Thomas Coc, Nicholas Cok, Nicholas Cok, Thomas Andr', Johannes le Veul, Hugo le

Cuppere, Willelmus le Paumer, Walterus filius Amicie, Willelmus filius Agathe, Willelmus le Smale, Ricardus le Haldare, Willelmus de Cimiterio, Nicholas le Brus, Robertus le Vaunter, Alanus Honning, Stephanus Canel, Johannes filius Galdine, Robertus Culvert, frat … [?], Willelmus de Arnicoth, Robertus Kiderel de Meriton', Bernardus Gareun, Willelmus le Say, Johannes Freman, Johannes filius Alicie, Ricardus le Seriant, Willelmus le Lung, Galfridus filius Gilberti Cole, Willelmus de Fonte, Willelmus Keming', Johannes le Neuman, Galfridus Burne, Nicholas Ketilbern', Thomas Norman, Ricardus Norman, Willelmus Gerard, Willelmus Gregor', Hugo le Canun, Hugo le Wilde, Ricardus filius Cecilie, Thomas le Gale, Galfridus frater eius, Nigellus Cupe, Walterus le Gale, Willelmus Cole, Thomas filius Galfridi Cupe, Nigellus filius Gilberti Thurbarn, Simon Coleman, Willelmus Noblet, Alexander le Parker, Hugo Sprot, Willelmus Scarbek, Willelmus Sweyn, Thomas le Hunte, Edward Scarbek, Ricardus Sachen, Willelmus Edward, Johannes Hereward, Hugo Cobbe, Nicholas filius Ran', Henricus filius Willelmi Herebert, Henricus Norman, Reginaldus atte Stile, Thomas de Aldrinton', Andreas Poleyn de Cherlton', Rogerus Cole, Rogerus filius Johannes le Poure, Ricardus Salomon, Willelmus Hervy, Robertus Pynel, summoniti fuerunt ad respondendum Ricardo Overea, de placito quare vi et armis venerunt ad gurgitem ipsius Ricardi in Islepe, et gurgitem illam fregerunt, et cleyas eiusdem gurgitis combusserunt, ad grave dampnum ipsius Ricardi, contra pacem etc. Etenim predictus Ricardus queritur quod predicti Johannes, Johannes, Robertus et alii, die Mercurii proxima post festum Sancti Jacobi Apostoli anno xliiij^{to}, venerunt ad quandam gurgitem ipsius Ricardi quam habet in aqua de Charwell in villa de Islepe vi et armis, scilicet hachiis, gladiis, furcisferreis, arcubus et sagittis et aliis armis diversis, et gurgitem illam fregerunt, et palos et cleyas extraxerunt, et illas combusserunt, et domum suam intraverunt in Islepe, et cleyas torell' sue ceperunt et asportaverunt et illas combusserunt, ad grave dampnum ipsius Ricardi contra pacem etc; unde dicit quod deterioratus est et dampnum habet ad valenciam x marcarum; et inde producit sectam etc.

Et predicti Johannes et alii venerunt [et] defendunt vim et iniuriam quando etc. Et bene defendunt quod numquam vi et armis nec contra pacem venerunt ad predictam gurgitem, nec predictas cleyas combusserunt, sicut eis inponit. Set re vera dicunt quod predictus Ricardus habuit quandam gurgitem in predicta aqua, in qua gurgite semper solebant esse in medio fili eiusdem aque xxv pedes absque obscuratione alicuius pali vel cleye, ubi inpetus aque cum inundaverit cursum suum habere consuevit. Et dicunt quod idem Ricardus postea iam uno anno elapso exaltavit predictam gurgitem, et filum eiusdem aque obstruxit palis et cleyis, ita quod apertum eiusdem aque non remansit nisi de quatuor pedibus. Dicunt enim quod circa predictum festum Sancti Jacobi, predicta aqua ita inundavit quod inebriavit omnes terras et prata iuxta eandem aquam fere circiter per iiij leucatas per predictam obscurationem. Et dicunt quod, propter predictam inundationem, palos et cleyas quas de novo apposuit extraxerunt et asportaverunt et ipsas super siccam terram posuerunt, et quod predictas cleyas non combusserunt, nec ibi vi et armis contra pacem venerunt, ponunt se super patriam, et predictus Ricardus similiter. Ideo fiat inde iurata.

Et iuratores de consensu partium electi dicunt super sacramentum suum quod predictus Ricardus a multo tempore transacto habuit quandam gurgitem in predicta aqua, in tali statu quod apertum eiusdem gurgitis solebat de xxv pedibus. Et dicunt quod idem Ricardus, uno anno elapso, obstruxit apertum eiusdem gurgitis per tres pedes, et eandem gurgitem exaltavit de altitudine trium pedum aliter quam esse consuevit, et ita spisse eandem gurgitem cleys munivit quod aqua illa redundabat ultra terras et prata vicina eidem aque, ad maximum dampnum totius patrie illius. Ob quod quamplures predictorum Johannis et aliorum accesserunt ad predictam gurgitem, et omnes palos et cleyas tam de veteri tempore quam de novo appositas extraxerunt. Set re vera dicunt quod predictas cleyas non combusserunt, nec vi et armis ibi venerunt, sicut eis imponit. Et iuratores, requisiti si omnes predicti fregerunt dictam gurgitem, dicunt quod Johannes ballivus, Johannes prepositus, Robertus le Lung, Hugo le Gay, Thomas Bonde, Willelmus de Cleydon', Radulfus de Cleydon', Willelmus de Hedindon', Johannes filius Thome, Thomas de Cleydon', Johannes le Frankeleyn, Willelmus le Keche, Ricardus filius Roberti de Cilere, Hugo Sweyn, Alanus Sweyn, Willelmus Gipcian, Johannes Coc, Nicholas Cok, Thomas Andreu, Johannes le Woul, Robertus Pinel, Ricardus

Salemon, Willelmus le Palmer, Walterus filius Amicie, Walterus de Cimiterio, Nicholas le Burs, Radulfus de Stowe, Stephanus le Foreyn, Stephanus Canel, Willelmus Cobbe, Thomas de Alcrinton', Henricus Noblet, Hugo Cobbe, interfuerunt predicte transgressioni. Ideo ipsi in misericordia, et satisfaciunt predicto Ricardo de dampnis, quod taxantur ad xij d. Et similiter predictus Ricardus in misericordia pro falso clamore versus alios, et similiter pro transgressione.

Glossary

barge-gutter a bypass stream around a weir, allowing the passage of boats and migrating fish.

blind lode   see *lode*

custumal   a document listing the services owed by tenants on a manor

eyre   an assize held in a locality by itinerant royal justices.

flash-lock   see *lock*

hythe   (OE *hȳð*) a landing place.

lake   (OE *lacu*) a small, slow-moving stream.

leet   a narrow artificial watercourse, usually serving a mill.

lock   a device allowing water on a river, with any craft on it, to be held back or released under more-or-less controlled conditions. A *flash-lock* (obsolete)—effectively a large sluice—is a gap in a weir closed by hurdles, boards, or paddles, the lifting of which allows a boat to pass downstream on the 'flash' of water thus released, or to be hauled upstream over the 'flash' by ropes. A *pound-lock* (the modern kind) is a section of a river shut off above and below by folding gates, which allow the water level between them to be raised or lowered.

lode   (OE *lād*) artificial watercourse. A *blind lode* is a lode which opens into a larger watercourse at one end only, usually terminating at the other end at a quay, manorial centre, or monastery where goods can be loaded and unloaded.

logboat   a small, narrow boat made by hollowing out a single half-log.

meander migration   the process by which a watercourse changes its configuration naturally through the erosion and deposition of material.

portage   the carrying of boats and their cargoes overland between waterways.

pound-lock   see *lock*

putts, putcher   fish-catching baskets, and weir made of them.

shout   (from Dutch *schuit*) a late medieval term for a flat-bottomed cargo vessel, like a barge but with pointed ends and a keel, used around the coasts and on rivers, especially the Thames.

sluice   an adjustable gate in a dam or weir which allows the water to be held back or released under controlled conditions.

staithe   (OE *stæð*, ON *stǫð*) a landing place.

trestle bridge   a timber bridge consisting of a planked causeway resting on a series of rigid trestle-like frames.

weir   (OE *wer*, Latin *gurges*) a barrier, usually of timber or wattle and sometimes associated with a more substantial dam, set across the flow of a watercourse to raise the water level for a mill, or to catch fish. (The Latin *gurges* often refers to the gap in the barrier—i.e. a sluice or flash-lock—rather than the barrier itself.)

Select Bibliography

This bibliography is designed as a starting point for further research. It includes works that are recurrently cited, or have a major general relevance to the themes of the book; it omits works only occasionally cited.

ASTON, M. (ed.), *Medieval Fish, Fisheries and Fishponds in England*, 2 vols., BAR British Ser. 182 (Oxford, 1988).

BLAIR, J., *Anglo-Saxon Oxfordshire* (Stroud, 1994).

___ 'The Minsters of the Thames', in J. Blair and B. Golding (eds.), *The Cloister and the World: Essays in Medieval History in Honour of Barbara Harvey* (Oxford, 1996), 5–28.

___ and RAMSAY, N. (eds.), *English Medieval Industries: Craftsmen, Techniques, Products* (London, 1991).

BOND, C. J. 'The Reconstruction of the Medieval Landscape: The Estates of Abingdon Abbey', *Landscape History*, 1 (1979), 59–75.

___ 'Monastic Water Management in Great Britain: A Review', in G. Keevill, M. Aston, and T. Hall (eds.), *Monastic Archaeology: Papers on the Study of Medieval Monasteries* (Oxford, 2001), 88–136.

BURWASH, D., *English Merchant Shipping 1460–1540* (Toronto, 1947; repr. Newton Abbot, 1969).

CAMPBELL, B. M. S., GALLOWAY, J. A., KEENE, D., and MURPHY, M., *A Medieval Capital and its Grain Supply: Agrarian Production and Distribution in the London Region c.1300* ([London], 1993).

CLARK, H. J. S., 'The Salmon Fishery and Weir at Wareham', *Proc. Dorset Nat. Hist. & Archaeol. Soc.* 72 (1950), 99–110.

CLARK, J. G. D., 'Report on Excavations on the Cambridgeshire Car Dyke, 1947', *Antiquaries Journal*, 29 (1949), 145–63.

COOK, H. F., and WILLIAMSON, T., *Water Management in the English Landscape: Field, Marsh and Meadow* (Edinburgh, 1999).

CRITCHLEY, R. H., 'The Lower Douglas and Ribble Valley Navigation', *Ribble Archaeology*, 6 (1973), 19–24.

CRUMLIN–PEDERSEN, O., 'The Boats and Ships of the Angles and Jutes', in S. McGrail, *Maritime Celts, Frisians and Saxons* (London, 1990), 98–116.

CURRIE, C. K., 'Saxon Charters and Landscape Evolution in the South-Central Hampshire Basin', *Proc. Hants Field Club and Archaeol. Soc.* 50 (1995), 103–25.

___ 'A Possible Ancient Water Channel around Woodmill and Gater's Mill in the Historic Manor of South Stoneham', *Proc. Hants Field Club and Archaeol. Soc.* 52 (1997), 89–106.

DARBY, H. C., *The Medieval Fenland* (Cambridge, 1940; repr. Newton Abbot, 1974).

___ *Domesday England* (Cambridge, 1977).

DAVIS, R. H. C., 'The Ford, the River and the City', *Oxoniensia*, 38 (1973), 258–67 (repr. in his *From Alfred the Great to Stephen* (London, 1991), 281–91).

Dodd, A. (ed.), *Oxford before the University* (Oxford, 2003).

Eddison, J., 'The Purpose, Construction and Operation of a Thirteenth-Century Watercourse: The Rhee, Romney Marsh, Kent', in A. Long, S. Hipkin, and H. Clarke (eds.), *Romney Marsh: Coastal and Landscape Change through the Ages* (Oxford, 2002), 127–39.

Edwards, J. F., 'The Transport System of Medieval England and Wales: A Geographical Synthesis' (University of Salford Ph.D. thesis, 1987).

——— and Hindle, B. P., 'The Transportation System of Medieval England and Wales', *JHG* 17 (1991), 123–34.

——— ——— 'Comment: Inland Water Transportation in Medieval England', *JHG* 19 (1993), 12–14.

Ekwall, E., *English River-Names* (Oxford, 1928).

——— *The Concise Oxford Dictionary of English Place-Names* (4th edn. Oxford, 1960).

Fenwick, V., *The Graveney Boat*, BAR British Ser. 53 (Oxford, 1978).

Fockema Andreae, S. J., 'The Canal Communications of Central Holland', *Jnl. of Transport History*, 4/3 (1960), 174–9.

Fowler, G., 'Fenland Waterways, Past and Present, South Levels District: I', *Proc. Cambridge Antiquarian Soc.* 33 (1933), 108–28.

——— 'Fenland Waterways, Past and Present, South Levels District: II', *Proc. Cambridge Antiquarian Soc.* 34 (1934), 17–33.

Gardiner, J. *Flatlands and Wetlands: Current Themes in East Anglian Archaeology*, EAA 50 (Cambridge, 1993).

Gardiner, M. F., 'Shipping and Trade between England and the Continent during the Eleventh Century', *Anglo-Norman Studies*, 22 (2000), 71–93.

Gaunt, G. D., 'The Artificial Nature of the River Don North of Thorne, Yorkshire', *YAJ* 47 (1975), 15–21.

Gelling, M., and Cole, A., *The Landscape of Place-Names* (Stamford, 2000).

Good, G. L., Jones, R. H., and Ponsford, M. W. (eds.), *Waterfront Archaeology: Proceedings of the Third International Conference on Waterfront Archaeology*, CBA Research Rep. 74 (London, 1991).

Griffiths, D., 'Coastal Trading Ports of the Irish Sea Region', in J. Graham-Campbell (ed.), *Viking Treasure from the North-West: The Cuerdale Hoard in its Context* (Liverpool, 1992), 63–72.

Guillerme, A. E., *The Age of Water: The Urban Environment in the North of France, AD 300–1800*, translation of *Les Temps de l'eau* (College Station, Tex. 1988).

Hall, D., *The Fenland Project 2: Fenland Landscapes and Settlement between Peterborough and March*, EAA 35 (Cambridge, 1987).

——— (1992). *The Fenland Project 6: The South-Western Cambridgeshire Fenlands*, EAA 56 (Cambridge, 1992).

——— *The Fenland Project 10: Cambridgeshire Survey, the Isle of Ely and Wisbech*, EAA 79 (Cambridge, 1996).

——— and Coles, J. M., *Fenland Survey: An Essay in Landscape and Persistence*, English Heritage Archaeological Rep. 1 (London, 1994).

Harrison, D., *The Bridges of Medieval England* (Oxford, 2004).

Haslam, J., *Anglo-Saxon Towns in Southern England* (Chichester, 1984).

Hayes, P. P., and Lane, T. W., *The Fenland Project 5: Lincolnshire Survey, the South-West Fens*, EAA 55 (Cambridge, 1992).

HILL, D., *An Atlas of Anglo-Saxon England* (Oxford, 1981).

HOLLINRAKE, C., and HOLLINRAKE, N., 'A Late Saxon Monastic Enclosure Ditch and Canal, Glastonbury, Somerset', *Antiquity*, 65 (1991), 117–18.

———— 'The Abbey Enclosure Ditch and a Late-Saxon Canal: Rescue Excavations at Glastonbury, 1984–1988', *Proc. Somerset Archaeol. and Nat. Hist. Soc.* 136 (1992), 73–94.

HOLT, R., 'Medieval England's Water-Related Technologies', in P. Squatriti (ed.), *Working with Water in Medieval Europe* (Leiden, 2000), 51–100.

HUGGINS, R., 'London and the River Lea', *London Archaeologist*, 8/9 (1998), 241–7.

JONES, E., 'River Navigation in Medieval England', *JHG* 26 (2000), 60–75.

KLÁPŠTĚ, J. (ed.), *Water Management in the Medieval Rural Economy/Les Usages de l'eau en milieu rural au Moyen Âge*, Ruralia 5 (Prague, 2005). [This work had not appeared at the time of going to press.]

KNIGHTON, D., *Fluvial Forms and Processes: A New Perspective* (London, 1998).

LANDERS, J., *The Field and the Forge* (Oxford, 2003).

LANGDON, J., 'Inland Water Transport in Medieval England', *JHG* 19 (1993), 1–11.

———— 'Inland Water Transport in Medieval England: The View from the Mills: A Response to Jones', *JHG* 26 (2000), 75–82.

———— *Mills in the Medieval Economy: England, 1300–1540* (Oxford, 2004).

LEIGHTON, A. C., *Transport and Communication in Early Medieval Europe, AD 500–1100* (Newton Abbot, 1972).

LOSCO-BRADLEY, P. M., and SALISBURY, C. R., 'A Saxon and a Norman Fish Weir at Colwick, Nottinghamshire', in M. Aston (ed.), *Medieval Fish, Fisheries and Fishponds in England* (Oxford, 1988), 329–52.

McGRAIL, S., *Logboats of England and Wales* BAR British Ser. 51 (Oxford, 1978).

———— (ed.), *The Archaeology of Medieval Ships and Harbours in Northern Europe*, BAR Internat. Ser. 66 (Oxford, 1979).

———— (ed.), *Maritime Celts, Frisians and Saxons*, CBA Research Rep. 71 (London, 1990).

———— *Boats of the World: From the Stone Age to Medieval Times* (Oxford, 2001).

MAGNUSSON, R. J., *Water Technology in the Middle Ages: Cities, Monasteries, and Waterworks after the Roman Empire* (Baltimore, 2001).

MARGARY, I. D., *Roman Roads in Britain* (3rd edn. London, 1973).

MARSDEN, P., *Ships of the Port of London: Twelfth to Seventeenth Centuries AD*, English Heritage Archaeological Rep. 5 (London, 1996).

MASSCHAELE, J., 'Transport Costs in Medieval England', *EcHR* 46 (1993), 266–79.

MILLS, A. D., *Dictionary of English Place-Names* (Oxford, 1991).

MILNE, G. and HOBLEY, B. (eds.), *Waterfront Archaeology in Britain and Northern Europe*, CBA Research Rep. 41 (London, 1981).

MOORE, F. G., 'Three Canal Projects, Roman and Byzantine', *American Journal of Archaeology*, 54 (1950), 97–111.

NAYLING, N., *The Magor Pill Medieval Wreck*, CBA Research Rep. 115 (York, 1998).

PANNETT, D. J., 'Fish Weirs of the River Severn with Particular Reference to Shropshire', in M. Aston (ed.), *Medieval Fish, Fisheries and Fishponds in England* (Oxford, 1988), 371–89.

PEBERDY, R., 'Navigation on the River Thames between London and Oxford in the Late Middle Ages: A Reconsideration', *Oxoniensia*, 61 (1996), 311–40.

PESTELL, T., and ULMSCHNEIDER, K. (eds.), *Markets in Early Medieval Europe: Trading and 'Productive' Sites, 650–850* (Macclesfield, 2003).

PHILLIPS, G., *Thames Crossings: Bridges, Tunnels and Ferries* (Newton Abbot, 1981).

PHYTHIAN–ADAMS, C., *Land of the Cumbrians: A Study in British Provincial Origins AD 400–1120* (Aldershot, 1996).

RAHTZ, P. A., and MEESON, R., *An Anglo-Saxon Watermill at Tamworth*, CBA Research Rep. 83 (London, 1992).

RAVENSDALE, J. R., *Liable to Floods: Village Landscape on the Edge of the Fens, AD 450–1850* (Cambridge, 1974).

RIPPON, S., *The Gwent Levels: the Evolution of a Wetland Landscape*, CBA Research Rep. 105 (York, 1996).

——*The Severn Estuary: Landscape Evolution and Wetland Reclamation* (London, 1997).

——*The Transformation of Coastal Wetlands: Exploitation and Management of Marshland Landscapes in North-West Europe during the Roman and Medieval Periods* (Oxford, 2000).

ROLT, L. T. C., *The Inland Waterways of England* (London, 1950).

RUSSETT, V. E. J., 'Hythes and Bows: Aspects of River Transport in Somerset', in G. L. Good, R. H. Jones, and M. W. Ponsford (eds.), *Waterfront Archaeology* (London, 1991), 60–6.

SALISBURY, C. R., 'Primitive British Fishweirs', in G. L. Good, R. H. Jones, and M. W. Ponsford (eds.), *Waterfront Archaeology* (London, 1991) 76–87.

SALZMAN, L. F., *Building in England down to 1540: A Documentary History* (Oxford, 1952).

SHERRATT, A., 'Why Wessex? The Avon Route and River Transport in Later British Prehistory', *Oxford Journal of Archaeology*, 15 (1996), 211–34.

SILVESTER, R. J., *The Fenland Project 3: Marshland and the Nar Valley, Norfolk* EAA 45 (Cambridge 1988).

SMITH, A. H., *English Place-Name Elements* (2 vols., 2nd impression Cambridge, 1970).

SMITH, N. A. F., 'Roman Canals', *Transactions of the Newcomen Society*, 49 (1978), 75–86.

SMYTH, A., *Scandinavian York and Dublin* (2 vols., Dublin, 1975–9).

SQUATRITI, P. (ed.) *Working with Water in Medieval Europe: Technology and Resource-Use* (Leiden, 2000).

——'Digging Ditches in Early Medieval Europe', *Past and Present*, 176 (2002), 11–65.

STENTON, F. M., 'The Road System of Medieval England', *EcHR*, 7 (1936), 1–21 (repr. in D. M. Stenton (ed.), *Preparatory to Anglo-Saxon England* (Oxford, 1970), 234–52).

SUMMERS, D., *The Great Ouse: The History of a River Navigation* (Newton Abbot, 1973).

THACKER, F. S., *The Thames Highway: A History of the Inland Navigation* (London, 1914 repr. New York, 1968).

——*The Thames Highway: A History of the Locks and Weirs* (London, 1920 repr. New York, 1968).

ULMSCHNEIDER, K., *Markets, Minsters and Metal-Detectors: The Archaeology of Middle Saxon Lincolnshire and Hampshire Compared* (Oxford, 2000).

WAITES, B., 'The Medieval Ports and Trade of North-Eastern Yorkshire', *Mariners Mirror*, 63 (1977), 137–49.

WILLAN, T. S., *River Navigation in England 1600–1750* (2nd impression London, 1964).

WILLIAMS, M., *The Draining of the Somerset Levels* (Cambridge, 1970).

WILSON, D. G., *The Thames: Record of a Working Waterway* (London, 1987).

WOODING, J. M., *Communication and Commerce along the Western Sealanes AD 400–800*, BAR Internat. Ser. 654 (Oxford, 1996).

Index

A book about waterways presents special indexing needs. Whereas most minor place-names can appropriately be indexed by parish, water-courses flow between parishes and even counties. Both their names and their configurations are often unstable, and they can be hard to locate in normal gazetteers.

Here, water-courses, crossings, weirs, locks and hythes are indexed in one sequence under their own names. County names (pre-1974) are given for minor water-courses, but not for large and well-known rivers. County names are also indicated for normal settlements, but riverine features can often be best identified in relation to their rivers (e.g. 'Newbridge, on upper Thames'), especially when those rivers are county boundaries. Place-names mentioned simply to indicate the locations of other features are not usually indexed.

Water-courses are identified as rivers ('r.'), artificial canals and other channels ('c.'), and other minor water-courses ('w-c').